Footprints of Regional Science

The Voice of Regional Science

The field of regional science continues to advance. The book series *The Voice of Regional Science* presents new scholarly thinking on important emerging issues and new scientific advances in the broad and multidisciplinary domain of regional science research. The contributions to the series invoke the aim of the Regional Science Academy to rethink "the spatial dynamics of people and socio-economic activities in the connected and complex spatial systems of our earth." The volumes in this series will serve the global regional science community by providing new insights on, or by calling attention to, recent developments that are rarely covered by conventional conferences, or are indicative of novel research directions.

The Voice of Regional Science welcomes both edited volumes and monographs, with works originating in the meetings of the Academy or proposed by external authors and volume editors. It features scholarly works that are retrospective only to the extent necessary to lay the groundwork for visions of future research in the area; all volumes have a forward-looking focus. Further, the series' emphasis on thematic publications is intended to promote a more future-oriented strategic focus for regional science research. Applied studies in regional science are also welcome, provided they reflect this forward-looking focus.

The two sub-series *The Voice of Regional Science* and *Great Minds in Regional Science* jointly form a series entitled *Footprints in Regional Science*. All three book series present work stemming from and related to the activities of The Regional Science Academy (www.regionalscienceacademy.org).

Louafi Bouzouina • Karima Kourtit
Peter Nijkamp
Editors

15-Minute Cities

Sustainable Mobility and Urban Livability

The Regional Science Academy

Springer

Editors
Louafi Bouzouina
LAET, ENTPE
University of Lyon
Vaulx-en-Velin, France

Karima Kourtit
Faculty of Management
Open University of the Netherlands
Heerlen, The Netherlands

Peter Nijkamp
University College Wrocław (WSKZ)
Wrocław, Poland

ISSN 2662-9623 ISSN 2662-9631 (electronic)
Footprints of Regional Science
ISSN 2662-9704 ISSN 2662-9712 (electronic)
The Voice of Regional Science
ISBN 978-3-032-20839-2 ISBN 978-3-032-20840-8 (eBook)
https://doi.org/10.1007/978-3-032-20840-8

This Springer imprint is published by the registered company Springer Nature Switzerland AG
The registered company address is: Gewerbestrasse 11, 6330 Cham, Switzerland

Contents

15-Minute Cities, Sustainable Mobility and Urban Livability: A New Urban Landscape Challenge

Louafi Bouzouina, Karima Kourtit, and Peter Nijkamp

1 Introduction

Climate and health challenges are currently prominent sources of concern at a local and global scale. A substantive part of scientific research addresses the short- and long-term effects of climate and health hazards on the sustainable development of cities and regions. On the other hand, an important share of research is also concerned with complex causality mechanisms, in particular how human choices and behaviour shape the geographic conditions for the occurrence of climate and health hazards. Urban land use, location of economic activities, new industrial technology, spatial mobility patterns and lifestyles, and urban transport and logistics are examples of determinants of unexpected or undesired hazardous consequences of urban life. The future of the liveability and functionality of cities is prone to new types of risks in many countries (see Fusco Girard et al., 2023). A broad and new research

L. Bouzouina (✉)
LAET, ENTPE, University of Lyon, Vaulx-en-Velin, France
e-mail: louafi.bouzouina@entpe.fr

K. Kourtit
Faculty of Management, Open University of the Netherlands, Heerlen, The Netherlands

EHERO – Erasmus University Rotterdam (EUR), Rotterdam, The Netherlands

University College Wrocław (WSKZ), Wrocław, Poland

P. Nijkamp
EHERO – Erasmus University Rotterdam (EUR), Rotterdam, The Netherlands

University College Wrocław (WSKZ), Wrocław, Poland

Tsinghua University, Beijing, China

L. Bouzouina et al. (eds.), *15-Minute Cities*, Footprints of Regional Science,
https://doi.org/10.1007/978-3-032-20840-8_1

theme is then centred on the question: "Will cities in the future continue to be sources of human wellbeing?"

In this context, the notion of the 15-minute city has in recent years gained much popularity in transportation science and urban planning (see e.g., Moreno, 2023; Moreno et al., 2021a, 2021b; Noworól et al., 2022; Zhang et al., 2023). This follows from the fact that the 15-minute city concept promotes the development of new local services, an increase in the density of the mixed-use neighbourhood and significant changes in urban design allowing access to work, services, education, and leisure within a short walk or bike ride. The 15-minute city offers a framework for addressing a range of urban challenges. Clearly, achieving the 15-minute goals may encounter fierce resistance from local residents (witness the response of some Not In My Backyard—NIMBY groups) who oppose the implementation of such changes in the city. To reach a consensus between the 15-minute city concept and the NIMBY concerns, complementary strategies need to be implemented. These strategies include the identification of local residents' concerns, the involvement of residents in the early participatory planning, and the use of incremental implementation, thus avoiding large-scale transformation that can trigger resistance from local residents. A broad consensus allows a balance between residents' concerns and collective urban sustainability goals. From this perspective, the chapters in this book elaborate on the links between urban form, mobility, social equity, and environmental outcomes, while combining conceptual, empirical, and policy perspectives for attaining liveable cities. The 15-minute city concept appears to be an ambitious and original vehicle for intense debates among scientists, professionals, and local stakeholders.

Recently, several initiatives have been launched to test the feasibility of this concept—for instance, the adoption of the concept in Paris or Brussels, whereas in other countries (e.g. Denmark and the Netherlands) it has been argued that the 15-minute city has already been long in existence as part of a long-term strategic policy to favour of "slow motion" or active mobility (e.g. cycling and walking) in combination with public transport (Bouzouina et al., 2024; Celbiş & Bouzouina, 2025). In any case, the 15-minute city notion opens up many new urban planning issues.

The 15-minute city is certainly not purely a transportation concept, nor is it a goal in itself. It positions urban liveability in the centre of citizen engagement regarding spatial mobility and quality of life (QoL) in a city. In other words, the well-being of residents and the attractiveness of local businesses are the key drivers of a modern liveable city (see Fusco Girard et al., 2023). The citizens owe the city and are responsible for its quality of life. In the contemporary "urban century" (the "New Urban World"; Kourtit, 2019), with a large-scale introduction of digital technology, the goal of data-driven decision-making for sustainable and climate-neutral development of the city is of prominent importance. The effectuation of climate-benign strategies makes the city also more robust and resilient in the face of future shocks and challenges, while favouring spatial accessibility and inclusivity for all citizens. In other words, the 15-minute city seeks to offer a new urban design model for enhanced quality of life and wellbeing in cities.

2 15-Minute City, Sustainable Mobility and Urban Liveability

Beyond its conceptual appeal, the 15-minute city also offers a practical framework for addressing several long-standing urban challenges that extend across spatial, social, and economic domains. One of its most compelling promises lies in its potential to address accessibility deficiencies and poverty-related disparities in large metropolitan areas. By ensuring that essential services, employment opportunities, education, healthcare, and green spaces are situated within short distances, the model reduces barriers faced by vulnerable populations who traditionally depend on costly or time-intensive mobility patterns. In this way, the 15-minute city can play a meaningful role in addressing structural inequalities and strengthening the social sustainability of urban environments. Achieving this ambition requires targeted investments in adapted transport infrastructure, particularly measures that promote safe, efficient, inclusive, active, and collective sustainable mobility.

Proximity-based living reframes transport not as a purely movement-oriented system, but as an essential enabler of short-distance accessibility. The development of high-quality cycling networks, pedestrian pathways, and integrated public transport hubs is critical for achieving spatial cohesion, preventing peripheral isolation, and enhancing accessibility for all urban residents. In many situations (peripheral urban form, lack of public transport, lack of local services, etc.), car use is considered a necessity. This car dependence reflects structural inequalities resulting from past planning models. The 15-minute city concept is a transition framework that aims to reduce forced car dependency by design rather than to oppose automobility, considering car dependency as a legacy constraint. This shift from long-travel distance and vehicle ownership to proximity-based living and diverse mobility access could be supported by car-sharing systems. Such a transition will reduce parking demand, which will progressively allow for reallocation of these spaces to public spaces and active mobility infrastructure without compromising the social inclusiveness of car-dependent populations. Without deliberate efforts to address the needs of disadvantaged groups—such as low-income households, migrants, seniors, or residents of underserved neighbourhoods—the concept risks reproducing segregation and sociospatial inequities (Bouzouina, 2008; Abbiasov et al., 2024). Ensuring that proximity does not translate into exclusion requires policies that balance local accessibility with equity-focused planning and investment priorities.

Central to this endeavour is citizen-oriented governance. The 15-minute city can only be effective when designed with and for its residents. Participatory planning, cocreation, digital engagement tools, and neighbourhood-level deliberation are essential mechanisms for aligning policy interventions with everyday mobility needs, lifestyle preferences, and community aspirations. A governance structure rooted in citizen input strengthens trust and social cohesion, while facilitating behavioural shifts toward sustainable mobility practices.

Moreover, the relationship between quality of life (QoL) and the 15-minute city requires more thorough analytical investigation. Proximity shapes daily rhythms, social interactions, mental health, physical activity, and environmental conditions. Despite the growing recognition of QoL as a guiding principle in proximity-based planning, empirical research exploring these interdependencies remains limited and calls for interdisciplinary analysis across urban geography, public health, transport studies, and behavioural sciences. Another important research gap concerns the role of public transport within the 15-minute city framework. Contrary to some popular interpretations, a proximity-oriented model does not diminish the relevance of public transport; rather, it reinforces its strategic importance for efficient metropolitan mobility, regional connectivity, and equitable access to opportunities. More detailed analytical work is clearly required to understand how public transport systems can complement proximity-oriented urban design.

The global expansion of the 15-minute city concept also calls for consideration of its applicability in developing countries, where rapid population growth, informal settlements, and infrastructural shortages complicate the implementation of proximity-based planning. This raises critical questions regarding the transferability of policy models and the conditions for successful adaptation in diverse socio--economic contexts. Here, a multifunctional perspective on cities—encompassing *Living, Working, Supplying, Caring, Learning, and Enjoying*—provides a comprehensive lens through which to conceptualise proximity not merely as spatial closeness, but as an integrated foundation for human development and urban wellbeing.

As the development of the concept of the 15-minute city gained momentum in recent years, particularly in the post-pandemic environment, the scientific literature has been facing challenges in keeping pace with real-world developments and novel challenges in relation to urban settings. The difficulty faced by the research landscape is particularly evident regarding the lack of thorough and operational examination of the transportation challenges of the 15-minute city in relation to urban form, health conditions, economic efficiency, social equity, and environmental sustainability. In this regard, it is of interest to explore whether the Sustainable Development Goals's—and in particular the New Urban Agenda—can be better realised through the presentation of a novel transportation perspective on cities or urban agglomerations that is inspired by the 15-minute city concept. While most publications on the 15-minute city are concerned with issues pertaining to urban planning, only a few put particular emphasis on the analysis of the urban-form transportation nexus. This volume predominantly addresses the opportunities and bottlenecks for the 15-minute city concept from a combined, evidence-based transportation-land use perspective, in relation to modal choice, traffic speed, activity radius, urban infrastructure demands, walkability, accessibility of amenities, environmental implications, and health benefits.

3 Overview of New Urban Landscape Challenges

The present book project aims to provide novel scientific and policy-relevant information with the potential to fill in the many gaps that exist in the research on the concept of the 15-minute city. In addition to intra-city accessibility issues, the volume seeks to offer a broad perspective on the liveability dimensions of the 15-minute city concept by addressing in particular quality of life aspects, land-use patterns, connectivity between urban districts and/or neighbourhoods, energy savings, social coherence and poverty problems, the use of public facilities (e.g. education, health care), and the economic implications for retailing or the real estate sector. A combined interdisciplinary and rigorous effort to explore and explain the implications of the 15-minute city on the urban population and their various, particularly vulnerable, subgroups will be presented as well in this volume. The book aims to present a comprehensive and integrated exploration of urban quality of life by providing pioneering empirical and theoretical perspectives that may both reinforce and oppose each other. Since urban indicators such as pollution, mobility patterns, infrastructure endowment, health outcomes, and transit mode usage cannot be studied without considering the behaviour and choices of individual urban dwellers and particular groups, the book also aims to bring together different theoretical and methodological perspectives to reinforce the improvement of post-pandemic urban liveability and welfare. The volume transcends individual case studies by seeking general sustainability perspectives on the future of the city.

The successful implementation of the 15-minute city model requires a robust conceptual foundation supported by empirical evidence on its acceptability and impact on quality of life. Such evidence-based research is essential for determining the conditions under which residents are willing to accept urban transitions and adjust their daily behaviour. The book addresses, therefore, also existing knowledge gaps by identifying the effect of the 15-minute city concept on the health and well-being of diverse social groups across various urban settings in European cities, as well as in developing countries. By focusing on these inherited urban and mobility challenges, the conducted research provides a practical roadmap for applying the model in heterogeneous global contexts.

To effectively mitigate opposition to urban densification, mixed-use, and significant infrastructure changes such as parking space reductions and road resizing, several chapters in this book also emphasise the importance of understanding governance capacity and encouraging active citizen participation. This participatory approach identifies specific population needs and the level of social acceptability for successful urban reorganisation and regeneration. Furthermore, the book also discusses the potential for resistance to massive transformation by advocating for incremental transition. By providing evidence-based policy recommendations that respect inherited urban structures, the presented research demonstrates possible approaches to facilitate a gradual land-use reform and adjustment toward more flexible and mixed urban functions.

The transition to a 15-minute city necessitates a novel assessment framework that incorporates various dimensions (spatial, sustainable, temporal, economic, and social). These frameworks are designed to support the equitable distribution of opportunities and mobility services to guarantee sociospatial equity. The studies in the book include assessments of service distribution in both high-density neighbourhoods and peripheral zones, with a particular focus on the mobility needs of socioeconomically disadvantaged groups, such as elderly or poor residents.

Furthermore, the presented contributions examine the relationship between urban form and mobility services, and their effect on health, economy, environmental sustainability, and social equity.

Finally, the book also examines the dual-natured role of digitalisation in the implementation of 15-minute city adjustments, noting that while it can support proximity, it also risks undermining it. For example, teleworking may support the proximity dimension, but could potentially encourage spatial dispersion and long-distance living patterns. The book addresses these complexities through observational and case studies, providing some evidence on how digital technologies can affect the objective of the 15-minute city concept. Advanced tools such as artificial intelligence are highlighted for their ability to provide a dynamic analysis of mobility demand and space occupancy. This allows for the identification of underused spaces and a deeper understanding of mobility behaviour, ultimately enabling a phased and data-driven reallocation of urban space as a catalyst for the 15-minute city transition. Various contributions in this book illustrate how these technologies can facilitate a progressive urban transition.

This book is organised into six parts and includes 18 chapters, beginning with this introductory chapter and followed by contributions by multiple authors from different countries. Bringing together conceptual, empirical, and methodological perspectives from diverse urban contexts, the volume examines how proximity-oriented planning can enhance accessibility, foster active mobility, promote social equity, and improve the well-being of urban residents.

The first part, "Conceptual Foundations", examines the 15-minute cites concept, how proximity-based thinking reshapes urban planning theory and how such ideas can be consistently evaluated. Second chapter, "15-Minute Cities in Uncertain Times: Rethinking Mobility, Proximity, and Circularity" by Carlos Moreno, Catherine Gall and Didier Chabaud underscores the emergence of the 15-Minute City model as a transformative framework by introducing the new concept of Proxilience, which emphasises the dynamic and adaptive capacity of the territories to withstand crises while preserving local proximity and quality of life. Third chapter, "Reimagining City Life: Key Concepts and Features of Proximity-Oriented Development" by Georgia Pozoukidou and Dimitra Plastara, explores the shift from mobility- to proximity-oriented planning through text mining and keyword frequency analysis of 277 academic works, proposing the Proximity-Oriented Development (POD) model to foster proximity-based living, where essential services, workplaces, and social hubs are accessible within short distances. Building on this conceptual framework, fourth chapter, "Defining and Measuring the 15-Minute City: Towards Consistent Methodologies for Urban Livability" by Efthymis Papadopoulos and Alexandros Sdoukopoulos, turns to the challenge of defining and measuring the 15-minute city, identifying methodological inconsistencies across studies and outlining directions for a unified framework. Together, these chapters establish the analytical and conceptual basis for understanding and assessing proximity-oriented urbanism.

The second part, "Policy and Public Support", focuses on how social attitudes, urban perceptions, and institutional practices shape the implementation of the

15-minute city. Fifth chapter, "Exploring People's Support of 15-Minute City Policy Implementation and Its Intersection with the Concept's Perceived Benefits and Drawbacks" by Jessica Wei-Lin Lam, Ariel Arosemena and Ehab Diab, investigates what drives citizens to support 15-minute city policies, using survey data from the University of Saskatchewan community in Saskatoon. Through multilevel mixed-effect logit modelling, the study finds that age, income, university position, commute mode, neighbourhood perception, and beliefs about the benefits and drawbacks of 15-minute city policies impact a person's likelihood of supporting 15-minute city policies in their neighbourhood. Extending this attention to public perception, sixth chapter, "Urban Factors and Walkability Perception in Medium-Sized Latin American Cities: The Case of Tepic, Mexico" by Luis Llamas-Rodriguez and Gabriela Ochoa-Covarrubias, assesses how residents' views on walkability align with planners' assessments. Developing a walkability assessment model, the authors show that the perceptions of the community and the knowledge of urban planners mostly coincide. Building on these insights, seventh chapter, "Transport Planning and the 15-Minute City: Opportunities and Challenges" by Clara Loubet, Francesco Bruzzone, Louafi Bouzouina, Cristina Bircu and Silvio Nocera, explores how cities can adapt the 15-minute city concept to improve urban mobility and sustainability. Using Milan as a case study, the authors examine how the city leveraged the COVID-19 crisis to shift its urban development policy toward the 15-minute city model.

The subsequent part, "Social and Spatial Equity", examines how the 15-minute city can promote inclusiveness across demographic and geographic divides. Eighth chapter, "Rethinking the Walkability and Accessibility of the Ageing Society: Milan 15-Minute City for Whom?" by Mina Akhavan, Fulvia Pinto, and Michael Rodrigues, investigates how Milan's proximity-based planning addresses the needs of older adults. Using Geographic Information System (GIS) and isochrone mapping across 88 neighbourhoods, the study underscores the importance of tailoring urban interventions to the unique requirements of different demographics, particularly seniors, to create more inclusive, age-friendly cities. Expanding from demographic to spatial equity, ninth chapter, "Cycling in Low-Density Areas and Outskirts: Enablers and Barriers to Achieve an Equitable 15-Minute City" by Isabel Cunha, assesses bicycle accessibility across the Lyon metropolitan region using the TIRE tool. The analysis reveals that high-density zones, specifically located in the core of the metropolitan region, have disproportionately more bicycle-accessible activities than low-density zones in the peripheral municipalities and discusses equity-oriented measures to improve the cycling conditions in zones with lower accessibility levels and higher concentrations of disadvantaged groups, as well as in low-density neighbourhoods on the outskirts. Shifting the regional lens to Africa, tenth chapter, "Everyday Mobility and the 15-Minute City in Sub-Saharan Africa: Insights from Dakar, Senegal" by Pascal Pochet and Gaële Lesteven, evaluates everyday mobility patterns in Dakar using household travel surveys and interviews. The findings show that while many residents already experience a de facto 15-minute city through constrained mobility, limited access to jobs and services exposes deep social inequalities.

Next, the fourth part, "Land Use and Urban Form", focuses on how the 15-minute city principle reshapes spatial organisation and development strategies within evolving urban systems. Eleventh chapter, "The City of Proximity as an Inspiration for New Concepts of Development of Kraków and Its Agglomeration" by Aleksander Noworól, Piotr Kopyciński, Jeremiasz Salamon, and Paweł Hałat, a proposal for a network of neighbourhood and suburban centres has been created, assuming the transformation of Krakow's highly compact downtown into a more polycentric system. The conclusions of the chapter formulate research postulates for further research, particularly concerning the implications of the polycentric urban system for the development of the metropolitan community. Twelfth chapter, "Chrono-Urbanism as a Driver of More Sustainable Post-Socialist Cities: Case Study Iași, Romania" by Alexandru Bănică, Cristian-Manuel Foșalău, Alexandra Sandu, and Lucian Roșu, extends the discussion to post-socialist urban contexts. Using GIS-based spatial analysis, the authors contrast the dense but polluted central districts of Iași with its greener yet service-poor peripheries. The findings of the chapter point towards viable planning solutions to address these challenges, which are vital for the highly mixed and full of contrasts post-socialist cities.

Fifth part, "Urban Accessibility", explores how proximity-based planning can be evaluated and operationalised through spatial analysis and accessibility metrics. Thirteenth chapter, titled "Assessing spatially the "15-minute city" in Spanish cities: the case of Granada" by Carme Miralles-Guash, Laia Mojica, Carmen Lizárraga and Alejandro Luis Grindlay applies a detailed spatial assessment to Spanish cities. The chapter provides a series of reflective insights and recommendations based on the findings that demonstrate the uneven distribution of everyday amenities and provide evidence to support more equitable proximity-based planning interventions. Building on a quantitative perspective, Nolwenn Sourimant, Peter Nijkamp, Karima Kourtit, and Louafi Bouzouina, in their chapter titled "Enhancing urban mobility: Analysis of the 15-minute city for an accessible and eco-friendly urban environment in Utrecht", analyse the duality of accessibility and quality of cycling in Utrecht, revealing spatial contrasts between central and peripheral districts and proposing a decision-support framework to address existing gaps. Extending the discussion to the health domain in fifteenth chapter titled "The 15-Minute City Concept and Public Health: Quantifying the Link between Pedestrian Accessibility and Population Wellbeing in Maastricht", Clémentine Ravereau, Karima Kourtit, Louafi Bouzouina and Peter Nijkamp use Maastricht as a case study to quantify how pedestrian accessibility correlates with physical and mental well-being, underscoring the public health value of walkable urban environments.

Finally, sixth part, "Mobility and the Digital City", examines how digitalisation reshapes proximity and mobility in contemporary urban life. Sixteenth chapter by Benjamin Motte Baumvol titled "Telework, Mobility, and the Paradox of Proximity in the 15-Minute City" highlights how telework challenges the traditional spatial logic of the 15-minute city: while it can cut commuting frequency, its environmental benefits depend on intensity and context, requiring policies that integrate digital flexibility with sustainable mobility and social equity. Seventeenth chapter titled "Teleworking and the 15-Minute City: Rethinking Lifestyles Between Proximity and Hypermobility" by Laura Hostettler Macias, Emmanuel Ravalet, and Patrick Rérat extends this reflection by exploring how teleworking reshapes the geography of lifestyles, examining both its potential to reinforce proximity-oriented urban models and the spatial complexities emerging from diverse teleworking practices impacting (im)mobility. Finally, in eighteenth chapter titled "Parking Availability Prediction using Deep Learning approaches: Temporal Convolutional Networks (TCN) and TimesNet", Hana Gharrad, Satria Bagus Wicaksonoa, Ansar Yasara, Iñaki Cejudo Fresnadillo, Josep Maria Salanova Grau and Evdokimos Konstantinidis explore how deep learning models can improve parking availability prediction, comparing the performance of two deep learning models for the prediction of parking availability based only on historical occupancy data.

This volume of varied and novel contributions from several authors seeks to highlight the significance of the 15-minute city concept for contemporaneous urban planning from a quality-of-life, mobility, and proximity perspective, based on place-based planning experiences and viewpoints from different cases from around the world. The various chapters in this book clearly demonstrate the powerful potential of this new idea, both conceptually and operationally. The successive studies included in this volume bring new views on the role of cities in human-oriented liveability, wellbeing, accessibility and climate change debates, both locally and globally, into the spotlight. Particular beacons for effective urban sustainability policy distilled from the various parts and chapters in this volume will now be presented concisely.

Each part introduces novel approaches into the extant literature. For example, first part introduces "Proxilience" which links proximity with territorial resilience, and proposes the Proximity-Oriented Development (POD) model grounded in large-scale text analysis. These contributions enable this section to advance an analytical and unified framework for consistently defining and measuring 15-minute cities. Second part identifies the drivers of public support, elaborating on the alignment of the perspectives of both residents and planners, and analyses policy strategies for implementing 15-minute cities. Third part highlights subsequent planning with respect to age and demographic attributes, and reveals spatial inequities in access to mobility and cycling, while uncovering how everyday mobility constraints reveal social inequalities in 15-minute city contexts. Fourth part addresses polycentric urban forms that link suburban and neighbourhood centres. It also proposes chrono-urbanism to guide sustainable planning in the context of post-Soviet cities. Then, fifth part demonstrates how spatial analysis and accessibility metrics can be used to assess progress toward a 15-minute city form while linking proximity and walkability to public health outcomes. Finally, sixth part shows how remote working is related to mobility patterns in a 15-minute city framework from a critical perspective. The section also presents a novel approach by implementing big data AI methods to analyse urban parking availability.

It goes without saying that the 15-minute city is not the only recipe for sound and liveable urban planning. It forms a complement to previous urban planning concepts and fits well into earlier urban planning and architecture ideas, like the garden city, the green city, the smart city, the happy city, the compact city, the polycentric city, the walkable city, or the liveable city. The main difference of the 15-minute city with earlier city planning concepts is that this notion has a normative and measurable target (i.e., a 15-minute action radius for daily activities), which makes this concept very suitable for concrete planning proposals and empirical tests on the achievements of quality-of-life goals of cities. In addition, its flexibility (in the form of X-minute cities and urban regions) makes it also appealing for more generic spatial planning applications in other contexts.

In conclusion, this volume brings together a variety of original perspectives on the 15-minute city, spanning conceptual frameworks, policy support, social equity, land use, accessibility, and digital mobility. By integrating theoretical insights, empirical analyses, and practical case studies from cities across the world, the chapters collectively illustrate both the opportunities and challenges of creating proximity-oriented, inclusive, and sustainable urban environments. The book highlights that realising the 15-minute city requires not only careful spatial planning but also attention to social, technological, and behavioural dimensions, providing a comprehensive foundation for scholars, planners, and policymakers committed to enhancing urban liveability.

References

Abbiasov, T., Heine, C., Sabouri, S., Salazar-Miranda, A., Santi, P., Glaeser, E., & Ratti, C. (2024). The 15-minute city quantified using human mobility data. *Nature Human Behaviour, 8*, 445–455. https://doi.org/10.1038/s41562-023-01770-y

Bouzouina, L., Kourtit, K., & Nijkamp, P. (2024). Modeling active mobility choices in urban spaces: A preface. *Journal of Transport and Land Use, 17*(1), 849–854. https://doi.org/10.5198/jtlu.2024.2620

Bouzouina, L. (2008). *Ségrégation Spatiale et Dynamiques Métropolitaines*. Thèse de Doctorat en Economie, Université Lumière Lyon, 2, 326.

Celbiş, M. G., & Bouzouina, L. (2025). To what extent walking and biking are substitutes or complements to public transport? Interpretable machine learning findings from the University of Lyon, France. *Journal of Transport Geography, 123*, 104071. https://doi.org/10.1016/j.jtrangeo.2024.104071

Fusco Girard, L., Kourtit, K., & Nijkamp, P. (Eds.). (2023). *The future of livable cities*. Springer.

Kourtit, K. (2019). *The new urban world – Economic-geographical studies on the performance of urban systems*. Shaker Verlag.

Moreno, C. (2023). *The 15-minute city: A solution for saving on time and our planet*. Wiley.

Moreno, C., Allam, Z., Chabaud, D., Gall, C., & Pratlong, F. (2021a). Introducing the "15-Minute City": Sustainability, resilience and place identity in future post-pandemic cities. *Smart Cities, 4*, 93–111. https://doi.org/10.3390/smartcities4010006

Moreno, C., Aguilar, A., Guerra, E., & Ortiz, R. (2021b). The 15-minute city: A conceptual analysis. *Sustainability, 13*(2), 498. https://doi.org/10.3390/su13020498

Noworól, A., Kopycinski, P., Halat, P., Salamon, J., & Holuj, A. (2022). The 15-minute city – The geographical proximity of services in Krakow. *Sustainability, 14*(12), 7103.

Zhang, S., Zhen, F., Kong, Y., Lobsang, T., & Zou, S. (2023). Towards a 15-minute city: A network-based evaluation framework. *Environment and Planning D, 50*(2), 500–514. https://doi.org/10.1177/23998083221111857

Conceptual Foundations

15-Minute Cities in Uncertain Times: Rethinking Mobility, Proximity, and Circularity

Carlos Moreno, Catherine Gall, and Didier Chabaud

1 Global Orientation

The 15-minute city concept has gained significant popularity in urban planning studies, and is increasingly playing a critical role in global sustainability studies and discussions. We will start our review with a global orientation.

The Intergovernmental Panel on Climate Change (IPCC, 2024) announced its decision to produce a special report on climate change and cities during its 61st plenary session, which took place in Sofia, Bulgaria, from July 27 to August 2, 2024. This report, scheduled for publication in March 2027, aims to thoroughly examine the intricate links between urbanization, climate change, and sustainable development. The IPCC recognizes the growing importance of cities for addressing climate change, both in terms of their vulnerability to its impacts and their significant potential for action.

The World Health Organization (WHO, 2024) has held its Second Global Conference on Air Pollution and Health in Cartagena, Colombia, from March 25 to 27, 2025. The event aims to address the health impacts of air pollution, promote clean energy solutions, and foster global collaboration to mitigate environmental health risks. It is an ecological imperative but a critical public health priority, as urban areas are home to most of the world's population and bear the brunt of toxic air exposure.

The COP16 on Biodiversity, held in Cali, Colombia (October 21–November 2, 2024), marked a significant milestone in adopting the "One Health" approach. Parties approved a Global Action Plan on Biodiversity and Health, emphasizing the deep interconnection between ecosystem, animal, and human health (COP16,

C. Moreno (✉) · C. Gall · D. Chabaud
ETI Chair - IAE Paris-Sorbonne, University Paris1 Panthéon-Sorbonne, Paris, France
e-mail: carlos.moreno@univ-paris1.fr

L. Bouzouina et al. (eds.), *15-Minute Cities*, Footprints of Regional Science,
https://doi.org/10.1007/978-3-032-20840-8_2

2024). The plan focuses on preventing zoonotic diseases, reducing noncommunicable diseases, and promoting sustainable ecosystems. It highlights the urgency of addressing shared threats such as deforestation, pollution, and climate change, all intensified in urban areas. Notably, the declaration calls for collaboration between health professionals, ecologists, and policymakers, advocating for integrating biodiversity and health considerations into urban planning and policy frameworks. This underscores cities' pivotal role as vulnerable hotspots and potential leaders in implementing health-conscious, biodiversity-friendly strategies for sustainable urban development.

The Call for Action from the 12th World Urban Forum in Cairo (WUF, 2024a) emphasizes that *"A holistic approach to the social, ecological, and economic functions of land, along with the continuum of land rights, is essential. Placing proximity and circularity at the center of urban development is necessary to ensure efficient resource consumption, environmental sustainability, economic prosperity, and social equity."* This statement underscores the importance of integrated urban planning strategies prioritizing proximity, enabling access to essential services and opportunities, and circularity, promoting resource efficiency and environmental responsibility. These principles aim to create resilient, inclusive, and sustainable cities that address contemporary urban challenges.

2 Navigating Uncertainty, Adaptation, and Transformation in Complex Cities

Nijkamp et al. (2024) emphasize that uncertainty and shocks are intrinsic to regional and urban development. Many interconnected factors shape these disruptions, including economic volatility, environmental crises, sociopolitical instability, and rapid technological advancements. These disruptions are not isolated events but systemic phenomena that ripple across spatial, temporal, and sectoral dimensions, amplifying their impact on urban ecosystems. As cities become increasingly complex and interdependent, the capacity to anticipate, absorb, and adapt to such shocks becomes a critical determinant of their long-term sustainability and livability.

In this regard, adaptive and resilient strategies are not merely reactive measures but proactive frameworks that enable cities and regions to remain agile in the face of uncertainty. Adaptation involves continuously re-evaluating spatial configurations, infrastructure systems, and governance mechanisms to address emerging risks and opportunities. Resilience, on the other hand, extends beyond mere recovery; it refers to a city's ability to transform itself in response to disruptions, building more substantial and more equitable foundations for future growth.

Moreover, spatial transformations are inherently dynamic and context-specific, influenced by local geography, governance structures, cultural values, and resource availability. By contrast, conventional urban planning approaches, often characterized by rigidity and siloed thinking, are insufficient to address the fluidity and complexity of contemporary urban challenges. We advocate for integrated, cross-disciplinary approaches that combine insights from urban economics, environmental science, social geography, and public policy to develop holistic solutions tailored to diverse urban contexts.

At the heart of these strategies lies the need for spatial flexibility and systemic foresight. Cities must develop planning tools and methodologies capable of anticipating potential disruptions, whether from climate change, pandemics, migration patterns, or digital transformations. This requires investment in robust infrastructure, adaptive land-use policies, and a cultural shift towards governance models prioritizing experimentation, iterative learning, and citizen participation.

Recognizing spatial inequalities and vulnerabilities as central challenges in building adaptive and resilient cities is crucial. Shocks often disproportionately affect marginalized communities, exacerbating preexisting social and economic disparities. Therefore, resilience strategies must go beyond physical infrastructure and encompass social and economic dimensions, ensuring all population segments have equal access to resources, opportunities, and decision-making processes.

Global crises, such as climate change, economic instability, and public health emergencies, have amplified these challenges and underscored the urgency of innovative urban models that can balance sustainability, social equity, and resilience.

3 The Emergence of the 15-Minute City Concept

In this context, the 15-minute city model emerges as a transformative framework supported by interdisciplinary contributions from urban studies, climate science, environmental research, economics, and social sciences (see, e.g., Moreno, 2021, 2023; Moreno et al., 2021; Noworól et al., 2022; Zhang et al., 2023). Each discipline offers a unique perspective on how proximity, accessibility, and spatial equity can be leveraged to create cities that are functional and resilient to systemic shocks. For instance, studies have consistently demonstrated that reducing motorized travel can significantly decrease greenhouse gas emissions and improve urban air quality (Brand et al., 2021). The shift toward active mobility—such as walking and cycling—also addresses public health concerns by encouraging physical activity and reducing sedentary lifestyles.

Moreover, increased accessibility to healthcare services, green spaces, and recreational facilities is pivotal in enhancing physical and mental well-being (WHO, 2021). Proximity to essential services fosters healthier lifestyles, reduces stress related to long commutes, and strengthens community bonds, contributing directly to the objectives of the New Urban Agenda (UN-Habitat, 2022a).

However, successfully adopting 15-minute city principles requires more than spatial reorganization; it necessitates integrated policies, strategic investments, and cross-sectoral collaboration. Investments in green infrastructure, including urban parks, permeable surfaces, and biodiversity corridors, are essential for environmental sustainability and for creating inclusive spaces that cater to diverse urban populations. Similarly, establishing robust active mobility systems—such as pedestrian zones, cycling networks, and seamless public transportation connections—must be prioritized to reduce car dependency and enhance urban connectivity.

Local governments and urban planners are central to this transformation. Their role extends beyond infrastructure delivery to fostering participatory governance frameworks involving local communities in the design and planning processes. Collaborative approaches ensure that urban spaces are designed for residents and *with* them, reflecting diverse needs, priorities, and cultural contexts (UN-Habitat, 2022b). Inclusive governance mechanisms, transparent decision-making processes, and continuous stakeholder engagement prevent social exclusion and unintended consequences, such as green gentrification or unequal access to urban services.

Additionally, technological innovations and data-driven urban management tools offer opportunities to monitor and evaluate the impact of 15-minute city initiatives in real time. Innovative city technologies can enhance service delivery efficiency, optimize mobility networks, and ensure resource allocation aligns with urban sustainability goals.

The 15-minute city model, when supported by adaptive, resilient strategies and a multidisciplinary approach, holds significant promise in addressing contemporary urban challenges. However, its successful realization depends on an integrated governance structure, equitable resource distribution, and proactive community engagement. As cities around the world continue to face uncertainty and shocks, the lessons drawn from these strategies provide a blueprint for building urban environments that are not only accessible and efficient but also resilient, inclusive, and sustainable.

4 A Brief Exploration of the Origins and Transformations of Territorial Proximity

The scientific understanding of proximity has continually evolved, shaped by technological innovations, social transformations, and economic changes. A brief exploration of the origins and transformations of territorial proximity offers valuable insights into the challenges and opportunities posed by contemporary paradigms, such as the 15-minute city, which is the subject of this book. These reflections provide a crucial foundation for reimagining our relationship with space and proximity to address the current and future needs.

In the preindustrial era, societies relied heavily on nature to survive, organizing themselves around local resources such as water and fertile land. In his seminal work *The Urban Revolution* (Childe, 1950), one of the most influential archaeologists of the twentieth century, Gordon Childe, highlights the idea of an "urban revolution," a transformative shift in production and social organization that underscores the importance of proximity in societies closely linked to the rhythms of nature.

Mesopotamia, often called the cradle of civilization, provides some of the earliest examples of organized urban life. In Ur, founded around 3800 BCE in present-day Iraq, we find evidence of planned infrastructure, including ziggurats, irrigation systems, and trade networks that connected the city to neighboring regions. Uruk, another prominent city in the region, is often recognized as the world's first proper city, with a population exceeding 50,000 by 3000 BCE and innovations such as cuneiform writing and monumental architecture.

These cities exemplified a societal shift from nomadic lifestyles to settled communities reliant on proximity to water resources like the Tigris and Euphrates rivers. They laid the foundation for governance structures, economic specialization, and cultural expressions that would influence urban organization for millennia.

More recently, the Amazon region—today considered one of the planet's lungs due to its critical bio-system—has been the subject of groundbreaking discoveries. In 2022, an international team of researchers led by French archaeologist Stéphen Rostain, under the auspices of the National Centre for Scientific Research, revealed the presence of a vast network of interconnected cities dating back approximately 2000 years (Rostain et al., 2024). These findings showcased complex urban planning, monumental architecture, and advanced socioenvironmental systems, reshaping our understanding of pre-Columbian civilizations in this region.

The eighteenth and nineteenth centuries Industrial Revolution transformed production, society, and urban organization, driving rural populations to cities for industrial jobs. Urban centers were densified around factories, with challenges like pollution and poor housing. In response, urbanists proposed solutions such as Ebenezer Howard's Garden City, which integrated green spaces with industrial zones. Berg and Hudson (1992) highlight how industrialization altered social proximity, making urban relationships more anonymous while centralizing the economy around industries and reshaping employment and professional dynamics.

In the twentieth century, the automobile emerged as a transformative force in urban planning and development, profoundly reshaping the spatial organization of cities and redefining the concept of distance (Mumford, 1961). The widespread adoption of personal vehicles enabled significant suburban expansion, allowing individuals to live farther from their workplaces while maintaining access to urban centers (Fishman, 1987). This suburbanization was often characterized by low-density, car-dependent communities, offering residential tranquility paired with city life's economic and cultural opportunities (Jackson, 1985).

However, this increased reliance on automobiles brought unintended consequences (Norton, 2011). The prioritization of road infrastructure and highways often came at the expense of public spaces and pedestrian-friendly environments (Jacobs, 1961). Cities became fragmented, with physical barriers such as highways dividing neighborhoods and disrupting traditional community ties (Appleyard, 1981). This fragmentation weakened social cohesion and contributed to spatial and economic inequalities, as wealthier populations moved to well-connected suburban areas. At the same time, marginalized communities remained in less accessible urban zones (Harvey, 1996).

Furthermore, car dependency exacerbated environmental challenges, including air pollution, greenhouse gas emissions, and urban heat island effects (Newman & Kenworthy, 1999). The sprawling urban form also increased the cost of public services, such as water, electricity, and waste management, due to the dispersed nature of suburban developments (Bruegmann, 2005).

Urban planners and scholars have advocated for alternative urban models in response to these issues. Jane Jacobs (1961) emphasized the importance of mixed-use neighborhoods and human-scale urban design to restore social cohesion and reduce automobile dependency.

Several urban thinkers and movements emerged to propose alternative paradigms for city development. Christopher Alexander's *"human-scale cities"* (Alexander et al., 1977) emphasized the importance of creating urban environments prioritizing human needs over automobile-centric design. Alexander argued for neighborhoods designed around natural patterns of human interaction, fostering community ties and walkable spaces. Similarly, concepts such as the *compact city* (Neuman, 2005) and *transit-oriented development* (Cervero & Kockelman, 1997) have been proposed to counteract suburban sprawl and prioritize walkability, public transit, and localized economies. Likewise, Jan Gehl's *"Cities for People"* (Gehl, 2010) advocated a shift away from car-dominated streets to cities designed at a pedestrian and cyclist scale. Gehl emphasized the importance of public spaces that invite social interaction, physical activity, and vibrant urban life. Both Alexander and Gehl's ideas reinforced the importance of human-centric design in counteracting the harmful effects of automobile dominance.

Another influential response came from the *New Urbanism* movement, which gained momentum in the late twentieth century. Spearheaded by architects and planners such as Duany et al. (1991), New Urbanism promoted compact, walkable, and mixed-use neighborhoods. Inspired by preautomobile cities, the movement sought to reduce sprawl, reconnect communities, and create spaces where daily needs could be met without car dependency. The Charter of the New Urbanism (1996) outlined principles for designing more sustainable, livable cities by focusing on walkability, transit accessibility, and architectural diversity.

5 The Rise of the 15-Minute City

The late twentieth and early twenty-first centuries, marked by the digital revolution, transformed interactions and economies through advances in information and communication technologies. The COVID-19 pandemic accelerated the adoption of telework, reshaping home-office dynamics and promoting hybrid spaces like coworking hubs (OECD, 2021). Digital tools also blur urban-rural divides, but challenges like digital inequality, privacy concerns, and impacts on social connections persist.

The 15-minute city interconnects two fundamental aspects of urban life that are considered deeply interdependent: sustainable mobility and the optimized use of urban space. These two dimensions are not isolated variables but rather two sides of the same coin, shaping how urban systems function, how citizens interact with their environments, and how cities can respond to the pressing challenges of the twenty-first century, including climate change, social inequality, and public health crises. Together, they form the backbone of a vision for cities that are efficient in their spatial and infrastructural organization, inclusive in their social outcomes, and livable in their everyday experience.

The first dimension, sustainable mobility, challenges the deeply ingrained reliance on car-centric urban planning that has dominated cities for much of the twentieth century. Research has repeatedly shown that excessive car dependency exacerbates a range of urban problems, including traffic congestion, air pollution, greenhouse gas emissions, and sedentary lifestyles (Desjardins, 2017; Newman & Kenworthy, 2015). By prioritizing walking, cycling, and efficient public transportation, the 15-minute city seeks to redefine the relationship between citizens and urban mobility, emphasizing human-scale transport systems rather than large-scale automotive infrastructure. This approach reduces the environmental footprint of transportation and enhances urban equity, as active mobility modes are generally more affordable and accessible to a larger segment of the population (Huré, 2017). Additionally, sustainable mobility directly influences public health outcomes. Numerous studies highlight the positive correlation between active mobility and improved physical and mental health, underscoring the multifaceted benefits of reducing car reliance (Lau et al., 2021).

However, sustainable mobility cannot function effectively without optimized urban space, the second key dimension of the 15-minute city. In traditional urban models, monofunctional zoning has often characterized land use, where residential, commercial, industrial, and recreational spaces are spatially segregated. This approach has historically reinforced spatial fragmentation, increasing reliance on motorized transport to bridge the physical distances between different urban functions (Jacobs, 1961; Fijalkow, 2018). The 15-minute city, by contrast, advocates for multifunctional urban spaces that blend residential, commercial, cultural, and recreational activities within the same neighborhoods. This principle is not merely about proximity but also about spatial diversity—ensuring that neighborhoods are vibrant, flexible, and capable of adapting to diverse social and economic needs over time (Gehl, 2010; Yurrita et al., 2021).

Furthermore, the optimized use of urban space goes hand in hand with spatial justice, ensuring that improvements in public infrastructure and services do not inadvertently displace vulnerable populations or exacerbate existing inequalities (Florida, 2017; Harvey, 2012). Gentrification remains a persistent risk in implementing spatial optimization policies, as enhanced accessibility and improved urban environments often attract higher-income residents, driving up property values and displacing lower-income communities (Raimbault, 2021). Therefore, the success of spatial optimization relies heavily on governance mechanisms prioritizing social equity and community participation in decision-making processes.

The interdependence between sustainable mobility and the optimized use of urban space becomes particularly evident when considering how they mutually reinforce each other. Effective sustainable mobility systems depend on spatial planning prioritizing proximity and accessibility. At the same time, optimized urban spaces only reach their full potential when supported by mobility networks that allow people to move efficiently within and between neighborhoods (Batty, 2018). Agent-based simulations of urban dynamics, such as those conducted by Yurrita et al. (2021), illustrate how interventions in one domain inevitably affect the other. For example, introducing a well-connected cycling network in a dense, mixed-use urban area can drastically reduce traffic congestion while enhancing the attractiveness of public spaces and increasing foot traffic for local businesses.

Moreover, digital technologies are emerging as key enablers of this interdependence. Integrating smart mobility systems, such as shared mobility platforms, real-time traffic data, and adaptive public transport schedules, bridges the gap between sustainable transport and urban space optimization (Raimbault, 2021). At the same time, geospatial analytics and urban data platforms provide policymakers with the tools to understand mobility patterns and land-use dynamics at unprecedented levels of granularity, enabling evidence-based decision-making that aligns with the principles of the 15-minute city.

6 Integrated Mobility and Participatory Governance and Urban Proximity

Implementing this dual framework—where mobility and spatial organization are treated as interconnected rather than siloed concerns—requires a multidisciplinary approach. Urban planners, transportation engineers, digital technology specialists, and social scientists must collaborate to design integrated strategies considering environmental sustainability, social inclusion, and economic feasibility (Desjardins, 2017; Huré, 2017). Furthermore, the active involvement of citizens in urban governance processes is essential to ensure that proposed solutions align with local needs and cultural contexts (Batty, 2018; Lau et al., 2021). The convergence of participatory governance and proximity represents a transformative approach to urban planning, offering substantial potential for creating sustainable, resilient, and human-centered cities. The success largely depends on the capacity of policymakers and urban planners to adapt it to local realities and address implementation challenges. Staricco (2022) highlights the importance of integrating citizen participation into the design and decision-making processes to ensure urban services reflect local needs. Participation is not merely consultative but must involve cocreation and shared responsibility. Similarly, Murgante et al. (2024) emphasize the critical role of localized governance structures that mediate between top-down policies and bottom-up citizen initiatives. This dual-level governance is essential to avoid mismatches between policy design and on-the-ground realities. Furthermore, proximity strengthens neighborhood ties, enhances social cohesion, and contributes to a greater sense of belonging (Mouratidis, 2018). Shokeen and Kaur (2024) highlight the importance of considering social interaction as a fundamental element of urban design, advocating for planning strategies that prioritize creating and maintaining social spaces to enhance the social sustainability of urban environments.

However, the equitable distribution of proximity remains a significant concern. Marquet et al. (2024) warn about the risk of gentrification associated with implementing the 15-minute city concept. The authors emphasize that the concentration of urban improvements in specific areas, already under significant real estate pressure, can make these spaces the exclusive domain of urban elites and investors. This disparity often reflects broader structural inequalities within urban systems. Scholars such as Anguelovski et al. (2022) have also raised alarms about "green gentrification" where improvements in urban livability inadvertently drive up housing costs, displacing vulnerable populations. These risks necessitate proactive antidisplacement strategies integrated into the implementation phase of the 15-minute city.

Adaptability emerges as another key factor in the academic discourse surrounding the 15-minute city. Moreno et al. (2021) stress that while the model's core principles are universally applicable, their practical realization is deeply contingent on local socioeconomic and spatial contexts. Markuske's comparative study (Markuske, 2024) of planning policies in Ottawa, Melbourne, and Portland reveals diverse approaches to applying the 15-minute city concept in suburban and semidense areas. The research finds that while cities adapt the concept to their unique contexts, many planning documents prioritize city centers over peripheries. Common strategies include improving public transit and cycling infrastructure, exploring mixed-use zoning in traditionally residential suburbs, and repurposing existing structures to create local amenities. The study highlights the challenges of balancing increased density with maintaining suburban character and emphasizes the importance of community engagement in successful implementation. Markuske concludes that while the 15-minute city concept shows promise for transforming urban peripheries, significant challenges remain in adapting it to less dense areas, necessitating tailored approaches considering local context and community needs.

7 Towards the x-Minute Urban Agglomeration

Büttner et al. (2024) provide an in-depth analysis of the global implementation of the 15-minute city concept. The study highlights key findings, including the growing geographic adoption of the model across diverse regions, terminological variations such as "20-minute neighborhoods" or "10-minute communities" adapted to local contexts, and a significant acceleration of initiatives since 2016, peaking in 2020–2021. Key objectives include promoting walkability, reducing car dependency, mitigating air pollution, and enhancing urban quality of life. However, the study also identifies challenges, such as adapting the concept to varying urban contexts and the need for integrated planning approaches. This research offers a valuable overview of the current state and emerging trends in the global adoption of the 15-minute city model and provides a detailed overview of various European cities' diverse strategies and policies to implement the 15-minute city concept. Through a comparative analysis of 100 different projects, the research examines how cities tailor the model to their specific local contexts, shedding light on the challenges and opportunities encountered during this transformative urban process. Each case study reveals unique adaptations, reflecting varying socioeconomic, cultural, and spatial realities while highlighting common trends, such as prioritizing sustainable mobility, equitable access to services, and fostering vibrant, inclusive neighborhoods.

Successfully implementing complex urban concepts like the 15-minute city demands a new form of governance. The Council of Europe's (2023) *Report on Multi-Level Governance* offers valuable insights into this issue. It explores the potential benefits of governance approaches that involve multiple levels of authority—local, regional, and national—while identifying key elements necessary for their success. Among these are transparency, stakeholder collaboration, and clearly defined roles and responsibilities across different levels of governance.

The report also emphasizes the inherent challenges of coordinating diverse actors, agencies, and jurisdictions—a critical factor for delivering ambitious urban projects like the 15-minute city. Fragmented governance structures, conflicting priorities, and resource disparities often hinder the seamless execution of urban policies. Therefore, the report advocates for integrated governance frameworks that are adaptable, inclusive, and capable of balancing top-down policy directives with bottom-up citizen engagement.

Together, these findings underscore the need for strategic alignment between urban planning objectives and governance mechanisms, ensuring that the principles of the 15-minute city—proximity, accessibility, and sustainability—are effectively embedded into urban systems, as is also illustrated in various scientific contributions.

In a study published by Allam et al. (2024), the evolution of the 15-Minute City concept reveals key trends and thematic focuses that define its global adoption. Bibliometric analysis highlights five recurring keywords—accessibility, walkability, mobility, sustainability, and proximity—as central to academic discussions. Accessibility emphasizes ease of access to essential services and opportunities, while walkability focuses on creating safe and user-friendly pedestrian environments. Mobility extends to efficient transport systems that reduce car dependency and foster healthier, more connected communities. Sustainability is an overarching theme, aligning urban planning practices with environmental, social, and economic objectives. Proximity ties these elements together, promoting mixed-use development, social interaction, and equitable service distribution.

In parallel, the historical timeline of 15-minute City initiatives reveals a significant growth trajectory. Initially sporadic, the concept gained traction after 2014, experiencing a sharp acceleration from 2019 to 2022, which continues today. This surge reflects growing global awareness of the concept's potential benefits for urban resilience and sustainability. Implementation varies across cities, with diverse approaches adapted to local socio-economic and spatial contexts. Despite progress, challenges like governance coordination, funding limitations, and sociospatial inequalities remain critical barriers.

The thematic and temporal analyses underscore the importance of integrated planning strategies, multilevel governance frameworks, and community engagement in ensuring the practical realization of 15-Minute City principles in diverse urban contexts.

8 Proximity and Resilience: Proxilience as a New Departure

In Nijkamp et al. (2024), "prosilience" is described as the capacity of regions and territories to adapt and remain resilient in the face of challenges and uncertainties, drawing on the principles of the New Economic Geography proposed by Krugman (1991) and adapting them to the evolving conditions of contemporary spatial economics. Building on this research, the concept of *Proxilience* (WUF12, 2024b)—prosilience as applied to proximity—enriches the 15-Minute City model by introducing a dynamic and adaptive dimension to the notion of proximity. While the 15-minute city focuses on local accessibility, Proxilience emphasizes the importance of cities being able to withstand crises while preserving the quality of life promised by proximity.

At the Global Forum on Human Settlements (GFHS, 2024), held at the United Nations headquarters in New York, Carlos Moreno described "proxilience" as a resilience strategy based on polycentric proximity that can create more livable and prosperous cities in the face of climate emergencies, economic crises, and social tensions in our increasingly urbanized world. This approach opens new avenues for modeling resilient cities and territories capable of addressing the increasingly frequent and severe disruptions observed in urban environments.

References

Alexander, C., Ishikawa, S., & Silverstein, M. (1977). *A pattern language: Towns, buildings, construction*. Oxford University Press.

Allam, Z., Khavarian-G, R., Lassaube, U., Chabaud, D., & Moreno, C. (2024). Mapping the implementation practices of the 15-minute city. *Smart Cities, 7*(4), 2094–2109. https://doi.org/10.3390/smartcities7040083

Anguelovski, I., Connolly, J. J., Cole, H., Garcia-Lamarca, M., & Triguero-Mas, M. (2022). Green gentrification in European and north American cities. *Nature Communications, 13*(1), 3816.

Appleyard, D. (1981). *Livable streets*. University of California Press.

Batty, M. (2018). *Inventing future cities*. MIT Press.

Berg, M., & Hudson, P. (1992). Rehabilitating the industrial revolution. *The Economic History Review, 45*(1), 24–50.

Brand, C., Dekker, H.-J., & Behrendt, F. (2021). Cycling, climate change and air pollution. *Advances in Transport Policy and Planning*.

Bruegmann, R. (2005). *Sprawl: A compact history*. University of Chicago Press.

Büttner, B., Seisenberger, S., McCormick, B., Silva, C., Papa, E., Cao, M., & Teixeira, J. F. (2024). *Mapping of 15-minute city practices: Overview on strategies, policies and implementation in Europe and beyond*. DUT Driving Urban Transition.

Cervero, R., & Kockelman, K. (1997). Travel demand and the 3Ds: Density, diversity, and design. *Transportation Research Part D: Transport and Environment, 2*(3), 199–219.

Childe, G. (1950). The urban revolution. *The Town Planning Review, 21*(1), 3.
Congress for the New Urbanism. (1996). Charter of the new urbanism. https://www.cnu.org/who-we-are/charter-new-urbanism
Convention on Biological Diversity COP 16. (2024). *Biodiversity COP 16: Important agreements reached towards goal of making peace with nature*. United Nations Sustainable Development. https://www.un.org/sustainabledevelopment/blog/2024/11/biodiversity-cop-16-important-agreement-reached-towards-goal-of-making-peace-with-nature-2/
Council of Europe. (2023). Report on multilevel governance. https://rm.coe.int/report-on-multilevel-governance-final-2768-6653-0568-v-1/1680ad9120
Desjardins, X. (2017). *Urbanisme et mobilité. De nouvelles pistes pour l'action*. Editions de La Sorbonne.
Duany, A., Plater-Zyberk, E., & Krieger, A. (1991). *Towns and town-making principles*. Rizzoli and Harvard University Graduate School of Design.
Fijalkow, Y. (2018). Quartiers, rues et espaces publics: éléments pour une histoire des Études Urbaines francophones. *Environnement urbain, 13*.
Fishman, R. (1987). *Bourgeois utopias: The rise and fall of suburbia*. Basic Books.
Florida, R. (2017). *The new urban crisis*. Basic Books.
Gehl, J. (2010). *Cities for people*. Island Press.
Global Forum on Human Settlements (GFHS). (2024, October 24). 19th Annual Session of Global Forum on Human Settlements. United Nations Web TV. https://webtv.un.org/en/asset/k1r/k1r2gb2zms
Harvey, D. (1996). *Justice, nature and the geography of difference*. Blackwell Publishing.
Harvey, D. (2012). *Rebel cities: From the right to the city to the urban revolution*. Verso Books.
Huré, M. (2017). *Les mobilités partagées. Nouveau capitalisme urbain*. Publications de la Sorbonne, Coll. Mobilités et Sociétés.
Intergovernmental Panel on Climate Change (IPCC). (2024). Special report on climate change and cities. https://www.ipcc.ch/report/special-report-on-climate-change-and-cities/
Jackson, K. T. (1985). *Crabgrass frontier: The suburbanization of the United States*. Oxford University Press.
Jacobs, J. (1961). *Death and Life of Great American Cities*. New York: Random House.
Krugman, P. (1991). Increasing returns and economic geography. *Journal of Political Economy, 99*(3), 483–499.
Lau, B. P. L., et al. (2021). The study of urban residential's public space activeness using space-centric approach. *IEEE Internet of Things Journal*.
Markuske, F. (2024). Applying the 15-minute city concept to urban peripheries: A comparative study of planning policies. Paper presented at the AESOP Annual Congress, Paris, France.
Marquet, O., Mojica, L., Fernández-Núñez, M. B., & Maciejewska, M. (2024). Pathways to 15-minute city adoption: Can our understanding of climate policies' acceptability explain the backlash towards x-minute city programs? *Cities, 148*, 104878.
Moreno, C. (2021). Living in proximity in a living city. *Glocalism, 3*. https://riviste.unimi.it/index.php/glocalism/article/view/20593
Moreno, C. (2023). *The 15-minute city: A solution for saving on time and our planet*. John Wiley.
Moreno, C., Allam, Z., Chabaud, D., Gall, C., & Pratlong, F. (2021). Introducing the "15-minute city": Sustainability, resilience and place identity in future post-pandemic cities. *Smart Cities, 4*, 93–111. https://doi.org/10.3390/smartcities4010006
Mouratidis, K. (2018). Built environment and social well-being: How does urban form affect social life and personal relationships? *Cities, 74*, 7–20.
Mumford, L. (1961). *The city in history: Its origins, its transformations, and its prospects*. Harcourt, Brace & World.
Murgante, B., Annunziata, A., Valluzzi, R., & Ferrer-Ortiz, C. (2024). Developing a 15-minute city: Evaluating urban quality using configurational analysis. The case study of Terni and Matera, Italy. *Applied Geography, 103171*.
Neuman, M. (2005). The compact city fallacy. *Journal of Planning Education and Research, 25*(1), 11–26.

Newman, P., & Kenworthy, J. (1999). *Sustainability and cities: Overcoming automobile dependence*. Island Press.

Newman, P., & Kenworthy, J. (2015). The end of automobile dependence: How cities are moving beyond car-based planning.

Nijkamp, P., Kourtit, K., Krugman, P., & Moreno, C. (2024). Old wisdom and the new economic geography: Managing uncertainty in 21st century regional and urban development. *Regional Science Policy & Practice, 16*(10).

Norton, P. (2011). *Fighting traffic? The dawn of the motor age in the American city (inside technology)*. MIT Press.

Noworól, A., Kopycinski, P., Halat, P., Salamon, J., & Holuj, A. (2022). The 15-minute city—The geographical proximity of services in Krakow. *Sustainability, 14*(12), 7103.

Organisation for Economic Co-operation and Development (OECD). (2021). *Teleworking in the COVID-19 pandemic: Trends and prospects*. OECD Publishing. https://doi.org/10.1787/72a416b6-en

Raimbault, J. (2021). Simulating urban dynamics and international governance of transportation infrastructure projects (arXiv preprint arXiv:2108.13915).

Rostain, S., Dorison, A., De Saulieu, G., Prümers, H., Le Pennec, J.-L., Mejia, F., Pagan-Jimenez, J. R., & Desla, P. (2024). Two thousand years of garden urbanism in the Upper Amazon. *Science, 383*(6679), 183–189. https://doi.org/10.1126/science.adi6317

Shokeen, S., & Kaur, R. (2024). Revitalizing urban social fabric through thoughtful street and built environment design. *International Research Journal on Advanced Science Hub, 6*(4).

Staricco, L. (2022). Participatory governance in urban planning: Lessons from the 15-minute city model. *Journal of Urban Affairs, 44*(3), 456–472.

UN-Habitat. (2022a). Action framework for the implementation of the new urban agenda. https://unhabitat.org/sites/default/files/2022/04/afinua_digital_a.pdf

UN-Habitat. (2022b). Our city plans: An incremental and participatory toolbox for urban planning. https://unhabitat.org/our-city-plans-an-incremental-and-participatory-toolbox-for-urban-planning

World Health Organization (WHO). (2021). Urban green spaces and health: A review of evidence. https://iris.who.int/handle/10665/345751

World Health Organization (WHO). (2024). Second global conference on air pollution and health. https://www.who.int/news-room/events/detail/2025/03/25/default-calendar/second-global-conference-on-air-pollution-and-health

World Urban Forum (WUF 12). (2024a). Cairo call to action (Twelfth Session of the World Urban Forum). UN-Habitat. https://unhabitat.org/sites/default/files/2024/11/cairo_call_to_action_wuf12.pdf

World Urban Forum (WUF 12). (2024b, November 11). *Earth negotiations bulletin (Twelfth Session of the World Urban Forum)*. International Institute for Sustainable Development. https://enb.iisd.org/world-urban-forum-wuf12-summary.

Yurrita, M., et al. (2021). Dynamic urban planning: An agent-based model coupling mobility mode and housing choice. In *Intelligent computing: Proceedings of the 2021 computing conference* (Vol. 2).

Zhang, S., Zhen, F., Kong, Y., Lobsang, T., & Zou, S. (2023). Towards a 15-minute city: A network-based evaluation framework. *Environment and Planning D, 50*(2), 500–514.

Reimagining City Life: Key Concepts and Features of Proximity-Oriented Development

Georgia Pozoukidou and Dimitra Plastara

1 Introduction

Nowadays, cities are increasingly confronting a range of challenges, including health, environmental, and economic crises. The lack of prompt and effective management of global shocks at the local level underscores urban areas' vulnerabilities and the structural deficiencies in their organization, operation, and governance. Both the European Environment Agency and the Intergovernmental Panel on Climate Change (IPCC) emphasize the importance of spatial organization in human activities as a key factor in reducing energy consumption and improving the adaptability of urban infrastructure and systems (European Environment Agency, 2016; IPPC, 2022).

Cities are interconnected entities, and urban sustainability depends on a complex network of interdependent subsystems, with varied temporal and spatial requirements (Patnaik, 2021). To build resilience against future crises and ensure a high quality of life for their inhabitants, urban planning approaches must be redefined. Key principles of sustainable urban development include: equitable access to urban services; provision of basic amenities at the neighborhood level; efficient land and resource use; increasing green and open spaces; reducing commuting needs; facilitating digital transitions; and, lastly, recognizing neighborhoods as vital community hubs (European Commission, 2020; Greenfield, 2023; Patnaik, 2021; UN-Habitat, 2018; United Nations, 2015).

Acknowledging cities' limited capacity to address current challenges, the global scientific community is advocating for a new urban model: the 15-minute city (15mC). With benefits for sustainability, health and social cohesion, the 15mC has

G. Pozoukidou (✉) · D. Plastara
School of Spatial Planning and Development, Faculty of Engineering, Aristotle University of Thessaloniki, Thessaloniki, Greece
e-mail: gpozoukid@plandevel.auth.gr

L. Bouzouina et al. (eds.), *15-Minute Cities*, Footprints of Regional Science,
https://doi.org/10.1007/978-3-032-20840-8_3

gained significant attention in international urban agendas and policy documents. The recent United Nations Habitat (United Nations Habitat, 2022) report on the future of cities emphasizes urban's planning regulatory role, addressing issues intensified by the COVID-19 pandemic, such as the vulnerability of socioeconomically disadvantaged groups, the need for green space investment, the benefits of compact city planning, and the complex urban-rural dynamics. The 15mC model is presented as a neighborhood-based approach with mixed land use and compact structure, allowing residents to meet their daily needs within a 15-minute walk, bike ride or transit stop. The report underscores the need to ensure that 15 mC initiatives will not exacerbate spatial inequalities but strengthen urban resilience through a polycentric city form.

The IPCC (Intergovernmental Panel on Climate Change, 2022) report stresses the urgent need for rapid adaptation and mitigation within urban systems to address the climate crisis. It highlights how urban planning and infrastructure can enhance mobility options and local service access, driving sociobehavioral change to reduce cities' climate impact. The report recognizes that implementing the 15mC model could significantly reduce urban carbon footprint by promoting accessible neighborhoods that offer diverse housing, local services, and a healthier lifestyle. The report acknowledges the model's rising popularity as a means to cut per capita travel distances and associated emissions. Additional perspectives are highlighted in report from C40 Cities and ARUP (C40 Cities Climate Leadership Group, 2021), entitled "Green and Prosperous Neighborhoods: A pathway to net zero, featuring the «15-minute city»" which outline the pathway towards the transition to net-zero emissions incorporating people-centric practices. The neighborhood is advocated as the most appropriate spatial unit due to its human scale for the promotion of a healthy lifestyle, as well as a more rational and efficient urban structure. The strategy is built on two pillars: "Green: zero emissions," and "Thriving neighborhoods: resilience and a people-centered approach." The latter encompasses 10 approaches, derived from the study of best practices across C40 cities initiative (2020), which collectively constitute the strategy's guiding principles. These approaches include complete neighborhoods, people-centered mobility, place-based inclusivity, green energy, green and circular economy, green infrastructure, nature-based solutions and sustainable lifestyle.

Similarly, Deloitte's (2021) report "Urban Future with a Purpose" identifies 12 key trends expected to shape cities by 2030. These trends emerge from the need for immediate mitigation strategies for low emission (resilient cities), the integration of big data and technology to enhance urban functionality (smart cities), a people-centered approach to achieve the UN's 17 Sustainable Development Goals, and the development of inclusive communities. Among these trends, the 15-minute city (15 mC) is introduced as a model for local living that supports neighborhood sustainability and resilience. The report emphasizes self-sufficiency in goods and services at the neighborhood level, the shift in work patterns and the resulting impacts on neighborhood and city center dynamics, while also acknowledging potential gentrification challenges. The Driving Urban Transition (DUT) Partnership (2024) position paper "Navigating the 15-minute City" outlines the perspectives regarding the 15 mC at three levels, European, national, and city. Through in-depth research and innovation, the DUT partnership empowers local authorities, municipalities, businesses, and citizens to effectively translate global strategies into actionable local initiatives. Additionally, the paper provides a comprehensive examination of the 15mC concept, elucidating its strategic and thematic integration into various initiatives and refining positions on the 15mC model to guide future policy recommendations within the DUT framework.

Building on these insights and prevailing trends, this research introduces an innovative urban development strategy inspired by the 15-minute city (15 mC), focusing on proximity-based living. The proposed Proximity Oriented Development (POD) model, structured as a "four-helix framework," promotes a shift from mobility- to proximity-centered urban planning. Starting with an exploration of the neighborhood's role as a foundational urban planning unit, the study examines how accessibility shapes urban patterns. By employing a text mining analysis and keyword frequency analysis of 277 academic works, the research captures prevailing priorities, insights, and emerging trends in the 15 mC discourse. Based on this analysis, the POD model and its four essential components are introduced.

2 From "Mobility" to "Proximity" Oriented Urban Development

Historically, urban development policies prioritized mobility infrastructure, enabling widespread access but also promoting car dependency and urban sprawl (Jabareen, 2006; Newman & Kenworthy, 1999). In recent years, however, there has been a renewed focus on neighborhoods as key planning units, valued for their role in facilitating local access to services and amenities. This shift toward proximity-centered planning brings accessibility, sustainability, and community well-being to the forefront of urban planning. Building on this perspective, this section explores the evolution of urban planning, focusing on the transition from car-centric, mobility-driven to neighborhood-centric, proximity-driven urban development.

2.1 The Neighborhood Unit as a Foundational Element of Urban Planning

The neighborhood has historically been a foundational element in the spatial and functional organization of the city, and its importance has been highlighted since the previous century in urban planning theory and practice (Duany & Plater-Zyberk, 2008; Patricios, 2002). Today, the neighborhood is gaining renewed importance as a foundation for urban planning, particularly through the lens of citizen well-being, anthropocentric design and bottom-up planning approaches.

Over the years, as urban planners sought to define the ideal city, the concept of the neighborhood has evolved. Initially viewed simply as a space for essential urban services, the neighborhood is now seen as a complex entity that reflects social values,

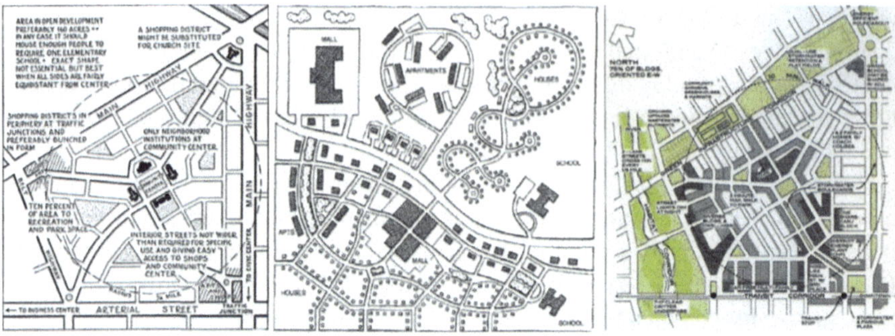

Fig. 1 The neighborhood—Perry (1929), functional separation and urban sprawl and Doug Farr's updated Sustainable Neighborhood. Source: Steuteville (2017)

motivations, and dynamics of its residents (Sanders, 2008). The idea of the neighborhood as a structural element in city planning was first introduced by Clarence Perry. In the metropolitan plan for New York, Perry proposed the neighborhood as a fundamental unit of urban planning, creating a hierarchy of services based on it (Perry, 1929). Perry developed specific design principles to organize the city life in the neighborhood. His model included urban services hubs with school, retail shops, services, a community center, a hierarchical road system, green spaces, public areas, and residential zones (Fig. 1). He also established quantitative guidelines for defining neighborhood boundaries, such as population, area, housing density, and a catchment area of about a ¼ of a mile radius from its center. This early attempt at systematic urban organization through neighborhood-based planning faced significant criticism from the late 1940s through the early 1960s. Critics argued that Perry's model encouraged racial, ethnic, and religious segregation. Others focused on design and methodology, challenging the reliance on quantitative criteria defining neighborhoods, as this approach overlooked the organic, inherent functional characteristics of existing structures and spaces (Bauer, 1945; Dyckman, 1959; Isaacs, 1948).

The post-World War II belief that cities should be radically transformed based on modernist principles of idealism and equality further fueled the criticism of traditional neighborhood-based planning (Kafkalas et al., 2016). Meanwhile, cities were rapidly expanding, with automobiles and buses emerging as the key transportation technologies shaping urban development, particularly in North America and Australia (Newman & Kenworthy, 1999). With the freedom provided by the car and bus to move in any direction, it became possible to expand cities first filling in between train lines (if existed) and then going out far for an average half-hour journey. Low-density housing became more feasible and separation of uses through zoning became the dominant planning practice. A sophisticated system of highways that moves people fast in and out of city centers has shaped urban development patterns. This further accelerated the decentralization of the city and the dispersion of urban activities in an unprecedented way. In this car-centric approach, cities are treated as destinations to be reached as quickly and efficiently as possible from the suburbs, making *mobility the primary focus* of urban planning. This mindset led to urban plans prioritizing seamless travel from point A (origin) to point B (destination), resulting in the widespread construction of highways and freeways through densely populated city centers. Highways, off-street, and multilevel parking facilities dominated the urban landscape, occupying valuable city space and disrupting both the social and economic vitality, with often irreversible effects on community cohesion. The uncontrolled urban sprawl driven by a *car-centric, mobility-oriented approach* has shaped the contemporary cities, leaving them with numerous problems: excessive land consumption, traffic congestion, inadequate transportation infrastructure, selective and limited accessibility, increased pollution, excessive energy consumption, and high greenhouse emissions (Hamidi et al., 2015; Jabareen, 2006; Newman & Kenworthy, 1999; Pozoukidou et al., 2018).

In response to the sprawling, low-density, and "anonymous" urban development of the 1960s, the neighborhood unit concept-originally proposed by Clarence Perry-was adapted to promote sustainable urban growth, in both dense dilapidated urban areas and suburban sprawl. This approach found a strong advocate in the New Urbanism movement, which emphasized physical design, drawing inspiration from early urban thinkers like Howards, Olmsted, and Geddes, who focused on creating cohesive human-centered communities (Kafkalas et al., 2016). New Urbanism reestablished the neighborhood as the core structural element, characterized by a defined size, clear boundaries and a core service area to ensure easy access to daily needs. This model has informed various urban development strategies, including "Transit Villages" (Bernick & Cervero, 1997), "Urban Villages" (Neal, 2003), "Traditional Neighborhood design" (Bothwell et al., 1998), and "Sustainable neighborhood Units" (Sharifi, 2016), all of which use the neighborhood unit to promote higher density, mixed land use, diversity, walkability, and connectivity.

A related approach, *Transit Oriented Development (TOD),* emerged as a popular strategy to limit urban sprawl and revitalize declining urban centers (Cervero et al., 2004). Inspired by the New Urbanism movement, Peter Calthorpe in his book «*The Next American Metropolis*» proposed a new approach to urban planning that promotes walkability and public transit use to reduce car dependency (Calthorpe, 1993; Schlossberg & Brown, 2004). He defined TOD as a mixed-use community within a 600-meter walking radius of a transit hub, such as a metro or commuter rail station, centered around a commercial core (Calthorpe, 1993). Since then, TOD has become an urban planning strategy aimed at shifting away from a car-centric development to a more human-centered approach that *prioritizes public transit accessibility*. TOD organizes urban development to maximize access to public transit and encourage nonmotorized transportation options, such as walking or other forms of micro mobility (Schlossberg & Brown, 2004). Thomas and Bertolini (2017) describe TOD as a planning approach that integrates land use and transportation to make nonmotorized transport and public transit both convenient and appealing. By focusing on development around transit stations, TOD enhances the efficiency of existing public transit and promotes multimodal options. To this end, some researchers propose TOD as a "reorientation" policy for sprawling, low-density areas around rail networks (Bertolini et al., 2012; Schlossberg & Brown, 2004). TOD emphasizes transit hub-centered organization while incorporating neighborhood-friendly features like pedestrian-friendly environments, safe walking routes, and public spaces (Bernick & Cervero, 1997). However, its primary focus remains on *enhancing urban mobility by leveraging public transit* for efficient movement across the city, rather than creating localized hubs of amenities and services.

2.2 Accessibility as an Urban Development Factor

Miller (2018) points out that the primary function of a transportation system—whether car-based or transit-based—is to provide access to various spatially dispersed activities. This capacity for connecting different activities defines accessibility. Hence, accessibility is a crucial feature of urban development, referring to the ease of reaching goods, services, activities, and destinations (Handy, 1993; Ibraeva et al., 2020). Literature suggests that from the 1940s onward, both car-centric development and the TOD approach primarily achieved *accessibility through mobility.*

Today, accessibility is a cornerstone of sustainable urban mobility policies, which aim to reduce reliance on private cars, encourage public nonmotorized transport, and ultimately improve transportation's environmental impact. Research also underscores that adequate accessibility promotes social inclusion and well-being (De Vos et al., 2014; Preston & Rajé, 2007; van Wee, 2016). Hence, transportation planning increasingly prioritizes accessibility goals over simply expanding mobility options (Boisjoly & El-Geneidy, 2017; European Environment Agency, 2016; Proffitt et al., 2019).

Grengs et al. (2010) suggest that accessibility can be achieved through mobility, proximity, and connectivity. Connectivity allows access to goods and services via the internet and digital communication, while mobility and proximity provide alternative or complementary ways to achieve physical access. Accessibility through mobility involves reaching urban amenities quickly and easily by public transit, driving or maybe cycling. In contrast, accessibility by proximity focuses on locating goods and services nearby, reducing the need for travel (Fig. 2).

The scheme above illustrates the relationship between these types of accessibility, highlighting the inverse relationship between proximity and mobility: as

Fig. 2 The various means of accessibility. Adjusted from Grengs et al. (2010)

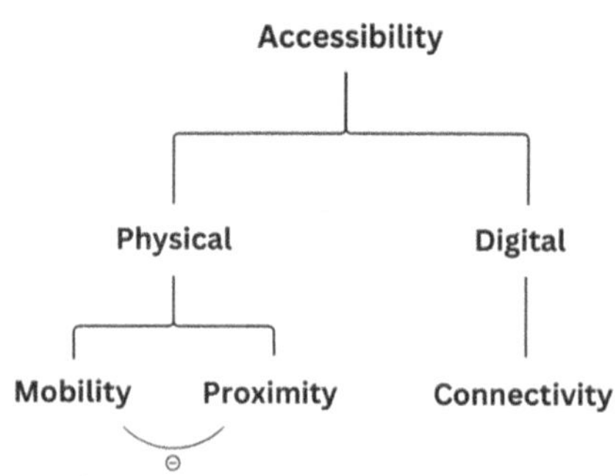

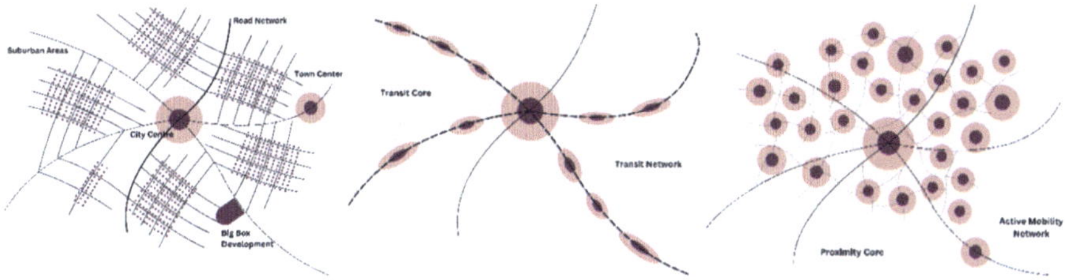

Fig. 3 The spatial imprint of mobility and proximity-oriented urban development (authors)

proximity increases by clustering urban functions closer together, the need for travel decreases, reducing mobility requirements, and vice versa. Examining accessibility further, it is clear that mobility-based and proximity-based approaches create fundamentally different urban layouts. Accessibility through mobility often relies on cars, transit, or other motorized means, shaping urban environment based on the chosen transport mode. For instance, the flexibility of cars or buses, which allow movement in any direction, typically results in an arterial grid layout, with linear development along highways, and infill with low-density, monofunctional areas like housing and big box development, e.g., shopping malls and office parks (Fig. 3).

In transit-focused infrastructure, the goal is to connect people to goods and services via reliable public transit, leading to a grid of transit-oriented nodes where most residents live and work near transit stations. Here, accessibility is achieved by linking nodes with concentrated employment, housing, retail, and more, though these functions *may not always coexist*. Fixed rail systems like metro or commuter rail—become the backbone of urban growth, with development organized around *"transit cores"* that act as hubs of high mobility. In contrast, accessibility by proximity focuses on enhancing local access to reduce the need for travel by locating goods and services nearby to form *"proximity cores."* In this *proximity-oriented development (POD)* approach, highways and fixed rail transit are less critical to the city's spatial and functional organization, as trips are typically short and made with nonmotorized means within these local hubs.

Overall, transitioning from a mobility- to a proximity-oriented approach offers a viable path for creating sustainable and resilient urban environments. By prioritizing accessibility within localized hubs, cities can reduce car dependency, enhance environmental outcomes, and strengthen community ties. This shift represents a pivotal step towards creating urban spaces that support equitable access, efficient land use, and the vitality of neighborhoods as essential community centers.

3 Methodology

This section details the research approach used to understand emerging themes in the 15-minute city discourse, focusing on text mining and quantitative content analysis (Qualtrics, 2024). In specific, it outlines the process of creating a corpus from

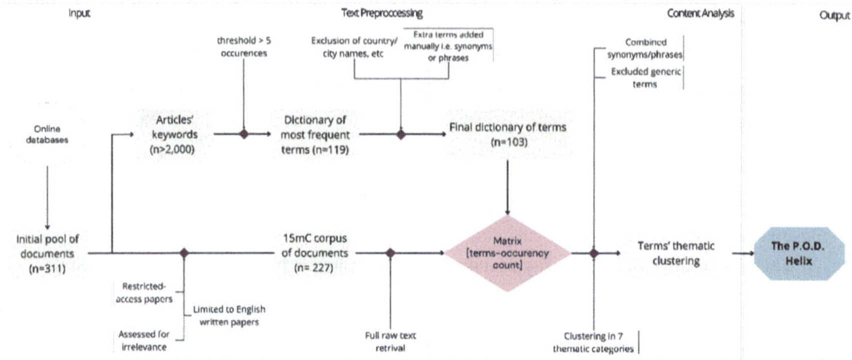

Fig. 4 Research methodology (authors)

academic literature, curating relevant keywords, and analyzing term frequencies to capture the prominence of various themes associated with proximity-oriented development. The proposed methodological approach (Fig. 4) provides a structured framework for identifying the key concepts and patterns shaping current research in proximity-based planning.

3.1 Corpus Formation and Keywords Dictionary

The study was conducted in August 2024. Scopus and Web of Science electronic databases were used to retrieve the scientific publications that compiled the corpus of the research. An initial search string ["minute cit*" OR "minute neigh*"] without exclusion criteria yielded 311 documents. Following a manual screening while retrieving the full texts from the publications, documents were excluded due to irrelevance (e.g., those not directly addressing the core subject) or due to restricted access. Non-English documents were also excluded since the analysis was conducted in English. The final corpus consisted of 227 English, peer-reviewed journal articles, book chapters, and conference papers.

To address the research scope, a dictionary of predefined terms was compiled by selecting the most frequent keywords from a pool of more than 2000 keywords, derived from the initial pool of 311 articles. Based on the frequency of occurrences, a threshold of at least five occurrences per keyword was imposed; thus, a total of 119 keywords were selected. From this set of keywords, terms related to country or city names, tools (e.g., GIS), methods (e.g., network analysis, composite indicator, etc.), and keywords including any form of the term "15-minute city" (e.g., 15 mC, 20-minute neighborhood, x-minute city, etc.), were excluded because they do not directly contribute to the purpose of the current analysis. In addition, to account for variations in terminology, preprocessing involved identifying synonyms or pluralization of the encountered terms, i.e., "amenit(?:y|ies)", which were consolidated in one single term. Finally, additional keywords that were not necessarily on the top of the list, were manually added, as overarching terms covering specific topics (e.g., "new working spaces" for coworking spaces and third places) or consolidate terms with multiple synonyms (e.g., "remote work" for work from home, telework, etc.) or cluster terms that compose complex concepts such as "wellbeing." The final keywords dictionary, containing 103 terms, was used for the analysis.

3.2 Content Analysis

The text analysis was conducted using R programming software and counted the frequency of keywords included in the dictionary throughout the corpus. The range of occurrences for these 103 terms was from 1 to 18.582. This wide range is due to high scores for general terms like "city", "urban," and "space," as expected, while more specific terms such as "urban metabolism" presented much lower frequency of occurrence. To narrow down the range, a process of combining keywords that were contextually or conceptually related was aggregated. For instance, the terms "active transport," "active transportation," and "active travel" were consolidated under the term "active mobility" with a total occurrence of 608 times. Then, general terms with very high occurrence values were left out as well, in an attempt to analyze the main terms in a more comparable manner. The consolidated list of terms totaled 51 with a range of frequencies from 6 to 11.254 occurrences. Finally, a structured frequency and contextual analysis of terms led to a clustering process that highlights the thematic priorities shaping contemporary urban planning discourse. These thematic priorities informed the proposed POD helix.

3.3 Limitations

There are certain limitations inherent in this methodology. By excluding non-English documents, there is potential bias, which also limits the representation of research from non-English-speaking countries. However, given that most of the peer-reviewed literature in this field is published in English, this decision was deemed reasonable for the purpose of this research. The analysis counted the raw frequency of terms or phrases but did not interpret their context within the text. This means that the total occurrence of the terms was recorded, while the variations in meaning or importance were not considered. Additionally, the analysis was limited to exact matches of the examined terms, meaning that alternative forms, such as synonyms or pluralization, may not be captured unless specifically accounted for during preprocessing. Future studies might address this by incorporating sentiment analysis or topic modeling to capture the significance of terms. Additionally, although efforts were made to include synonyms and pluralized forms, more advanced natural language processing techniques like word embeddings could improve this process in future analyses.

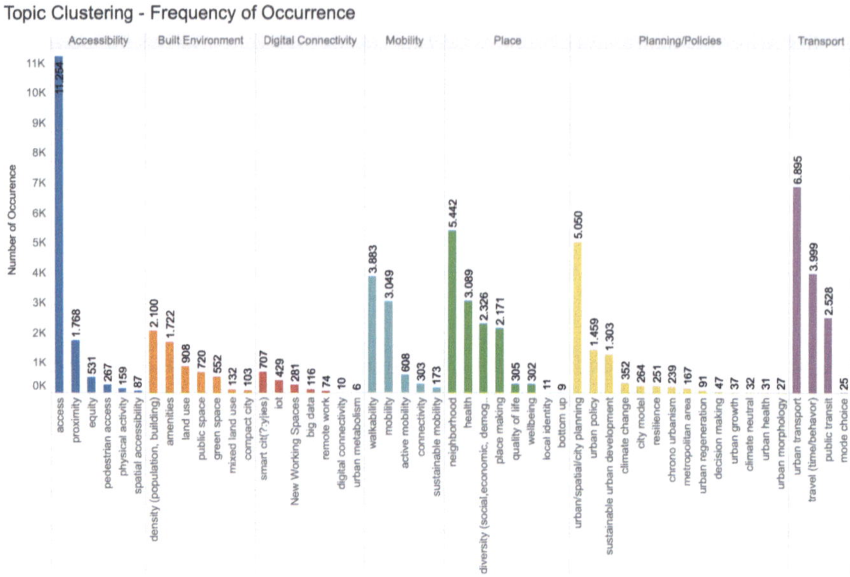

Fig. 5 Topic clustering—frequency of occurencies (authors)

4 Emerging Thematic Priorities in the 15 mC Discourse

This section delves into the specific language and terminology most frequently associated with the 15 mC discourse. By clustering key terms and concepts from selected literature, this analysis identifies the thematic priorities shaping contemporary urban planning conversations. Terms are grouped into clusters reflecting major areas of focus, including accessibility, built environment, digital connectivity, mobility, transport, place, and planning policies. This structured frequency analysis provides quantitative insights into the vocabulary and themes that most frequently arise, highlighting the essential concepts and priorities underpinning the shift toward proximity-based urban frameworks. Figure 5 shows the occurrences of the 51 most frequent key terms grouped into 7 thematic clusters. A detailed description per cluster is set out below.

1. *Accessibility:* The cluster of "Accessibility" has a range of words emphasizing the importance of equitable access and ease of movement in urban areas. The term "access" appears most frequently, with over 11,000 occurrences, suggesting a strong emphasis on general accessibility as a core theme in the 15-minute city discussions. "Equity" (1768 occurrences) and "pedestrian access" (1531 occurrences) are mentioned with notable frequency highlighting the importance of both fairness and pedestrian-friendly design in accessibility discussions. The relatively high frequency of "proximity" (531 occurrences) along with "spatial accessibility" (87 occurrences) indicates an emphasis on the importance of the allocation of resources, services, and amenities. Lower frequencies for terms

like "pedestrian access" (267 occurrences) and "physical activity" (159 occurrences) further highlight related yet less emphasized aspects of accessibility.

2. *Built environment:* The "Built Environment" cluster is diverse with several terms receiving considerable attention. Most of the terms are related to the physical and functional components of urban spaces. "Density," referring to population and building density, appears with high frequency (2100 occurrences), along with "amenities" (1722 occurrences) and "land use" (908 occurrences), reflecting the urban planning priorities. "Public spaces" and "green space" are also notable with 720 and 552 occurrences, respectively, showing that sustainable, resilient, and livable spaces are crucial in the build environment discussions. Finally, the terms "mixed land use," "compact city," and "urban morphology" are less frequent (132, 103, and 27 occurrences, respectively); however, their contextual relevance and interconnections with more general terms in the corpus like "land use" and "density" indicate their importance in the discussions regarding future urban development focus.
3. *Digital connectivity:* The "Digital Connectivity" cluster is marked by significant mentions of terms like "smart cities" (707 occurrences) and "IoT" (429 occurrences) highlighting the increasing importance of technological integration in urban planning. "New working spaces" (281 occurrences) along with "remote work" (74 occurrences) reflect the shift towards flexible working arrangements and spaces, a response to changing urban work dynamics with several spatial implications for urban areas. Terms like "big data," "digital connectivity," and "urban metabolism" also appear, though less frequently (116, 10, and 6 occurrences, respectively), suggesting that while they are part of the conversation, they have a more specialized focus within digital connectivity discussion. The category overall shows the critical role of technology and data-driven innovation in creating interconnected urban spaces.
4. *Mobility:* The "Mobility" cluster demonstrates high-frequency terms associated with sustainable and active transportation. The term "walkability" with 3883 occurrences reflects a core aspect of mobility by emphasizing pedestrian-friendly environments where essential services and amenities are within reach. "Active mobility" (608 occurrences) such as walking and cycling is promoted as healthy, environmentally friendly mode of transport. In addition, "connectivity" (303 occurrences) refers to the seamless facilitation of travel across the neighborhood, mostly related to the characteristics of the built environment (e.g., size of building block, etc.). Overall, the general term "mobility" with 3049 occurrences involves creating urban spaces where movement is efficient and convenient for all, regardless of transport modes, while "sustainable mobility" (173 occurrences) focuses on the importance of reducing reliance on private vehicles and minimizing environmental impacts through active transportation and eco-friendly transit options.
5. *Place:* The "Place" cluster highlights the quality and livability of urban neighborhoods. The leading term, "neighborhood", appears 5442 times, underscoring a strong focus on the neighborhood and community-centric planning within the 15-minute city discourse. "Health" (3089 occurrences) also emerges as a signifi-

cant element, with a focus on physical and mental wellness through well-designed accessible public spaces. Although with quite less frequent occurrences "quality of life" (305 occurrences) and "well-being" (302 occurrences), along with the "health" term underscore the significance in achieving healthy and thriving urban environments at the neighborhood scale. "Diversity" (2326 occurrences) encompassing social, economic, and design aspects accentuate the importance of creating spaces that cater to people from various backgrounds, encouraging inclusivity and interactions. The concept of "placemaking" itself, highly frequent in the discourse (2171 occurrences), emphasizes active community engagement and the transformation of public spaces into meaningful places that reflect local identity. "Local identity" (11 occurrences) and "bottom-up" approaches (9 occurrences), though less frequently mentioned, play critical roles in fostering spaces that are not only functional but also deeply connected to the lives and cultures of those who inhabit them.

6. *Planning policies:* The "Planning Policies" cluster includes all the related policies and approaches framing 15-minute city discourse. With 5050 occurrences, it is obvious that the 15-minute city is considered a "city," "urban," and "spatial" planning strategy focusing on reconsidering neighborhoods' layout in a city-wide level. "Chrono-urbanism" though with fewer occurrences (239 occurrences) indicates a recognition of time-sensitive urban dynamics. Other terms like "urban growth" (37 occurrences) indicate that the 15-minute city model is considered as urban growth management strategy that promotes sustainable urban development (through densification and urban sprawl containment) at the metropolitan scale (167 occurrences). Aiming for sustainable urban development (1303 occurrences), this urban model also addresses "climate change" (352 occurrences) by reducing emissions through proximity-based living and active transportation options. Furthermore, it is a "city model" (264 occurrences) with an emphasis on the notion of resilience (251 occurrences) by fostering self-sufficient, adaptable communities that respond to crisis without centralizing resources. Lesser-used terms like "urban regeneration" (91 occurrences), "climate neutral" (32 occurrences) and "urban health" (31 occurrences) indicate niche but significant aspects in urban planning. Finally, the moderate frequency of "urban policy" (1459 occurrences) along with "decision making" (47 occurrences) reflects a focus on strategic informed planning.
7. *Transport:* The "Transport" cluster reflects the focus on different modes of transportation in urban settings. The high occurrence of "urban transport" (6895 occurrences) emphasizes the focus of the 15-minute city model on designing transit systems that prioritize local accessibility, allowing residents to meet their daily needs close to home. Reducing "travel time" and changing "travel behavior" (3999 occurrences) are key to the model, as shorter commutes contribute to improved quality of life and reduce travel emissions. "Public transit" (2528 occurrences) also ranks high, underscoring the importance of accessible public transportation within walking distance. Finally, with moderate frequency "modal choice" (25 occurrences) underscores the importance of providing diverse and flexible transport options.

Overall, the word frequency analysis underscores the prominence of themes such as inclusive accessibility, sustainability of build environment, community well-being, and planning policies, in the 15mC discourse. The high frequency of specific terms reveals strong alignment with proximity-based living and sustainable urban planning objectives. This analysis not only confirms the importance of these themes but also establishes a foundation for understanding how proximity-oriented approaches are framed and discussed in the academic literature. Such insights are instrumental in guiding further development of POD approaches, emphasizing language that resonates with current urban planning priorities and future needs.

5 The POD Helix

The POD (Proximity Oriented Development) helix is proposed as a transformative urban planning approach, prioritizing local access to essential services and amenities. Organized around the concept of "proximity cores", the framework seeks to redefine neighborhood self-sufficiency and resilience. By focusing on four fundamental pillars—*Proximity, Urban Environment, Meta City and Placemaking*—the POD framework addresses the multifaceted needs of urban populations, supporting localized living, reduced environmental impacts, and enhanced community engagement.

POD aims to reconnect people to their neighborhood by focusing on two main aspects: proximity to key urban functions and the localization of primary urban functions. A key principle is to provide direct access to a wide range of essential services at the local level, ensuring neighborhood self-sufficiency in critical urban amenities that enhance quality of life (Büttner et al., 2022; Moreno et al., 2021; Pozoukidou & Chatziyiannaki, 2021). Such infrastructure and functions include health services, kindergartens, schools, social services, commercial activities, recreation, cultural and leisure activities, parks, natural areas, etc. Since each neighborhood and its residents may have different priorities and needs (Gil Solá & Vilhelmson, 2019), the above list is intended to be indicative rather than exhaustive.

As proximity is the main feature, the "proximity cores" are a critical structural element in the application of POD. Proximity cores can typologically describe clusters of urban forms unified by shared ground. Drawing from the concept of "archipelagos" as theorized by Ungers and Koolhaas in 1977and latter discussed by Shannon (Shannon, 2014), these proximity cores can be understood as integral parts of a broader urban struccture In this context, the archipelago serves as a metaphor for an urban entity composed of self-sufficient "islands" set within a city's landscape (Shannon, 2014). This idea of autonomy and self-sufficiency—illustrated in the *The City in the City: Berlin as a Green Archipelago*—envisions distinct assemblies connected by a shared landscape, with each island acting as a systematic, self-contained functional structure within larger urban areas. These islands are defined by structural elements such as building structure, axes, street systems, natural surroundings, etc. In POD, the notion of islands is enhanced, to create "urban condensers" and places with their own gravitational pull, attracting and fostering collective activity. Each proximity core is

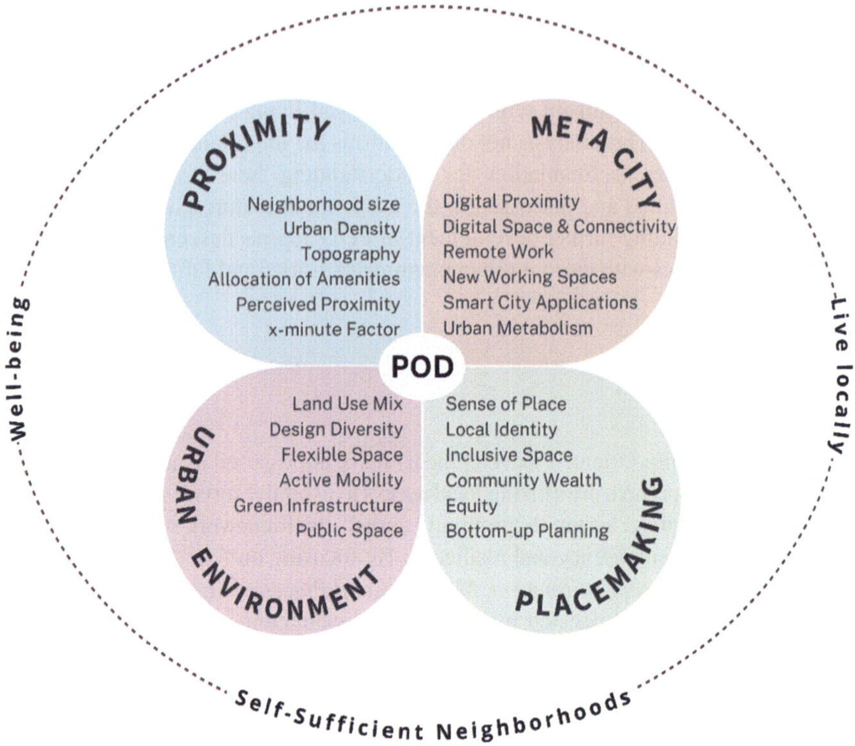

Fig. 6 The proximity-oriented development helix (authors)

defined by the activities it supports and the social relationships it cultivates, radiating its impact like ripples in a pond until it overlaps with the sphere of the next core. While each proximity core has a unique character, it remains part of a larger urban setting.

Building on these concepts, the corpus analysis, the frequency matrix, and the seven thematic categories outlined above, a framework for POD is presented below. This framework structures around four helices: proximity, urban environment, meta city, and placemaking. Each helix is described according to the seven thematic groups incorporating the most frequently occurring *concepts, notions, and attributes*. The four helices are closely interrelated, exhibiting multiple points of overlap and interaction that highlight the dynamic interplay among their respective dimensions (Fig. 6).

5.1 *Accessibly by Proximity*

Proximity, defined by several key attributes, sets this urban development approach apart from past models. In physical space, proximity is shaped by factors such as neighborhood size, urban density, topography, allocation of amenities, and perceived proximity.

POD perceives the *neighborhood* as a unit of spatial and functional organization on a citywide scale. Bibliography indicates that within the 15 mC model, there are many variations in the minimum walking time, which can range from 5 to 20 min, with the most common being 15 or 20 min (Logan et al., 2022; Teixeira et al., 2024). Setting the minimum walking time and therefore distance to amenities is related to the *topography* of places as well as other geographical factors that can limit or enhance accessibility. These include physical and artificial barriers, e.g., rivers, railway lines, freeways or any other features that disrupt the urban continuum. Apart from the geographical features, the size of the neighborhood unit and the *x-minute factor* is heavily affected by the *urban density*. Several studies emphasize the significance of high-density urban environments in the *allocation of urban amenities* as well as generating the agglomeration forces to form clusters of amenities that could act as the core of an x-minute neighborhood. Urban density is correlated to population density and population size, the latter being a crucial factor in the distribution of urban amenities. Earlier work of the authors (Plastara & Pozoukidou, 2023, 2024; Pozoukidou, 2024; Pozoukidou & Chatziyiannaki, 2021) and other research (Delventhal et al., 2022; Macharia et al., 2021) highlight the importance of the urban density as a factor in the process of the definition of the catchment area for each amenity. The catchment area is a critical metric for robust and reliable estimation of the population denominator used in city planning, enabling the estimation of amenities, commodity needs, and their spatial distribution.

Hence, POD presents an opportunity to rethink resource allocation on a citywide scale based on a hierarchical order of primary, secondary, tertiary, … z-iary urban amenities. Each type of amenity requires a different catchment area and population size to ensure its efficient operation, so the combination of population size and population-density thresholds is necessary to capture the full hierarchy of urban amenities. Thus, similar catchment areas allow the grouping of different amenities and services with a similar range and critical size, contributing to the creation of multiple and hierarchically graded "service nodes". This inevitably creates a hierarchy of urban services or else proximity cores that contribute to the creation of multiple and hierarchically graded city layers. Therefore, proximity cores are part of a system and not mere hierarchies or single entities (Pozoukidou & Chatziyiannaki, 2021) in a larger urban—x-minute—area of a monocentric or polycentric urban system (Duranton & Puga, 2020; Kamruzzaman, 2022; Rappaport, 2008). The citywide value of *x* is related to and defined by the overall spatial and functional order of the urban territory (Fig. 7). Thus, effective implementation of a POD strategy relies on precise metrics of service access and catchment areas based on the population served and the complexity of the allocated amenities with clear implications for urban planning. This is especially critical when planning essential public services such as education, healthcare, and social welfare.

Additionally, few but countable research points out that when studying geographic proximity, individual's functional capabilities should be acknowledged. This is crucial, considering that the main objective of the POD approach is to promote and encourage active travel, albeit to varying degrees (Pozoukidou & Angelidou, 2022). Hence, *perceived proximity* becomes a critical element in the

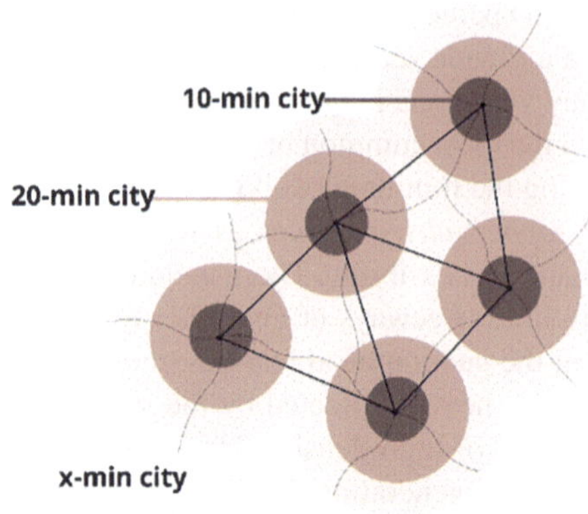

Fig. 7 The hierarchically graded city layers (authors)

design and creation of proximity cores. Perceived proximity refers to the subjective experience of distance and accessibility within a city or neighborhood. It shapes desirable proximity—the distance and accessibility that individuals or groups consider necessary, adequate, acceptable or even optimal from their perspective (Phillips et al., 2023). Perceived and desirable proximity, although related to, are distinct from objective proximity, which is a measure of physical distance or travel time (Phillips et al., 2023). Perceived and desirable proximity are influenced by a variety of attributes, including physical/build features, individual characteristics and behavior, and sociocultural factors (Gil Solá & Vilhelmson, 2019; Struwe, 2021). Therefore, urban qualities such as the internal and external connectivity, characteristics of the built environment (urban density, etc.) and allocation of essential services and amenities (Stojanovski, 2020), acquire different dimensions and significance under the notion of perceived proximity. Additionally, factors like age, gender, physical impairment, economic status, movement and travel behavior, sociability, social interaction, and sociocultural values and norms (e.g., attributes toward personal space and social distance) are crucial in studying proximity. These factors influence individuals' willingness to adapt their behavior and lifestyles to achieve the "proximity" effectiveness (Stojanovski, 2020).

5.2 *Quality of Urban Environment*

The notions of geographic and perceived proximity highlight the importance of the attributes that define the quality of the urban landscape. Most of these attributes have been used as design flagships in the past, namely land use mix and design diversity; however, pairing them with contemporary approaches regarding flexible use of space, multifunctional green spaces, and active mobility within the proximity context, set the basis for new approaches in urban planning.

Land use mix is a key feature in the POD approach. According to Shatu and Kamruzzaman (2021) and Grengs et al. (2010), a mixed-use urban environment is crucial in achieving the proximity objective. Appropriate land use mix reduces the need for long commutes, promotes active transport and enhances self-sufficiency, by localizing a variety of urban functions such as residential, commercial, recreational, and institutional. A well-balanced land use mix is aligned with the distribution of urban services and amenities at the neighborhood scale, based on the spatial and functional order of the urban territory as well as the catchment area and critical population size per amenity (see proximity helix).

The mix of land uses allows for *flexibility in space*, particularly in the POD approach, which advocates for adaptable, multifunctional spaces that respond dynamically to shifting urban needs. According to the basic principles of chrono-urbanism, the mixing of land uses can occur not only in space but also in time (Gwiazdzinski, 2014). POD supports the decoupling of space from time, with each person and function having its own rhythm, giving place the quality of multi-temporality (Crang, 2001). This means that spaces can be used differently according to varying schedules that can last from 1 day to longer periods of time. This flexibility enables urban areas to maximize land use efficiency by accommodating various activities throughout the day and evolving in response to community needs (Abdelfattah et al., 2022). For example, a single space might serve as a coworking space area, social gathering spot, or an event venue at different times, catering to diverse user needs without requiring extensive new infrastructure. Such adaptability is particularly valuable in densely populated areas where space is scarce, as it supports efficient use of land and reduces the urban footprint. Moreover, flexible spaces foster resilience by allowing neighborhoods to pivot quickly during disruptions, such as health and economic crisis, offering residents continual access to essential services and community spaces. This approach not only enhances functionality and sustainability but also enriches urban life by creating dynamic, community-centered environment that supports varied and inclusive uses. Such approaches have already been applied in the context of temporal land uses, giving the flexibility needed for cities to respond and adapt quickly to crises such as health, climate, etc. Thus, traditional urban planning is being gradually replaced by flexible spatiotemporal urban planning (Pozoukidou, 2026; Teller, 2021).

5.2.1 Design Diversity and Inclusivity

The importance of design diversity in urban environments is rooted in its capacity to make cities more resilient, inclusive, and sustainable addressing both current and emerging challenges. Design diversity refers to the inclusion of a variety of architectural styles, building forms, public spaces, landscapes within a single area or neighborhood (Steuteville, 2017). Diversity aims to create more visually engaging, functional, and *inclusive urban environments* that cater different community needs and lifestyles. Design diversity often enhances the character of an area, making it more resilient and adaptable to social and economic changes over time, contributing to a sense of place and identity. However, design diversity aims not only to promote more adaptable urban environments but to ensure socioeconomic diversity as well. Diverse urban design can combat sociospatial inequalities, ensuring more equitable access to services, green spaces, and affordable housing options across neighborhoods. *Housing affordability* is a critical component and has been pointed out as a drawback in long-standing urban development and revitalization strategies such as TOD and New Urbanism. In the POD approach, affordable housing initiatives are seen as integral to urban inclusivity providing a buffer against sociospatial segregation, helping to stabilize communities by reducing displacement pressures and supporting diverse socioeconomic representation. Moreover, policy frameworks that prioritize affordability alongside land use diversity such as zoning reforms and incentives for low-cost housing developments, are vital for creating equitable, thriving urban places where all residents can maintain a stable connected livelihood. Apart from housing affordability, by localizing amenities and job opportunities, urban design can dismantle barriers that often segregate communities by income or race, thereby promoting a more inclusive urban fabric. This approach aligns with the principle of the "right of the city," enabling residents to participate in shaping the urban environment and ensuring marginalized groups are not overlooked.

5.2.2 Green Infrastructure, Public and Green Spaces

POD incorporates the concept of green infrastructure (GI) which is extensively used in spatial planning to mitigate the effects of climate change and prevent natural disasters such as floods, heat waves, fires, etc. (Matthews et al., 2015; Olcina, 2022). GI is defined as a strategically planned network of natural and seminatural areas with diverse environmental characteristics, designed and managed to provide a wide range of ecosystem services. It includes public and private, terrestrial and aquatic areas, and its effective implementation and management is based on increased *connectivity, multifunctionality* and *planning at multiple scales* (from neighborhood to city and periphery) (ESPON EGTC, 2018; Pozoukidou, 2020). In relation to natural disasters, GI contributes to reducing flood risk, mitigating the urban heat island (UHI) effect, and in many cases is used as buffer zone to shield urban areas from fires (ESPON, 2019; Foster et al., 2011; Sebesvari et al., 2019). Paired with proximity, the properties of connectivity and multifunctionality, are essential in achieving an accessible and multifunctional hierarchical system of parks, green and open spaces, that contribute to the creation of healthy environments and lifestyles for all ages and abilities. In this network of accessible and high-quality local open spaces, other types of spaces, such as schoolyards, can be included to offer space for multiple activities, i.e., sports and vice versa.

POD values the importance of physical activity by ensuring an urban environment that induces walkability, improves air quality, and reduces city noise. The benefits and therapeutic power of nature are widely documented, and a well-connected system of green spaces can play a crucial role in reducing health disparities (World Health Organization. Regional Office for Europe, 2016). To fully harness these benefits, green and open spaces need to be integrated as part of a city-wide network, with emphasis on multifunctionality. Finally, POD's prioritization of pedestrian mobility minimizes the need for extensive road networks, parking areas, and impermeable surfaces which usually impede the natural flow of water or even intensify the phenomenon of UHI (Akbar & Edelenbos, 2021; Lefebvre, 1991) (Heymans et al., 2019), contributing to the effective implementation of the GI and enhancing urban resilience.

To this end, urban planning has a pivotal role in supporting the implementation of green infrastructure, managing trade-offs, and addressing conflicts (Pozoukidou, 2020). Viewing cities as shared habitat through multipiece approach is essential, and green and blue infrastructure are increasingly seen as key tools for delivering ecosystem services in urban areas (Bush & Doyon, 2019; Heymans et al., 2019).

5.2.3 Active and Equitable Mobility

Active mobility is more of an outcome of the structural and functional layout of urban space as defined by the land use mix, design diversity, and allocation of open space. By fostering proximity between residences, workplace, and essential services, active mobility reduces dependency on cars, thereby lowering emission and traffic congestion (Logan et al., 2022; Teixeira et al., 2024). POD promotes urban layouts that prioritize pedestrian-friendly infrastructure—such as wide sidewalks, protected bike lanes, and well-connected pathways—that enhance both accessibility and safety, making active travel a viable option for all age groups and abilities. Furthermore, the integration of green spaces and public areas along these routes promotes health and well-being, providing residents with safe and attractive spaces for physical activity. Importantly, within the POD paradigm, active mobility supports social equity by ensuring equitable access to resources and services without the economic burden of car ownership.

As urban *transport* is a vital component of contemporary cities, POD aims at reducing reliance on private vehicles promoting *accessible and efficient alternatives* that enhance connectivity and reduce environmental impact. It integrates multimodal transport solutions, such as active mobility options and well-coordinated public transit, to ensure residents can easily reach essential services without long commutes. As previously mentioned, POD promotes high-density, multitemporal, mixed-use neighborhoods that present adequate self-sufficiency, encouraging short-distance travel. However, situating public transport hubs within or near neighborhoods improves accessibility and encourages the use of transit rather than car. This in turn has several implications in reducing emissions and energy consumption but also supports equitable mobility by providing affordable reliable travel options for all residents. By making transport options sustainable and accessible POD promotes healthier cities that are resilient and inclusive, meeting the needs of diverse populations.

5.3 Placemaking

Placemaking is a community-centered approach aiming at enhancing the identity, functionality, and inclusiveness in public spaces. It involves efforts by local communities, planners, and other stakeholders to design and manage spaces that reflect the unique cultural, social, and economic characteristics of an area. As Istoriou and Pozoukidou (2024) point out, placemaking practices are perceived as experimental, nonplanned approaches with great flexibility, consisting of various processes, and tied to the local community. Placemaking's main drive is to get things done, to learn by doing, and to establish a transformative learning and collective understanding of *place* versus *space*. Indeed, the works of Lefebvre, Soja, Massey, Entrikin, and others accentuate the significance of "place" versus "space" as a criticism of modernism (Saar & Palang, 2009), which shaped the foundations for placemaking (Legge, 2016).

Placemaking brings the focus to public space and human activity while *community involvement* rises as an imperative ingredient. It is a concept that prioritizes human-centric design and active community engagement, recognizing public spaces as dynamic environments shaped by social and cultural interactions of those who use them (Akbar & Edelenbos, 2021; Lefebvre, 1991). In the POD concept, placemaking is perceived as a process and practice with an emphasis on active resident participation in shaping their neighborhoods, making planning processes more inclusive, adaptive, and aligned with the local needs. Unlike top-down planning practices where decisions are made at a central level, POD embraces *bottom-up planning* to empower residents and engage them in decision making. This includes identifying local priorities to contributing to design choices and implementation. This approach builds a sense of ownership and accountability among community members, fostering trust and ensuring that urban development reflects diverse perspectives, especially those of traditionally marginalized groups (Chinis et al., 2024). Placemaking practices value *local identity* by emphasizing the unique cultural, historical, and social characteristics of neighborhoods, fostering a sense of belonging and pride among residents. The POD approach prioritizes the integration of local identity by designing spaces that reflect and support community traditions, values, and daily needs. In fact, by encouraging neighborhood-based amenities, green spaces, and gathering points, POD strengthens community bonds, preserves unique local features, resists urban homogenization and promotes diverse, vibrant local hubs. Local identity thus contributes to social resilience, allowing residents to feel connected to their environment while promoting sustainable, inclusive urban growth that respects cultural heritage. By engaging community members in the design and stewardship of public spaces, placemaking encourages local investment, stimulates small business growth and revitalizes underutilized areas, which in turn generates economic opportunities. As Papastergiou et al. point out when defining a neighborhood's identity, its commercial character is equally, or even more important to its residential qualities. Being the "visible face of the neighborhood," local businesses have a fundamental role in driving the place's identity, but also reflecting it

(Papastergiou et al., 2024). Local businesses contribute to residents' feeling of ownership and generate a sense of familiarity and belonging (Martucci, 2019), building *collective identity* and *community wealth.*

Finally, POD approach focuses on tailoring access to resources, services, and places to meet the diverse needs of community members, ensuring fair outcomes rather than uniform solutions. Unlike equality, which offers the same resources to all, *equity* recognizes that different groups—such as low-income families, elderly residents, and people with disabilities often face unique barriers. Through the process of placemaking, the POD approach goes beyond basic accessibility by integrating community voices into the design and implementation process to create *inclusive spaces* that reflect local identity, enhancing affordable transit and housing options in underserved areas and foster social cohesion.

5.4 Meta City

In the context of POD, *digital proximity and connectivity* create another layer of intangible environment above the built environment, the virtual one. By utilizing technology and data, *smart city* applications can optimize the use of local services, ensuring efficient use of resources and data-driven planning (Hui et al., 2023). Such technologies provide the capability for real-time adjustments in regulating public transportation, energy consumption, and infrastructure. To achieve this, adopting Internet of Things (IoT) and digital twin applications is essential (Pozoukidou & Angelidou, 2022). This approach enhances *urban metabolism* and real-time adjustment of urban functions (Teller, 2021).

The so-called *Meta city* aligns with the vision of POD for reducing the need for physical travel by offering residents the *digital space,* where remote access to services, personalized user experiences, and virtual interactions complement the physical world. The implementation of metaverse-driven urban planning strategies has the potential to foster economic growth and innovation while improving the quality of life for residents (I.S.I.—Industrial Systems Institute, 2023).

At this edge, remote working arrangements have gained popularity particularly after the COVID-19 outbreak. The rise of *remote work* has resulted in more people moving to suburban and rural areas, as well as to smaller cities. This trend has also spurred the development of *new working spaces* (NWS), which differ from traditional workplaces by supporting "near working"—working closer to home (Biagetti et al., 2024). NWSs such as coworking spaces, third places or neighborhood-based working hubs like cafes and libraries, provide an equipped working environment and facilitate social interaction and community engagement. As more people embrace hybrid work routines, many prefer using third places near their houses, reducing the need for long daily commutes to a traditional office (Biagetti et al., 2024; Li et al., 2024).

Hölzel et al. (2023) highlight that coworking spaces providing amenities beyond a simple workspace encourage active mobility within the "city of proximity." The growth of remote work is a significant factor in the transition towards more localized and decentralized workplaces. The reallocation of workers to regional areas might increase the demand for associated changes in urban infrastructure planning (Yu et al., 2019). Integrating NWS in both central and peripheral areas represents a strategy for maintaining economic activity in cities and serves as an effective strategy for advancing POD. The importance of *digitization* and digital skills for economic growth has increasingly become a priority for urban planners aiming to reorganize urban spaces to better support the digital economy (Pajević, 2021). In the context of remote work, POD which emphasizes proximity to essential services, is particularly important. With fewer people commuting daily, the demand for local amenities and services within walking or biking distance increases (Elldér, 2022; Pozoukidou & Chatziyiannaki, 2021). By localizing workspaces, POD supports sustainable living and reduces long commutes, and the associated environmental costs while enabling individuals to work in closer proximity to their place of residence (Pozoukidou & Angelidou, 2022). To fully embrace the POD model, cities need to promote hybrid and flexible work environments, ensuring that local workplaces contribute to both social interaction and economic vitality.

A precise digital representation of urban space enables the recording and evaluation of infrastructure adequacy, mobility options, travel durations, accessibility to public open spaces, the environmental impact of vehicles, and patterns of energy consumption. *Digital twins*—virtual replicas of cities—can significantly enhance the efficiency and sustainability of urban environments. They have practical applications in key sectors such as energy management, public safety, healthcare, logistics, urban planning, construction, and transportation (Deng et al., 2021). When combined with citizen feedback, these systems can improve city planning and offer valuable insights into how residents perceive and experience their environment. Moreover, digital twins support data-driven decision-making, helping authorities anticipate problems, optimize resource allocation, and design more resilient and livable cities (Botín-Sanabria et al., 2022; Lehtola et al., 2022).

In relation to transport systems and mobility, digital twin technology can both support and be enhanced by intelligent transportation systems (Rudskoy et al., 2021). A city-scale digital twin can model road networks and their environment, simulate scenarios, and provide real-time data connectivity for effective traffic management. Reliable connectivity ensures timely responses to emergencies (Feng et al., 2023), addressing uncertainties such as accidents or adverse weather. As transportation networks grow more complex, digital twins combined with artificial intelligence offer predictive insights, enabling smart mobility solutions, car-free urban models, and the integration of innovative modes like autonomous vehicles (van Arem et al., 2022).

As a final note, the POD framework offers a holistic approach to proximity-oriented planning, addressing accessibility, environmental sustainability, digital integration, and placemaking. By establishing proximity cores as foundational urban units, the framework aligns with current trends towards local decentralization of crucial urban amenities and self-sufficiency in urban areas. This approach not only supports more resilient and adaptable cities but also fosters inclusive, community-centered development, presenting a scalable model for achieving sustainable and equitable urban growth.

6 Conclusions

This research underscores the importance of proximity-based living as an effective response to contemporary urban challenges, aligning closely with the 15 mC principles. The transition from a mobility—to a proximity-oriented development approach represents a paradigm shift in urban planning. This shift prioritizes local accessibility over long commutes, aiming to create self-sufficient neighborhoods that reduce car dependency, enhance environmental sustainability, and improve quality of life. The POD framework builds on this perspective, advocating for the reorganization of urban spaces around proximity cores that promote walkability, active travel, and localized access to essential services and amenities.

The methodology employed in this study combines text mining and keyword frequency analysis to examine emerging themes and priorities within the discourse of 15-minute city and proximity-oriented urban planning. Drawing on a corpus of peer-reviewed literature, this approach identified key thematic clusters that resonate with sustainable urban priorities, including accessibility, built environment, transport, mobility, digital connectivity, placemaking, and strategic planning, which informed the development of the POD helix.

The POD helix is proposed as a framework structured around four foundational pillars: proximity, urban environment, meta city, and placemaking. These pillars address various dimensions of sustainable urban development:

- *Proximity* emphasizes accessibility to essential urban services, leveraging geographic and perceived proximity to foster self-sufficient neighborhoods.
- *Urban environment* advocates for diverse land use, flexible and multifunctional spaces, and the integration of green infrastructure to enhance sustainability and resilience.
- *Meta city* focuses on digital proximity, utilizing technologies to optimize urban systems and support hybrid working arrangements that reduce long and frequent commutes.
- *Placemaking* underscores the importance of community-driven planning processes, local identity, and equity in creating vibrant, inclusive urban spaces.

Overall, the POD approach offers a scalable, adaptable model for cities aiming to balance growth with quality of life, ensuring urban areas that are sustainable, resilient, and inclusive. This proximity-centered approach holds significant potential as a guiding model for urban planners seeking to foster dynamic, self-sustaining, and equitable urban environments that support the well-being of diverse communities.

Future research should explore the dynamic interactions between four pillars, particularly the integration of digital and physical proximity in urban systems. Equally vital is the weighting of each foundational dimension, a step that would substantially strengthen the analytical precision of the framework. Establishing their relative significance—grounded in empirical evidence, urban typologies, or policy priorities—would allow for a more accurate recording and assessment of transformative processes toward the POD paradigm. Such an approach would also foster context-sensitive adaptations of the model, acknowledging that cities may differently value proximity, environmental quality, digital integration, or community engagement according to their unique sociospatial and economic realities. In terms of the methodology, Natural Language Processing techniques could further refine thematic analysis, while longitudinal studies on the implementation of proximity cores could provide insights into the real-world impact on sustainability, equity, and community cohesion. Expanding research to include diverse cultural and geographical contexts will also be essential in adapting the POD framework to varying urban realities.

References

Abdelfattah, L., Deponte, D., & Fossa, G. (2022). The 15-minute city: Interpreting the model to bring out urban resiliencies. In M. Tira & G. Maternini (Eds.), *Presented at the transportation research procedia* (pp. 330–337). Elsevier B.V.. https://doi.org/10.1016/j.trpro.2021.12.043

Akbar, P. N. G., & Edelenbos, J. (2021). Positioning place-making as a social process: A systematic literature review. *Cogent Social Sciences, 7*, 1905920. https://doi.org/10.1080/23311886.2021.1905920

Bauer, C. (1945). *Good neighborhoods - Catherine Bauer, 1945* (1st ed.). The ANNALS of the American Academy of Political and Social Science.

Bernick, M., & Cervero, R. (1997). *Transit villages in the 21st century*. McGraw-Hill.

Bertolini, L., Curtis, C., & Renne, J. (2012). Station area projects in Europe and beyond: Towards transit oriented development? *Built Environment, 38*, 31–50. https://doi.org/10.2148/benv.38.1.31

Biagetti, M., Croce, G., Mariotti, I., Rossi, F., & Scicchitano, S. (2024). The call of nature. Three post-pandemic scenarios about remote working in Milan. *Futures*. https://doi.org/10.1016/j.futures.2024.103337

Boisjoly, G., & El-Geneidy, A. M. (2017). How to get there? A critical assessment of accessibility objectives and indicators in metropolitan transportation plans. *Transport Policy, 55*, 38–50.

Bothwell, S. E., Gindroz, R., & Lang, R. E. (1998). Restoring community through traditional neighborhood design: A case study of Diggs town public housing. *Housing Policy Debate, 9*, 89–114. https://doi.org/10.1080/10511482.1998.9521287

Botín-Sanabria, D. M., et al. (2022). Digital twin technology challenges and applications: A comprehensive review. *Remote Sensing*. https://doi.org/10.3390/rs14061335

Bush, J., & Doyon, A. (2019). Building urban resilience with nature-based solutions: How can urban planning contribute? *Cities, 95*, 102483. https://doi.org/10.1016/j.cities.2019.102483

Büttner, B., Seisenberger, S., Baquero Larriva, M., Gante, A., Ramírez, A., & Haxhija, S. (2022). *±15-Minute City: Human-centred planning in action. Mobility for more liveable urban spaces*. EIT Urban Mobility.

C40 Cities Climate Leadership Group, ARUP. (2021). Green and Thriving Neighbourhoods: A pathway to net zero, featuring the '15-minute city' [WWW Document]. https://www.c40knowledgehub.org/s/article/Green-and-Thriving-Neighbourhoods-A-pathway-to-net-zero-featuring-the-15-minute-city?language=en_US (Accessed 13 Nov 24).

C40 Global Mayors. (2020). *Green & Just Recovery Agenda [WWW document]*. C40 Cities. https://www.c40.org/what-we-do/raising-climate-ambition/green-just-recovery-agenda/ (Accessed 13 Nov 24)

Calthorpe, P. (1993). *The next American metropolis: Ecology, community, and the American dream* (3rd ed.). Princeton Architectural Press.

Cervero, R., Murphy, S., Ferrell, C., Goguts, N., & Tsai, Y.-H. (2004). *Transit-oriented development the United States: Experiences, challenges, and prospects*. Transportation Research Board. https://doi.org/10.17226/23360

Chinis, I., Pozoukidou, G., & Istoriou, T. (2024). Renegotiating spatial planning practices: The role of collective initiatives and informal networks. *European Planning Studies, 32*, 263–280. https://doi.org/10.1080/09654313.2021.1903400

Crang, M. (2001). Rhythms of the city: Temporalised space and motion.

De Vos, J., Van Acker, V., & Witlox, F. (2014). The influence of attitudes on Transit-Oriented Development: An explorative analysis. *Transport Policy, 35*, 326–329. https://doi.org/10.1016/j.tranpol.2014.04.004

Deloitte. (2021). Urban future with a purpose 12 trends shaping the future of cities by 2030 [WWW Document]. https://www.deloitte.com/global/en/Industries/government-public/perspectives/urban-future-with-a-purpose/15-minute-city.html (Accessed 12 Nov 24).

Delventhal, M. J., Kwon, E., & Parkhomenko, A. (2022). JUE insight: How do cities change when we work from home? *Journal of Urban Economics, JUE Insights: COVID-19 and Cities, 127*, 103331. https://doi.org/10.1016/j.jue.2021.103331

Deng, T., Zhang, K., & Shen, Z. J. (2021). A systematic review of a digital twin city: A new pattern of urban governance toward smart cities. *Journal of Management Science and Engineering, 6*(2), 125–134. https://doi.org/10.1016/j.jmse.2021.03.003

Duany, A., & Plater-Zyberk, E. (2008). The traditional neighborhood and urban sprawl. In *New urbanism and beyond designing cities for the future* (pp. 64–69). Rizzoli.

Duranton, G., & Puga, D. (2020). The economics of urban density. *Journal of Economic Perspectives, 34*, 3–26. https://doi.org/10.1257/jep.34.3.3

DUT. (2024). *Navigating the 15-minute city first learnings and discussions from the driving urban transitions partnership*. Driving Urban Transitions Partnership.

Dyckman, J. W. (1959). Comment on glazer's school proposals. *Journal of the American Institute of Planners, 25*, 196–199.

Elldér, E. (2022). Active travel and telework in Sweden: Teleworkers walk more, but cycle less. *Transportation Research Part D: Transport and Environment, 109*, 103362. https://doi.org/10.1016/j.trd.2022.103362

ESPON. (2019). *GRETA - "GReen infrastructure: Enhancing biodiversity and ecosysTem services for territoriAl development"*. European Union.

ESPON EGTC. (2018). *Territorial potentials for green infrastructure (working paper)*. ESPON European Grouping on Territorial Cooperation.

European Commission. (2020). *Handbook of sustainable urban development strategies [WWW document]*. European Commission. https://urban.ec.europa.eu/strategies/handbook-sud/ (Accessed 14 Nov 24)

European Environment Agency. (2016). *Urban sprawl in Europe, Urban sprawl in Europe*. Joint EEA-FOEN report, Publications Office.

Feng, H., Lv, H., & Lv, Z. (2023). Resilience towarded Digital Twins to improve the adaptability of transportation systems. *Transportation Research Part A: Policy and Practice, 173*, 103686. https://doi.org/10.1016/j.tra.2023.103686

Foster, J., Lowe, A., & Winkelman, S. (2011). *The value of green infrastructure for urban climate adaptation*. Center for Clean Air Policy.

Gil Solá, A., & Vilhelmson, B. (2019). Negotiating proximity in sustainable urban planning: A Swedish case. *Sustainability, 11*, 31. https://doi.org/10.3390/su11010031

Greenfield, D. E. (2023). *What are the principles of sustainable urban planning? [WWW document]*. Sigma Earth. https://sigmaearth.com/what-are-the-principles-of-sustainable-urban-planning/ (Accessed 13 Nov 24)

Grengs, J., Levine, J., Shen, Q., & Shen, Q. (2010). Intermetropolitan comparison of transportation accessibility: Sorting out mobility and proximity in San Francisco and Washington, D.C. *Journal of Planning Education and Research, 29*, 427–443. https://doi.org/10.1177/0739456X10363278

Gwiazdzinski, L. (2014). Adaptable cities and chrono-urbanism. In *Presented at the adaptable cities and chrono-urbanism.*

Hamidi, S., Ewing, R., Preuss, I., & Dodds, A. (2015). Measuring sprawl and its impacts: An update. *Journal of Planning Education and Research, 35*, 35–50. https://doi.org/10.1177/0739456X14565247

Handy, S. (1993). Regional versus local accessibility: Implications for nonwork travel. *Transportation Research Record.*

Heymans, A., Breadsell, J., Morrison, G. M., Byrne, J. J., & Eon, C. (2019). Ecological urban planning and design: A systematic literature review. *Sustainability, 11*, 3723. https://doi.org/10.3390/su11133723

Hölzel, M., Akhavan, M., & Leducq, D. (2023). Concluding remarks: European narratives about the effects of the COVID-19 pandemic on coworking. *SpringerBriefs in Applied Sciences and Technology*. https://doi.org/10.1007/978-3-031-26018-6_15

Hui, C. X., Dan, G., Alamri, S., & Toghraie, D. (2023). Greening smart cities: An investigation of the integration of urban natural resources and smart city technologies for promoting environmental sustainability. *Sustainable Cities and Society, 99*, 104985. https://doi.org/10.1016/j.scs.2023.104985

Ibraeva, A., Correia, G. H. d. A., Silva, C., & Antunes, A. P. (2020). Transit-oriented development: A review of research achievements and challenges. *Transportation Research Part A: Policy and Practice, 132*, 110–130. https://doi.org/10.1016/j.tra.2019.10.018

Industrial Systems Institute (ISI). (2023). Digital Intelligent Urban Services in the Metaverse—a solution for increased competitiveness of small and median cities | ISI [WWW Document]. https://www.isi.gr/node/1096 (Accessed 13 Nov 24).

Intergovernmental Pannel on Climate Change (IPCC). (2022). Climate Change 2022—Impacts, Adaptation and Vulnerability. Contribution of Working Group II to the Sixth Assessment Report of the Intergovernmental Panel on Climate Change. [H.-O. Pörtner, D.C. Roberts, M. Tignor, E.S. Poloczanska, K. Mintenbeck, A. Alegría, M. Craig, S. Langsdorf, S. Löschke, V. Möller, A. Okem, B. Rama (eds.)]. Cambridge University Press, p. 3056. https://doi.org/10.1017/9781009325844.

Isaacs, R. R. (1948). The neighborhood theory: An analysis of its adequacy. *Journal of the American Institute of Planners*. https://doi.org/10.1080/01944364808978605

Istoriou, T., & Pozoukidou, G. (2024). Understanding placemaking in the context of spatial planning: Insights from a literature review. *Journal of Urbanism: International Research on Placemaking and Urban Sustainability, 0*, 1–18. https://doi.org/10.1080/17549175.2024.2415632

Jabareen, Y. R. (2006). Sustainable urban forms: Their typologies, models, and concepts. *Journal of Planning Education and Research, 26*, 38–52. https://doi.org/10.1177/0739456X05285119

Kafkalas, G., Vitopoulou, A., Gemenetzi, G., Giannakou, A., & Tasopoulou, A. (2016). The City as an evolutionary process. In *Sustainable cities: Adaptation and resilience in times of crisis*. Hellenic Academic Ebooks.

Kamruzzaman, M. (2022). Key technologies, applications and trends of internet of things for energy-efficient 6G wireless communication in smart cities. *Energies, 15*. https://doi.org/10.3390/en15155608

Lefebvre, H. (1991). *The production of space*. Blackwell.

Legge, K., 2016. The evolution of place making - What is next?' Place Leaders - Asia Pacific. https://www.placeleadersmedia.com/the-evolution-of-placemaking

Lehtola, V. V., et al. (2022). Digital twin of a city: Review of technology serving city needs. *International Journal of Applied Earth Observation and Geoinformation, 114*(July), 102915. https://doi.org/10.1016/j.jag.2022.102915

Li, W., Zhang, E., & Long, Y. (2024). Unveiling fine-scale urban third places for remote work using mobile phone big data. *Sustainable Cities and Society*. https://doi.org/10.1016/j.scs.2024.105258

Logan, T. M., Hobbs, M. H., Conrow, L. C., Reid, N. L., Young, R. A., & Anderson, M. J. (2022). The x-minute city: Measuring the 10, 15, 20-minute city and an evaluation of its use for sustainable urban design. *Cities, 131*, 103924. https://doi.org/10.1016/j.cities.2022.103924

Macharia, P. M., Ray, N., Giorgi, E., Okiro, E. A., & Snow, R. W. (2021). Defining service catchment areas in low-resource settings. *BMJ Global Health, 6*, e006381. https://doi.org/10.1136/bmjgh-2021-006381

Martucci, S. (2019). Shopping streets and neighborhood identity: Retail theming as symbolic ownership in New York. *City & Community, 18*, 1123–1141. https://doi.org/10.1111/cico.12465

Matthews, T., Lo, A. Y., & Byrne, J. A. (2015). Reconceptualizing green infrastructure for climate change adaptation: Barriers to adoption and drivers for uptake by spatial planners. *Landscape and Urban Planning, 138*, 155–163. https://doi.org/10.1016/j.landurbplan.2015.02.010

Miller, E. J. (2018). Accessibility: Measurement and application in transportation planning. *Transport Reviews, 38*, 551–555.

Moreno, C., Allam, Z., Chabaud, D., Gall, C., & Pratlong, F. (2021). Introducing the "15-minute city": Sustainability, resilience and place identity in future post-pandemic cities. *Smart Cities, 4*, 93–111. https://doi.org/10.3390/smartcities4010006

Neal, P. (2003). *Urban villages and the making of communities* (1st ed.). Taylor & Francis.

Newman, P., & Kenworthy, J. (1999). *Sustainability and cities: Overcoming automobile dependence*. Island Press.

Olcina, J. (2022). Land use planning and green infrastructure: Tools for natural hazards reduction. In S. Eslamian & F. Eslamian (Eds.), *Disaster risk reduction for resilience: Disaster risk management strategies* (pp. 129–146). Springer International Publishing. https://doi.org/10.1007/978-3-030-72196-1_6

Pajević, F. (2021). The Tetris office: Flexwork, real estate and city planning in Silicon Valley North, Canada. *Cities*. https://doi.org/10.1016/j.cities.2020.103060

Papastergiou, E., Kalogeresis, A., & Pozoukidou, G. (2024). Subjective (social) well-being in the neighbourhood: A conceptual exploration. In *Presented at the Game Changer? Planning of just and sustainable regions*. AESOP.

Patnaik, M. (2021). *What are the principles of urban sustainability [WWW document]*. RTF | Rethinking The Future. https://www.re-thinkingthefuture.com/sustainable-architecture/a4249-what-are-the-principles-of-urban-sustainability/ (Accessed 12 Nov 24)

Patricios, N. N. (2002). Urban design principles of the original neighbourhood concepts. *Urban Morphology, 6*, 21–32. https://doi.org/10.51347/jum.v6i1.3900

Perry, C. A. (1929). The neighborhood unit-a scheme of arrangement for the family-life community, regional survey of New York and its environments. In *Neighborhood and community planning*. Committee of Regional Plan of New York and Its Environs.

Phillips, A., Plastara, D., Khan, A. Z., & Canters, F. (2023). Integrating public perceptions of proximity and quality in the modelling of urban green space access. *Landscape and Urban Planning, 240*, 104875. https://doi.org/10.1016/j.landurbplan.2023.104875

Plastara, D., & Pozoukidou, G. (2023). 15-minute cities: A literature overview exploring international planning strategies. In *18th international conference on environmental science and technology (CEST2023)*. ISSN: 2944-9820.9820. https://doi.org/10.30955/gnc2023.00391

Plastara, D., & Pozoukidou, G. (2024). Advancing proximity oriented development in the 15mC paradigm: A bibliometric analysis. In *Proximity planning international congress. Universidad Politécnica de Madrid. June 26th–28th 2024 Madrid.*

Pozoukidou, G. (2020). Designing a green infrastructure network for metropolitan areas: A spatial planning approach. *Euro-Mediterranean Journal for Environmental Integration, 5*, 40. https://doi.org/10.1007/s41207-020-00178-8

Pozoukidou, G. (2024). Urban resilience and new patterns of urban organization: The case of the city of 15 minutes. In *Spatial resilience: Conceptual approaches, policies, applications* (pp. 227–256). Kleidarithmos.

Pozoukidou, G. (2026). Temporal transformations and prospects in spatial planning: From neutrality to adaptability and flexibility. In D. Melissas & A. Tasopoulou (Eds.), *Participation and transparency in spatial planning processes: Social, technological and institutional innovation*. National Technical University of Athens. (in press).

Pozoukidou, G., & Angelidou, M. (2022). Urban planning in the 15-minute city: Revisited under sustainable and smart city developments until 2030. *Smart Cities, 5*, 1356–1375. https://doi.org/10.3390/smartcities5040069

Pozoukidou, G., & Chatziyiannaki, Z. (2021). 15-minute city: Decomposing the new urban planning Eutopia. *Sustainability, 13*, 928. https://doi.org/10.3390/su13020928

Pozoukidou, G., Evdou, M., & Valatidou, E. (2018). The use of soil sealing indicators in identifying peri urban development patterns. *RELAND: International Journal of Real Estate & Land Planning, 1*, 409–416. https://doi.org/10.26262/reland.v1i0.6506

Preston, J., & Rajé, F. (2007). Accessibility, mobility and transport-related social exclusion. *Journal of Transport Geography, 15*, 151–160. https://doi.org/10.1016/j.jtrangeo.2006.05.002

Proffitt, D. G., Bartholomew, K., Ewing, R., & Miller, H. J. (2019). Accessibility planning in American metropolitan areas: Are we there yet? *Urban Studies, 56*, 167–192.

Qualtrics (2024). The definitive guide to text analysis [WWW Document]. Qualtrics. https://www.qualtrics.com/experience-management/research/text-analysis/ (Accessed 13 Nov 24).

Rappaport, J. (2008). Consumption amenities and city population density. *Regional Science and Urban Economics, 38*, 533–552.

Rudskoy, A., Ilin, I., & Prokhorov, A. (2021). Digital twins in the intelligent transport systems. In *Transportation research procedia* (pp. 927–935). Elsevier B.V. https://doi.org/10.1016/j.trpro.2021.02.152

Saar, M., & Palang, H. (2009). The dimensions of place meanings. *Living Reviews in Landscape Research, 3*. https://doi.org/10.12942/lrlr-2009-3

Sanders, T. I. (2008). Complex systems thinking and new urbanism. In *New urbanism and beyond: Designing cities for the future* (pp. 275–279). Rizzoli.

Schlossberg, M., & Brown, N. (2004). Comparing transit-oriented development sites by walkability indicators. *Transportation Research Record, 1887*, 34–42. https://doi.org/10.3141/1887-05

Sebesvari, Z., Woelki, J., Walz, Y., Sudmeier-Rieux, K., Sandholz, S., Tol, S., Ruíz García, V., Blackwood, K., & Renaud, F. G. (2019). Opportunities for considering green infrastructure and ecosystems in the Sendai Framework Monitor. *Progress in Disaster Science, 2*, 100021. https://doi.org/10.1016/j.pdisas.2019.100021

Shannon, K. (2014). The city in the city. Berlin: A green archipelago—Oswald Mathias Ungers and Rem Koolhaas with Peter Riemann, Hans Kollhoff and Arthur Ovaska. A critical edition by Florian Hertweck and Sébastien Marot. *Journal of Landscape Architecture, 9*, 77–77. https://doi.org/10.1080/18626033.2014.931723

Sharifi, A. (2016). From garden city to eco-urbanism: The quest for sustainable neighborhood development. *Sustainable Cities and Society, 20*, 1–16. https://doi.org/10.1016/j.scs.2015.09.002

Shatu, F., & Kamruzzaman, M. (2021). Determining optimum design density for 20-minute neighbourhoods. *Findings*. https://doi.org/10.32866/001c.27391

Steuteville, R. (2017). *25 great ideas of the new urbanism [WWW document]*. CNU. https://doi.org/10/31/25-great-ideas-new-urbanism

Stojanovski, T. (2020). Urban design and public transportation—Public spaces, visual proximity and Transit-Oriented Development (TOD). *Journal of Urban Design*.

Struwe, L. (2021). How can Perceived Proximity support sustainable urban planning for health equality on the neighbourhood level? (Bachelor).

Teixeira, J. F., Silva, C., Seisenberger, S., Büttner, B., McCormick, B., Papa, E., & Cao, M. (2024). Classifying 15-minute cities: A review of worldwide practices. *Transportation Research Part A: Policy and Practice, 189*, 104234. https://doi.org/10.1016/j.tra.2024.104234

Teller, J. (2021). Urban density and Covid-19: Towards an adaptive approach. *Buildings and Cities, 2*, 150–165. https://doi.org/10.5334/bc.89

Thomas, R., & Bertolini, L. (2017). Defining critical success factors in TOD implementation using rough set analysis. *Journal of Transport and Land Use, 10*. https://doi.org/10.5198/jtlu.2015.513

UN-Habitat. (2018). SDG Indicator 11.3.1 Training Module: Land Use Efficiency.

UN-Habitat. (2022). *World cities report 2022: Envisaging the future of cities | UN-Habitat*. United Nations Human Settlements Programme.

United Nations. (2015). *Transforming our world: The 2030 agenda for sustainable development*. (No. A/RES/70/1). United Nations.

van Arem, B., et al. (2022). Accessibility of urban regions on a low car diet—A research agenda for digital twins. *Communications in Transportation Research, 2*, 100077. https://doi.org/10.1016/j.commtr.2022.100077

van Wee, B. (2016). Accessible accessibility research challenges. *Journal of Transport Geography, 51*, 9–16. https://doi.org/10.1016/j.jtrangeo.2015.10.018

World Health Organization. Regional Office for Europe. (2016). *Urban green spaces and health*. (No. WHO/EURO:2016-3352-43111-60341). World Health Organization.

Yu, R., Burke, M., & Raad, N. (2019). Exploring impact of future flexible working model evolution on urban environment, economy and planning. *Journal of Urban Management*. https://doi.org/10.1016/j.jum.2019.05.002

Defining and Measuring the 15-Minute City: Towards Consistent Methodologies for Urban Livability

Efthymis Papadopoulos and Alexandros Sdoukopoulos

1 Introduction

The 15-minute city, hailed as the next urban planning paradigm, aims to transform cities into sustainable, healthy environments where residents can access most of their daily needs through active mobility. While this concept may seem to be a new planning paradigm, its core ideas can be traced back to proximity-based models from the early twentieth century (Wang et al., 2024), including Ebenezer Howard's "Garden City Movement" of 1898, Clarence Perry's "Neighbourhood Unit" concept from the 1920s (Perry, 1929), as well as the "Compact City" (Dantzig & Saaty, 1973), and "The City Within the City" (Krier, 1984).

The 15-minute city concept was shaped and popularized by Carlos Moreno, a French-Colombian Professor at the IAE—Paris1 Sorbonne University. Moreno introduced this novel perspective of "chrono-urbanism" in 2016 as a response to current global and local challenges such as climate change, social inequality, and urban sprawl (Moreno et al., 2021). He advocated for decentralized cities, where residents can easily meet their needs for work, education, healthcare, shopping, and entertainment within a 15-minute walk or bike ride from their homes (C40 Cities Climate Leadership Group, 2020; Moreno et al., 2021; Pozoukidou & Chatziyiannaki, 2021). The 15-minute city concept aligns with the United Nations' Sustainable Development Goals, particularly those related to sustainable cities and communities, climate action, and good health and well-being (Ferrer-Ortiz et al., 2022). By reducing reliance on cars, it aims to promote urban sustainability, local resilience,

E. Papadopoulos (✉)
Laboratory of Transportation Engineering, Department of Civil Engineering, Aristotle University of Thessaloniki, Thessaloniki, Greece
e-mail: efthympg@civil.auth.gr

A. Sdoukopoulos
Mobilités, Research and Consulting, Thessaloniki, Greece

L. Bouzouina et al. (eds.), *15-Minute Cities*, Footprints of Regional Science,
https://doi.org/10.1007/978-3-032-20840-8_4

place identity, inclusivity, and quality of life (Moreno et al., 2021). To this end, the concept encapsulates four dimensions: proximity, density, diversity, and digitalization (Moreno et al., 2021).

The concept of the 15-minute city has the potential to significantly improve various aspects of urban life and provide solutions to some of the biggest challenges faced by cities today. One of its significant benefits is its ability to reduce environmental impact and fight against climate change. By designing cities around proximity and walkability, the 15-minute city minimizes the need for cars, which are responsible for significant levels of air pollution, carbon emissions, and traffic noise (Allam, Bibri, et al., 2022; Allam, Nieuwenhuijsen, et al., 2022; Khavarian-Garmsir et al., 2023). Another key advantage of the 15-minute city concept is its potential to improve public health. By encouraging walking and cycling as primary means of transport, this model promotes active lifestyles, reducing the risk of obesity, heart disease, and other health conditions linked to physical inactivity. Additionally, the proximity to parks, sports facilities, and outdoor public spaces enhances mental well-being, providing opportunities for relaxation, exercise, and socialization. Reducing long commutes also lowers stress levels and enhances work-life balance, allowing residents more time for leisure, family, and community activities. The 15-minute city also addresses social equity, a pressing issue in many urban areas. In traditional city layouts, access to essential services such as healthcare and green spaces is often unevenly distributed, with specific communities suffering from a lack of proximity to these resources. By ensuring that every neighborhood has access to such amenities within a short walk or bike ride, the 15-minute city can reduce these inequalities, creating more inclusive and equitable urban environments (Khavarian-Garmsir et al., 2023). Moreover, the 15-minute city fosters stronger local economies. The decentralization of urban functions encourages the growth of small businesses, local shops, and markets, as residents rely more on nearby services instead of distant commercial centers (Allam, Bibri, et al., 2022; Allam, Nieuwenhuijsen, et al., 2022; Khavarian-Garmsir et al., 2023). This can help revitalize local economies, creating jobs and opportunities within neighborhoods, and fostering a sense of community. Mixed-use developments, where residential areas coexist with commercial and public services, support local entrepreneurship and enhance economic resilience by reducing dependence on large, centralized business districts. Finally, by distributing businesses at the neighbourhood level and focusing on proximity to jobs, the 15-minute city concept contributes to lower energy consumption and enables residents to save on transport costs (Khavarian-Garmsir et al., 2023).

Despite its great potential benefits, it was only during the COVID-19 pandemic that interest in the 15-minute city accelerated significantly (Allam, Nieuwenhuijsen, et al., 2022; Ferrer-Ortiz et al., 2022). Global lockdowns and the shift to remote work made people reconsider their relationship with the urban environment. The pandemic highlighted the importance of having nearby accessible services, as many people experienced challenges relying on distant amenities and facing long commutes during periods of restriction. Cities like Paris, under the leadership of Mayor Anne Hidalgo, embraced the concept by redesigning streets to promote active mobility and localized living. At the same time, a growing body of scholars began exploring and contributing to the 15-minute city concept, resulting in an increasing volume of academic research on the subject.

Based on a brief bibliometric analysis using data from Scopus to explore the evolution and context of the relevant literature, approximately 300 research papers focused on the 15-minute city concept were published between 2020 and September 2024. As shown in Fig. 1, which presents key findings from this analysis, the annual number of publications has grown steadily, reflecting the increasing scholarly interest in the concept. This is further supported by the significant number of citations—around 3600—demonstrating the broad impact and engagement with this body of work.

Figures 2 and 3 depict the results of a co-occurrence analysis of author and index keywords from these papers, conducted using VOS Viewer software. The authors of these papers tend to associate the 15-minute city concept with a relatively small set of terms, the most frequent being proximity, accessibility, walkability, and urban planning. In contrast, the index keywords reveal a broader range of associations, with terms like human and climate change standing out as indicative of the concept's wider implications.

The popularity of the 15-minute city concept is evident in the growing number of media articles, social media posts, and policy discussions, making it a focal point in conversations about post-pandemic urban recovery as cities explore ways to redesign public spaces, repurpose streets, and support localized economies. According to Papadopoulos et al. (2023), who examined the concept's popularity using Google

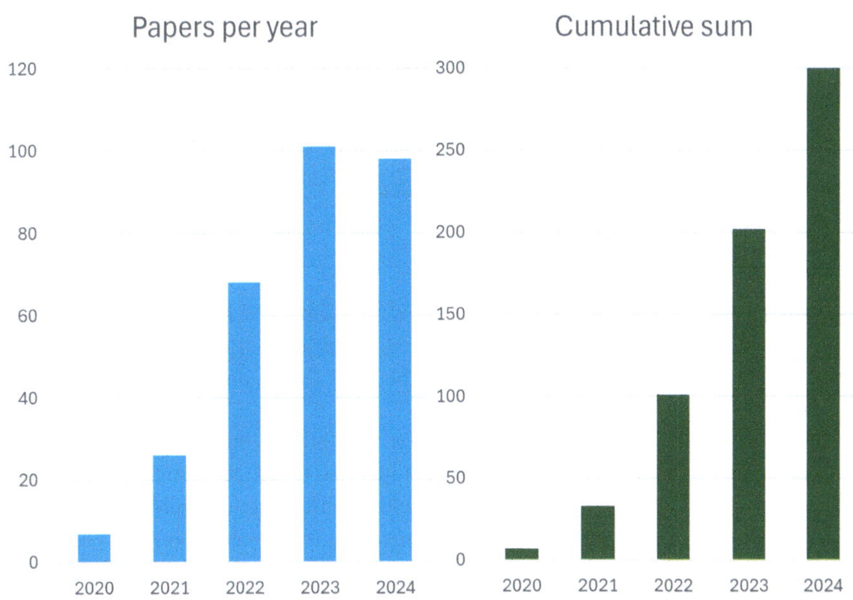

Fig. 1 Evolution over time of the relevant literature

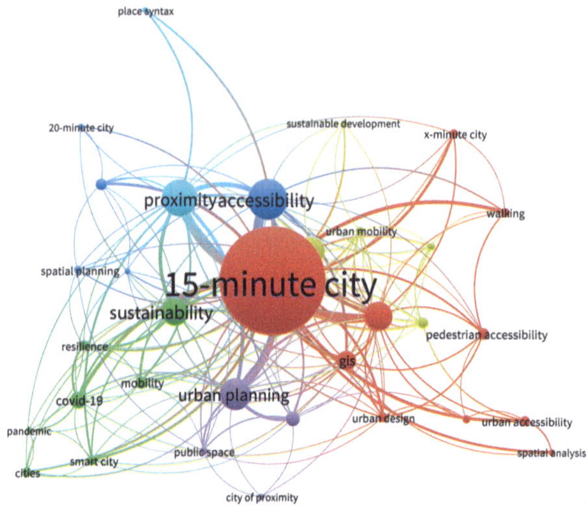

Fig. 2 Co-occurrence analysis of author keywords in the relevant literature

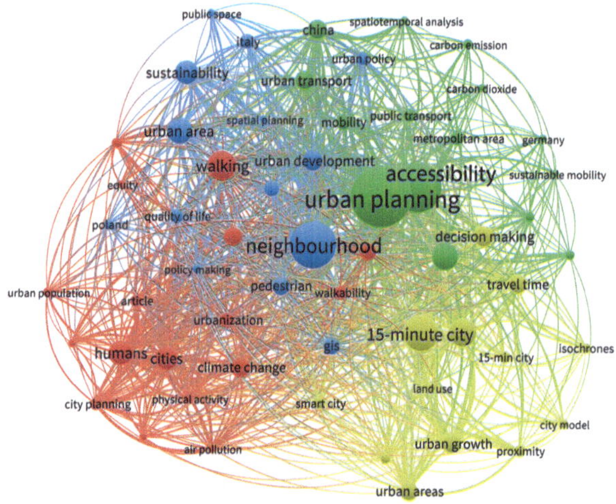

Fig. 3 Co-occurrence analysis of index keywords in the relevant literature

Trends data, interest peaked in March 2022, with a sustained high level of engagement since then. The concept has attracted significant global attention, with European countries like Ireland, France, the UK, Italy, and Germany leading the way.

However, despite its growing popularity, not all attention surrounding the 15-minute city has been positive. The concept has also faced criticism and backlash, with some arguing that its implementation is either unrealistic or could lead to unintended negative consequences such as gentrification or concerns about urban control and surveillance. Headlines reflecting this skepticism include: "The 15-Minute City Is a Distracting Utopia," "The 15-Minute City: An Extraordinarily Bad Idea," "The 15-Minute City Meets Human Needs but Leaves Desires Wanting–Here's

Why," and "Transformation or Gentrification? The Hazy Politics of the 15-Minute City" (Bertaud, 2024; Herbert, 2021; O'Toole, 2024; Ratti & Florida, 2021).

Part of the negative reaction to the 15-minute city stems from the fact that the concept includes many dimensions and aspects that require a clearer definition. For instance, while the core ideas of walkability and proximity are relatively straightforward, the specifics of implementing this model—such as which amenities should be prioritized or whether the 15-minute trip refers to a one-way or round trip—remain unclear. Additionally, questions surrounding the evaluation of other dimensions, such as how they should be measured, whether they are easily quantifiable, and the relationships between them, further complicate the concept. This lack of clarity has resulted in significant variations in the methodologies used to assess progress toward the concept's implementation.

In this context, the aim of this chapter is to discuss the ambiguous aspects of the 15-minute city concept as highlighted by the diverse methodologies developed so far. By doing so, it seeks to offer guidance in establishing a common framework for the assessment process. This will not only enable a more accurate evaluation of cities' current performance but also help refine the concept, facilitating its transition from theory to practice and promoting more livable and accessible urban neighborhoods.

The remainder of the chapter is structured as follows: Sect. 2 briefly presents how researchers have approached the dimensions of the 15-minute city concept in their attempt to assess urban performance relative to the concept's objectives, while also discussing the challenges that arise when measuring the 15-minute city at this higher level of its structure. Moving to a lower level of analysis, Sect. 3 shifts focus to the practical challenges encountered in measuring the 15-minute city, specifically addressing the various components that are inherently involved in the performance evaluation process. Finally, Sect. 4 summarizes the conclusions of this chapter.

2 Evaluating the 15-Minute City: Current Approaches and Conceptual Concerns at the Higher Level of Its Structure

The implementation of the 15-minute city concept is based on four key dimensions: proximity, density, diversity, and digitalization. As initially proposed by Moreno et al. (2021), the core framework of the 15-minute city concept comprises three physical dimensions (proximity, density, and diversity), with digitalization serving as an additional supportive component within a smart city ecosystem.

Proximity is widely regarded as the most critical dimension among these, having been considered in all pertinent studies that assess cities' performance in relation to the concept objectives. This underscores its indispensable role in shaping the 15-minute city idea, as it directly relates to the accessibility of essential amenities within a short distance or time. Density emerges as the second most frequently considered dimension, predominantly contextualized in terms of population density. In this context, studies that accounted for the density dimension typically measured population density at each spatial scale of analysis (Papadopoulos et al., 2023).

Diversity has only been considered in a few studies when measuring compliance with the 15-minute city principles. Often, it is addressed indirectly through proximity metrics, such as measuring the variety of amenities accessible within a specified distance or time threshold (e.g., Abdelfattah et al., 2022; Fazio et al., 2023; Knap et al., 2023). Some studies, however, have explored diversity more explicitly by employing metrics like Shannon Entropy or Simpson's Diversity Index to evaluate land use mix (Di Marino et al., 2022; Graells-Garrido et al., 2021). Notably, as proposed by Moreno et al. (2021), diversity is twofold and extends beyond land use mix to encompass multiculturalism, i.e., diversity in culture and people. While a few studies approached the 15-minute city concept from an equity-focused perspective (examining socioeconomic, spatial, and age inequalities; e.g., Knap et al., 2023; Calafiore et al., 2022; Ma, 2020; Weng et al., 2019), none have attempted to measure diversity in terms of cultural or demographic variation.

Similarly, digitalization has been largely overlooked in the existing literature, with no studies specifically addressing this dimension. This gap highlights the lack of well-defined methods and metrics to capture and quantify digitalization at the neighborhood level, despite its importance in supporting the actualization of the other three dimensions (Moreno et al., 2021).

Lessons from the relevant literature indicate that, at a more aggregated or higher level of analysis, specific dimensions of the 15-minute city concept are consistently overlooked when evaluating compliance with its guidelines. This trend is particularly evident with dimensions such as digitalization and diversity, specifically in terms of multiculturalism. The consistent neglect of these aspects gives rise to concerns about the extent to which all dimensions of the 15-minute city concept can be genuinely quantified. Are some dimensions more inherently measurable than others, or does the lack of standardized methods and metrics contribute to their omission? This issue points to the need for the development of more robust evaluation tools than can adequately capture the complexity of these often-ignored dimensions.

Even when all dimensions are quantified, interpreting their outcomes as simply "good" or "bad" is not always straightforward. For instance, proximity can be easily evaluated using metrics like the percentage of population within a 15-minute walk of core destinations or the number of amenities accessible in that timeframe. In such cases, a higher value generally signifies better performance. However, this clarity does not extend to dimensions like density: although higher density is frequently regarded as an advantageous factor in urban efficiency, the assumption that "more is better" does not apply indefinitely. Beyond a certain threshold, which is not easily determined, increased density may lead to overcrowding or a decline in quality of life, introducing a more nuanced—and sometimes negative—interpretation of performance.

Similarly, the absence of universally accepted thresholds for evaluating compliance with the 15-minute city concept necessitates that the final determination be based on the subjective interpretations of researchers. For example, if the population density of a study area is calculated to be 16,000 people/km^2, does this figure meet the density objectives of the 15-minute city concept, or should a higher or lower density be targeted? Likewise, if a proximity-related metric indicates that 75% of city parcels have access to core amenities within a 15-minute walk, is this sufficient to be considered compliant with proximity goals, or should a threshold of 90% be met? If the latter is true, would an area with 88% accessibility not be considered compliant, while one with 90% would? These questions highlight the need for standardized benchmarks to ensure consistent evaluations of progress toward the concept's objectives.

Perhaps the most crucial insight derived from the literature on measuring progress towards the 15-minute city goals is that, almost universally, the concept dimensions are treated as separate, individual aspects of the 15-minute city framework, with only a few exceptions such as Sdoukopoulos et al. (2024). In other words, researchers have tended to quantify and measure each dimension independently, employing specific indicators to assess each one. This fragmented approach limits the understanding of the interplay between the concept's dimensions and precludes a comprehensive view of the "big picture," diminishing the ability to effectively capture and summarize the complex, multi-dimensional realities of the 15-minute city concept. At the same time, some researchers conceptualize certain dimensions as being interrelated and embedded within others. For instance, they may evaluate diversity by utilizing proximity metrics, such as counting the number of different amenities or amenity categories reachable within the 15-minute threshold.

The above raises a conceptual concern regarding the overarching structure of the 15-minute city concept, particularly in terms of how its dimensions are treated and understood: are the dimensions of the 15-minute city distinct entities, or do they converge towards a common underlying principle, that is, proximity? Perhaps it all comes down to proximity, which is why it is regarded as the fundamental principle underpinning the 15-minute city concept? Can a city or urban area excel in proximity while performing poorly in diversity or density? Alternatively, might the dimensions of the 15-minute city concept be strongly interrelated, such that high performance in one aspect implies or facilitates performance in others? Are all dimensions equally important? Or is digitalization supplementary? (Fig. 4).

What is practically implied here is that urban and economic dynamics largely determine a region's performance in a uniformly positive or negative manner across the different dimensions of the 15-minute city concept, particularly with regard to private sector services. The private sector often favors locations with larger populations, as this increases the likelihood of patronage for various services and businesses. Conversely, a location with many and varied amenities within a short distance typically attracts a higher population density. This is, of course, not the case with public land uses, such as schools, hospitals, parks, and other public

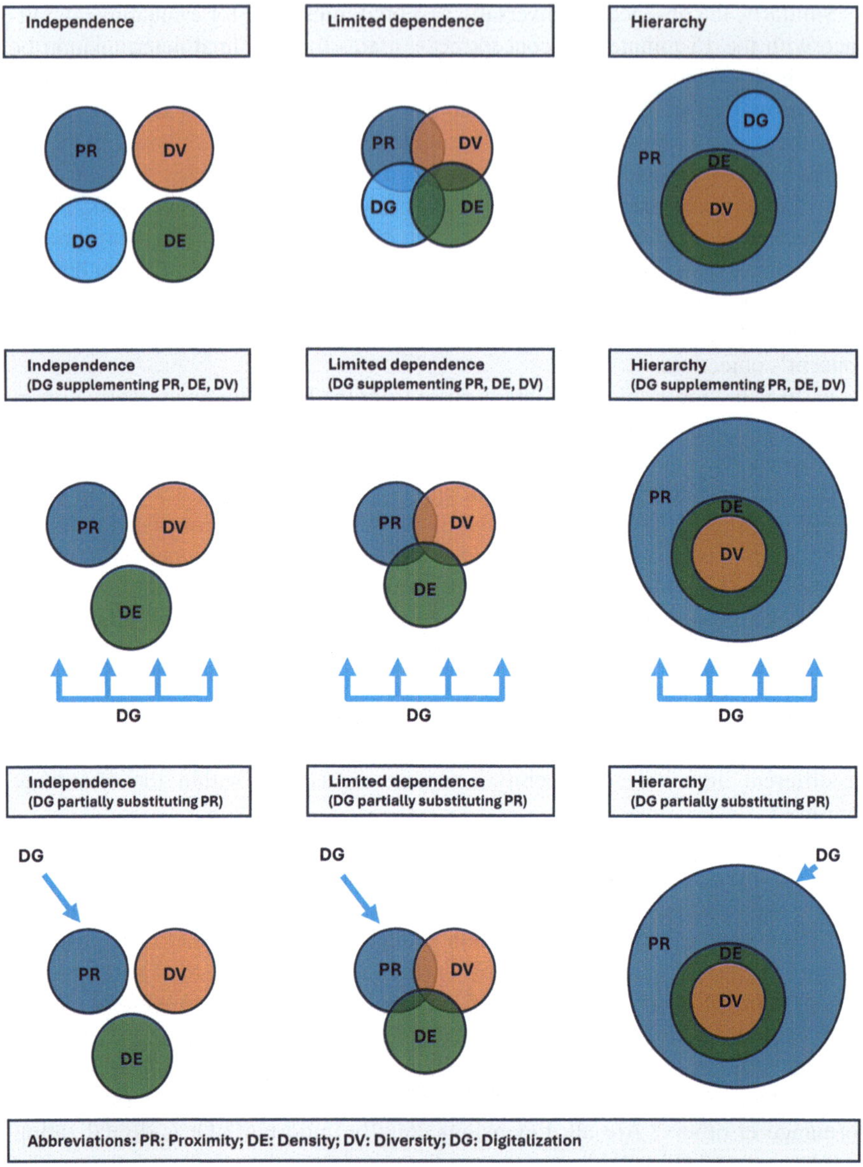

Fig. 4 Different versions of the overarching structure of the 15-minute city concept

services, as the allocation of such services and spaces is often driven by community needs, equity considerations, and public policy, rather than direct profitability.

In some cases, the dimensions of the 15-minute city concept may even be contextualized as substitutes for each other. This may be particularly true for proximity and digitalization. To illustrate this point, one might examine an extreme scenario in which residents of an urban neighborhood lack access to any grocery store, supermarket, or restaurant within a 15-minute walking distance. However, they are able to utilize real-time delivery services, such as online grocery shopping, to receive food and essentials within that same timeframe. This situation raises an intriguing question of whether this urban neighborhood still meets the 15-minute city proximity objectives. While the absence of physical amenities suggests noncompliance, the availability of digital solutions might challenge traditional interpretations of proximity. Nevertheless, this potential substitution of proximity with digitalization could only apply to a limited set of amenities, and not to the majority of essential services. Furthermore, it is at odds with the core values that underpin the 15-minute city concept, including inclusivity, resident participation, and fostering a sense of community, which are considered integral design attributes.

A broader understanding of the relationships and interconnections between the different concept dimensions, and their examination in a more holistic and comprehensive manner when assessing city performance in relation to its objectives, would not only facilitate a more precise and integrated evaluation process but also encourage a reconsideration of the concept's basic principles, at this higher level of its structure. Although the 15-minute city represents an inspiring urban planning concept for fostering more livable and well-connected urban environments, the challenge lies in quantifying and measuring the current level of compliance with its objectives. This is the first and decisive step towards its wide adoption and implementation. Therefore, developing robust, integrated, and holistic methodologies is imperative to truly grasp the multifaceted nature of the 15-minute city concept.

3 Current Approaches and Measurement Challenges in the 15-Minute City: Navigating Complexities beyond the Concept

In addition to the variations identified in the pertinent literature, particularly regarding the higher-level structure of the 15-minute city concept (i.e., the concept dimensions and how they have been interpreted and contextualized when measuring city performance in relation to the concept's goals), there exists significant heterogeneity in the individual, intricate components also involved in the evaluation process (Papadopoulos et al., 2023). This aspect can be considered a lower, more disaggregated level of analysis. These components also reflect the diverse ways in which the 15-minute city concept has been interpreted and applied in different research studies—this time at a lower level—and relate to defined trip origins, selected urban amenities (trip destinations), considered transport modes and travel speeds, the network chosen for analysis, time or distance thresholds used, and the consideration of one-way or round trips.

3.1 Defining Trip Origins and Selecting Trip Destinations

Determining trip origins is a logical initial step in the subsequent analysis of how well cities and urban areas perform in relation to the 15-minute city model requirements. Typically, residential addresses serve as the actual origins of the journeys that people make to access different urban amenities. However, the relevant literature largely relies on more aggregated spatial units—mainly census tracts and city blocks—due to a lack of more granular population data (Papadopoulos et al., 2023). While a few studies have commendably used individual residences as trip origins (Birkenfeld et al., 2023; Guzman et al., 2021; Zhang et al., 2023), these cases are the exception rather than the rule, highlighting a potential limitation and raising concerns about whether the commonly selected spatial units accurately reflect the true starting points of residents' trips.

Accurately defining trip origins cannot be overstated, since it directly influences the precision of distance calculations and consequently, the exact number of residents that have access to core amenities within the defined time threshold. In cases where disaggregated population data is unavailable, the use of centroids of barycenters for aggregated spatial units offers a reasonable approximation of distance from trip origins to destinations. While this option can mitigate some of the inaccuracies introduced by using broader spatial units, it still does not fully account for the finer details of population distribution within those units. More sophisticated methodologies, such as population-weighted approaches, could significantly enhance precision by factoring in the actual distribution of residents within spatial units, leading to a more accurate measurement and representation of accessibility.

Selecting trip destinations, or urban amenities expected to be accessible within the defined time or distance threshold, represents another important component of the performance evaluation process that naturally follows the selection of trip origins. This step is crucial since it determines which amenities will be evaluated for their proximity and accessibility from the chosen trip origins. The relevant literature shows great variation in this component, with a wide range of urban amenities designated as trip destinations (Papadopoulos et al., 2023; Sdoukopoulos et al., 2024). Even without rigid selection criteria, certain essential amenities, including kindergartens, primary schools, high schools, grocery stores, supermarkets, parks, and pharmacies, have been consistently selected across the relevant studies, owing to their indispensable and non-substitutable role in everyday life. Transport-related urban amenities, such as bus and metro stations/stops, as well as metropolitan-scale amenities, including hospitals and universities, have also occasionally been selected as trip destinations (Burke et al., 2022; Gorrini et al., 2023; Rhoads et al., 2023; Weng et al., 2019). In contrast, job opportunities have been largely overlooked.

Meanwhile, it is common for the selected urban amenities to be grouped into broader categories, typically referred to as amenity categories. Although the specifics of the classification may vary, the literature generally converges on several key categories, including food/grocery stores, education, and health services. It should be noted here that this classification can be particularly important in certain cases, as the metrics employed to assess city performance in relation to the 15-minute city objectives are often linked directly to these broader amenity categories, rather than to the individual amenities themselves (Papadopoulos et al., 2023).

The inclusion of transport-related amenities among the destinations that should be reachable within a 15-minute walk can be viewed as an extension of the 15-minute city concept from a within-neighborhood framework to a between-neighborhoods framework. Even if these amenities typically serve as intermediate points in a trip chain rather than final destinations, public transport is considered crucial to access spatially distributed opportunities beyond the neighborhood or city scale. Therefore, considering these amenities as potential trip destinations can provide valuable insights into connectivity and accessibility at a spatial level that extends beyond the neighborhood.

In contrast, certain urban amenities inherently reference the city scale rather than the neighborhood scale. However, the amenities considered in the performance evaluation process are anticipated to be reachable at the neighborhood level. Consequently, amenities such as hospitals and universities should not be considered potential trip destinations, since they are typically offered at a metropolitan or city-wide level, not at the local or neighborhood level. Ultimately, access to such services is strongly affected by the availability of transportation facilities within neighborhoods, with the inclusion of such amenities as potential trip destinations have already been highly encouraged.

Localizing employment opportunities represents a key ambition of the 15-minute city idea, yet the existing literature on evaluating compliance with its goals largely ignores to include job opportunities as potential trip destinations (Papadopoulos et al., 2023). The localization of employment could be understood in two ways: indirectly, as a driver of local economic growth, and directly, by ensuring that residents are able to access employment within a short, reasonable timeframe. Notably, Moreno et al. (2021) identified "working" as one of the six essential urban social functions that residents should be able to effectively fulfil to sustain a decent quality of life. Furthermore, time spent on daily commuting could instead be used for activities that foster a sense of community, which is largely in line with the concept aspirations. The systematic omission of job opportunities from the list of potential trip destinations may be attributed to data availability challenges, as comprehensive employment data at granular spatial levels is frequently hard to obtain.

At the same time, relevant studies have shown that individuals are typically willing to commute to work for extended periods, frequently up to 30 minutes, regardless of the transport mode they select (Logan et al., 2022; Wu et al., 2021). This is in stark contrast to the shorter times generally deemed acceptable for other activities. This suggests that commuting is a distinctly different transport function from local traveling. However, incorporating employment opportunities as potential trip destinations could provide valuable insights into the potential of citizens to live and work in the same urban neighborhood. To this end, disaggregated employment data are required, which could be further combined with transport accessibility measures to assess proximity and ease of access to local jobs.

Given the usual grouping of amenities into broader categories, careful consideration should be given to the selection of metrics used for evaluating city's performance in relation to the 15-minute city goals. To prevent misleading outcomes, it is essential to account for the lack of substitutability between different amenities that fall within the same category. For example, while primary and secondary schools are usually included in the same broader category, that of educational facilities, they serve distinct purposes and age groups and therefore both need to be present to meet local needs. Thus, metrics that fail to recognize this distinction—such as the percentage of residents that have access to at least one amenity per category—may lead to inaccurate assessments.

Finally, in the broader context of selecting trip origins and destinations, radical approaches that depart from the conventional method of selecting specific destinations, instead treating any point in space as a potential trip destination, are highly encouraged. Such approaches have already emerged in the relevant literature, recognizing other types of walking activities beyond strictly utilitarian, destination-based trips (Alawadi et al., 2022). Taking this a step further, if trip origins are similarly generalized, the 15-minute city framework could be extended beyond its traditional focus on home-based trips, opening up new possibilities for assessing accessibility across urban spaces.

3.2 *Selecting Transport Modes, Travel Speeds, and Network Specifications*

In relation to the transport modes used to access urban amenities, the existing literature predominantly emphasizes walking, with only a few studies incorporating cycling and public transport (Birkenfeld et al., 2023; Capasso Da Silva et al., 2020; Hosford et al., 2022). While the consideration of these alternatives is a valuable addition, prioritizing walking ensures that findings remain conservative and less optimistic, since the distances achievable on foot within the same time window are consistently shorter than those possible with other transport modes. Nonetheless, cycling and public transit are essential complements, particularly regarding destinations that lie beyond the 15-minute walking radius. Therefore, integrating cycling and public transport is highly encouraged to expand methodologies for assessing the X-minute city, although this would require additional data, some of which may be challenging to obtain (such as public transit frequency).

In terms of average travel speed, the existing literature on measuring compliance with the 15-minute city concept commonly assumes a walking speed between 4 km/h and 5 km/h, which is consistent with the definitions frequently found in relevant studies (Papadopoulos et al., 2023). Distinctively, a few approaches have employed varied walking speeds to account for different weather conditions and, most importantly, specific demographic groups often recognized as vulnerable road users, including children and older individuals (Hosford et al., 2022; Willberg et al.,

2023). However, these average walking speed values were often applied uniformly across populations to develop different scenarios, rather than reflecting the actual demographic composition within specific areas. The above highlights a gap in detailed demographic data, which is crucial for enhancing the accuracy of assessments. In cases where such detailed information is unavailable, selecting an average walking speed below 5 km/h might be prudent to accommodate all road users, particularly the more vulnerable groups.

Along with the selection of transport modes and the assignment of travel speeds, another critical component of the performance evaluation process is, of course, the choice of the network utilized in the analysis. The road network serves as the canvas for the subsequent analysis, shaping how distances are measured and how accessibility to facilities is assessed. The importance of network selection is further highlighted by the fact that the relevant literature almost universally agrees on using network distance rather than Euclidean distance as the distance measure for performing proximity analysis (Papadopoulos et al., 2023; Sdoukopoulos et al., 2024). Although the predominant approach involves the inclusion of the entire road network without applying any inclusion/exclusion criteria, this raises concerns about the representativeness of the selected network in capturing the true dynamics of urban mobility, particularly for vulnerable populations.

In response to this limitation, a few recent studies have taken network characteristics into account when selecting the network utilized for analysis. Relevant approaches include excluding road segments deemed unsuitable for walking or cycling, incorporating network characteristics into the calculation of travel impedance between origins and destinations—specifically when calculating the travel time along the street network—and, in some cases, focusing solely on the sidewalk network rather than the broader road network (Capasso Da Silva et al., 2020; Gaglione et al., 2022; Rhoads et al., 2023). These refined approaches offer a more precise depiction of the actual conditions influencing accessibility, especially for vulnerable road users, while emphasizing walkability as a central element in the proximity measurement process, making them highly recommended.

3.3 *Determining Time Thresholds*

The time threshold used for measuring city performance in terms of accessibility to local services is inextricably linked to the concept itself, as evidenced by its name and its chrono-urban essence. As such, determining the time threshold is undoubtedly one of the key components of the performance evaluation process, as it sets the time window within which selected urban amenities are anticipated to be accessible, either by walking or through other transport modes considered in the analysis. As demonstrated in the extant literature, the 15-minute threshold is the most prevalent, indicating a heightened popularity of the 15-minute city concept in comparison to its variants, such as the 20-minute or the 45-minute city.

It is crucial to acknowledge that the selection of the time threshold is often contingent upon the distinctive urban characteristics of the area under examination. In general, longer time thresholds—such as 20 minutes or more—are typically used when evaluating accessibility in cities with lower population densities, where sprawling urban landscapes result in longer travel times. Typical examples of such cities include American and Australian cities. In contrast, shorter time thresholds—especially the 15-minute threshold—are favored in high-density cities across Europe and Asia, where the compact urban structure facilitates easier access to services (Papadopoulos et al., 2023). This, of course, indicates that the 15-minute city is not a one-size-fits-all city model: findings from proximity analyses, when combined with the use of different time thresholds, suggest that the 15-minute proximity target should be re-evaluated and tailored to fit the unique characteristics of each individual case area. Besides, relevant findings reveal that many European cities are already largely aligned with the 15-minute city proximity objectives, which is not typically the case in more spread-out cities across America and the Middle East. Overall, cities should be assessed based on the city model they adopt.

Although the time thresholds employed in the literature vary, a common feature of the analyses is the uniform application of these thresholds across all selected urban amenities, without any differentiation between them. However, a few studies have deviated from this approach, using different time thresholds for the different amenities, based on the specific characteristics of each urban function. In these cases, factors such as the frequency of use, the essential role of amenities in everyday life, as well as the time residents are willing to spend accessing each service were taken into account when determining the varying time thresholds (Ferrer-Ortiz et al., 2022). Additionally, some studies used varied time thresholds for both amenity types and population groups, acknowledging that the distance decay effect may vary considerably not only across the different amenities, but also across the different populations (Weng et al., 2019).

This flexibility in defining time thresholds is undoubtedly an important enhancement in evaluating city performance against the 15-minute city objectives, as it enables more accurate assessments. By adjusting thresholds to accommodate the distinct characteristics of different urban amenities and the diverse needs of different population groups, the analysis better reflects real-world conditions, ultimately leading to more precise and meaningful insights. However, it is essential to consider a wide variety of criteria when defining appropriate, acceptable time thresholds for each type of amenity.

3.4 Clarifying Trip Types: One-Way or Round Trips?

A piece of information that is systematically missing from research on measuring city compliance from a 15-minute city perspective is whether the proximity analyses are performed for one-way trips or for round trips (aller-retour). This distinction is of great importance as the time threshold for accessing urban services is frequently framed in terms of a one-way journey. For example, a common assertion is that residents should be able to reach a specific destination, such as a grocery store or a park, within a 15-minute walk. However, this raises an important question: does this time threshold account for the return journey? A coarse assumption that a 15-minute travel time suffices for a round trip may lead to significant discrepancies in accessibility evaluations.

An analysis solely based on one-way trips is insufficient for accurately reflecting the full trip chain that residents have to undertake, as it fails to account for the return journey to their homes. To illustrate further, a resident may successfully reach their intended destination within the 15-minute threshold but could then face an additional 15 minutes on their way back, resulting in a total travel time of 30 minutes. This is contrary to the foundational premise of the 15-minute city model, which aims to enhance local accessibility and minimize travel times.

Consequently, ignoring this distinction may lead to misleading conclusions about how well cities and urban areas meet the 15-minute city proximity objectives. It may foster a false sense of achievement, suggesting that urban areas are more accessible than they genuinely are. Of course, the above could be reduced to a higher conceptual question: does the 15-minute proximity target pertain solely to one-way trips, i.e., journeys from residences to specific urban amenities, or does it also encompass round trips, including the return to home journeys? Given the fundamental objectives of the 15-minute city model, future research should focus on round trips when evaluating city performance in relation to the 15-minute city goals, thereby tightening the time window within which core amenities should be accessible.

4 Concluding Remarks: Towards a Widespread Implementation of the 15-Minute City Concept

In response to the challenges posed by the COVID-19 pandemic, the concept of a city that can be traversed in 15 minutes has emerged as a novel and insightful urban vision. By centering urban planning around proximity to core amenities and promoting local living to foster community-oriented lifestyles, the 15-minute city model aims to tackle the critical issues that modern cities face. Despite its great potential for reshaping urban environments and enhancing quality of life, its successful implementation and widespread adoption rely heavily on developing appropriate methodologies for measuring cities' current level of compliance, which in turn requires a deeper understanding of its fundamental elements.

As revealed by the pertinent literature, despite its seemingly simple and measurable premise, the 15-minute city concept has been interpreted and applied differently in the different research studies. These variations reflect the different conceptualizations that arise when attempting to quantify and evaluate the concept in practice, and derive from the concept's inherently broad and fluid definition. When it comes to measuring compliance with the 15-minute city objectives, several measurement challenges arise. These challenges can be grouped into two broad categories: those at the higher-level structure of the concept, which are directly related to the concept dimensions, and those at the lower level, which refer to the various, intricate components that are inherently involved in the performance evaluation process. These two categories can also be viewed as the conceptual and practical challenges of evaluating the 15-minute city, though this distinction is not always clearly defined.

Regarding the former, proximity, density, diversity, and digitalization—the four dimensions of the 15-minute city concept—are often treated in a fragmented manner, limiting a more comprehensive understanding of how these dimensions interact and relate to each other. Moreover, the observed inability to quantify specific dimensions, namely diversity and digitalization, and the absence of standardized benchmarks for all dimensions further complicate the evaluation process. Future research should go beyond fragmented approaches, shifting its focus on the development of robust methodologies that can quantitatively capture the multifaceted nature of the 15-minute city concept, including often-ignored dimensions.

As for the latter, several distinct approaches have already emerged in the literature that deviate from common practice and further enhance the performance evaluation process, by artfully addressing the practical challenges related to its various components. Regardless of which specific component these approaches focus on, they all share two main objectives: to further tighten up the performance evaluation process, which aligns with existing findings that show the 15-minute proximity objective is already largely met, at least in dense European cities, and to account for diverse population groups and their varying needs, giving a more equity-related twist to the 15-minute city concept.

Of course, many of these challenges that emerge when attempting to measure and evaluate the 15-minute city could not have been fully anticipated or precisely defined during the initial conceptualization phase. However, now that the concept has been thoroughly explored and debated, these challenges have to be addressed in order to move on from theory to practice, paving the way for the successful implementation and widespread adoption of this inspiring urban planning idea.

References

Abdelfattah, L., Deponte, D., & Fossa, G. (2022). The 15-minute city: Interpreting the model to bring out urban resiliencies. *Transportation Research Procedia, 60*, 330–337. https://doi.org/10.1016/j.trpro.2021.12.043

Alawadi, K., Khanal, A., Mouselly, A., & Aletaywi, A. B. (2022). Planning in the age of pandemics: Renewing suburban design. *Sustainable Cities and Society, 87*, 104261. https://doi.org/10.1016/j.scs.2022.104261

Allam, Z., Bibri, S. E., Chabaud, D., & Moreno, C. (2022). The '15-minute city' concept can shape a net-zero urban future. *Humanities and Social Sciences Communications, 9*(1), 1–5. https://doi.org/10.1057/s41599-022-01145-0

Allam, Z., Nieuwenhuijsen, M., Chabaud, D., & Moreno, C. (2022). The 15-minute city offers a new framework for sustainability, liveability, and health. *Lancet Planet Health, 6*(3), e181–e183. https://doi.org/10.1016/S2542-5196(22)00014-6

Bertaud, A. (2024). The 15-minute city is a distracting utopia. *Market Urbanism*. Retrieved from https://marketurbanism.com/2024/08/06/the-15-minute-city-is-a-distracting-utopia/. Accessed October 2, 2024.

Birkenfeld, C., Victoriano-Habit, R., Alousi-Jones, M., Soliz, A., & El-Geneidy, A. (2023). Who is living a local lifestyle? Towards a better understanding of the 15-minute-city and 30-minute-city concepts from a behavioural perspective in Montréal, Canada. *The Journal of Urban Mobility, 3*, 100048. https://doi.org/10.1016/j.urbmob.2023.100048

Burke, J., Gras Alomà, R., Yu, F., & Kruguer, J. (2022). Geospatial analysis framework for evaluating urban design typologies in relation with the 15-minute city standards. *Journal of Business Research, 151*, 651–667. https://doi.org/10.1016/j.jbusres.2022.06.024

C40. (2020). How to build back better with a 15-minute city. Retrieved from https://www.c40knowledgehub.org/s/article/How-to-build-back-better-with-a-15-minute-city?language=en_US. Accessed October 2, 2024.

Calafiore, A., Dunning, R., Nurse, A., & Singleton, A. (2022). The 20-minute city: An equity analysis of Liverpool City region. *Transportation Research Part D: Transport and Environment, 102*, 103111. https://doi.org/10.1016/j.trd.2021.103111

Capasso Da Silva, D., King, D. A., & Lemar, S. (2020). Accessibility in practice: 20-minute city as a sustainability planning goal. *Sustainability, 12*(1), 129. https://doi.org/10.3390/su12010129

Dantzig, G. B., & Saaty, T. L. (1973). Compact city: A plan for a liveable urban environment.

Di Marino, M., Tomaz, E., Henriques, C., & Chavoshi, H. S. (2022). The 15-minute city concept and new working spaces: A planning perspective from Oslo and Lisbon. *European Planning Studies*. https://doi.org/10.1080/09654313.2022.2082837

Fazio, M., Borghetti, F., Giuffrida, N., Le Pira, M., Longo, M., Ignaccolo, M., Inturri, G., & Maja, R. (2023). The "15-minutes station": A case study to evaluate the pedestrian accessibility of railway transport in Southern Italy. *Transportation Research Procedia, 69*, 536–543. https://doi.org/10.1016/j.trpro.2023.02.205

Ferrer-Ortiz, C., Marquet, O., Mojica, L., & Vich, G. (2022). Barcelona under the 15-minute city lens: Mapping the accessibility and proximity potential based on pedestrian travel times. *Smart Cities, 5*(1), 146–161. https://doi.org/10.3390/smartcities5010010

Gaglione, F., Gargiulo, C., Zucaro, F., & Cottrill, C. (2022). Urban accessibility in a 15-minute city: A measure in the city of Naples, Italy. *Transportation Research Procedia, 60*, 378–385. https://doi.org/10.1016/j.trpro.2021.12.049

Gorrini, A., Presicce, D., Messa, F., & Choubassi, R. (2023). Walkability for children in Bologna: Beyond the 15-minute city framework. *The Journal of Urban Mobility, 3*, 100052. https://doi.org/10.1016/j.urbmob.2023.100052

Graells-Garrido, E., Serra-Burriel, F., Rowe, F., Cucchietti, F. M., & Reyes, P. (2021). A city of cities: Measuring how 15-minutes urban accessibility shapes human mobility in Barcelona. *PLoS One, 16*(5), e0250080. https://doi.org/10.1371/journal.pone.0250080

Guzman, L. A., Arellana, J., Oviedo, D., & Moncada Aristizábal, C. A. (2021). COVID-19, activity and mobility patterns in Bogotá. Are we ready for a '15-minute city'? *Travel Behaviour and Society, 24*, 245–256. https://doi.org/10.1016/j.tbs.2021.04.008

Herbert, J. (2021). *Transformation or gentrification? The hazy politics of the 15-minute city*. Resilience. Retrieved from https://www.resilience.org/stories/2021-09-02/transformation-or-gentrification-the-hazy-politics-of-the-15-minute-city/. Accessed October 5, 2024

Hosford, K., Beairsto, J., & Winters, M. (2022). Is the 15-minute city within reach? Evaluating walking and cycling accessibility to grocery stores in Vancouver. *Transportation Research Interdisciplinary Perspectives, 14*, 100602. https://doi.org/10.1016/j.trip.2022.100602

Khavarian-Garmsir, A. R., Sharifi, A., & Sadeghi, A. (2023). The 15-minute city: Urban planning and design efforts toward creating sustainable neighborhoods. *Cities, 132*, 104101. https://doi.org/10.1016/j.cities.2022.104101

Knap, E., Baran Ulak, M., Geurs, K. T., Mulders, A., & van der Drift, S. (2023). A composite X-minute city cycling accessibility metric and its role in assessing spatial and socioeconomic inequalities—A case study in Utrecht, The Netherlands. *The Journal of Urban Mobility, 3*, 100043. https://doi.org/10.1016/j.urbmob.2022.100043

Krier, L. (1984). The city within the city, Les Quartiers. *Architectural Design, 54*(7/8).
Logan, T. M., Hobbs, M. H., Conrow, L. C., Reid, N. L., Young, R. A., & Anderson, M. J. (2022). The x-minute city: Measuring the 10, 15, 20-minute city and an evaluation of its use for sustainable urban design. *Cities, 131*, 103924. https://doi.org/10.1016/j.cities.2022.103924
Ma, F. (2020). Spatial equity analysis of urban green space based on spatial design network analysis (sDNA): A case study of Central Jinan, China. *Sustainable Cities and Society, 60*, 102256. https://doi.org/10.1016/j.scs.2020.102256
Moreno, C., Allam, Z., Chabaud, D., Gall, C., & Pratlong, F. (2021). Introducing the "15-minute city": Sustainability, resilience and place identity in future post-pandemic cities. *Smart Cities, 4*(1), 93–111. https://doi.org/10.3390/smartcities4010006
O'Toole, R. (2024). *The 15-minute city: An extraordinarily bad idea*. Frontier Centre for Public Policy. Retrieved from https://fcpp.org/2024/03/28/the-15-minute-city-an-extraordinarily-bad-idea/. Accessed October 5, 2024
Papadopoulos, E., Sdoukopoulos, A., & Politis, I. (2023). Measuring compliance with the 15-minute city concept: State-of-the-art, major components and further requirements. *Sustainable Cities and Society*, 104875. https://doi.org/10.1016/j.scs.2023.104875
Perry, C. A. (1929). The neighborhood unit: A scheme of arrangement for the family-life community. *Reg. Plan New York: Its Environment, 7*, 2–140.
Pozoukidou, G., & Chatziyiannaki, Z. (2021). 15-minute city: Decomposing the new urban planning Eutopia. *Sustainability, 13*(2), 928. https://doi.org/10.3390/su13020928
Ratti, C., & Florida, R. (2021). *The 15-minute city meets human needs but leaves desires wanting. Here's why*. World Economic Forum. Retrieved from https://www.weforum.org/agenda/2021/11/15minute-city-falls-short/. Accessed October 4, 2024
Rhoads, D., Solé-Ribalta, A., & Borge-Holthoefer, J. (2023). The inclusive 15-minute city: Walkability analysis with sidewalk networks. *Computers, Environment and Urban Systems, 100*, 101936. https://doi.org/10.1016/j.compenvurbsys.2022.101936
Sdoukopoulos, A., Papadopoulos, E., Verani, E., & Politis, I. (2024). Putting theory into practice: A novel methodological framework for assessing cities' compliance with the 15-min city concept. *Journal of Transport Geography, 114*, 103771. https://doi.org/10.1016/j.jtrangeo.2023.103771
Wang, T., Li, Y., Chuang, I. T., Qiao, W., Jiang, J., & Beattie, L. (2024). Evaluating the 15-minute city paradigm across urban districts: A mobility-based approach in Hamilton, New Zealand. *Cities, 151*, 105147. https://doi.org/10.1016/j.cities.2024.105147
Weng, M., Ding, N., Li, J., Jin, X., Xiao, H., He, Z., & Su, S. (2019). The 15-minute walkable neighborhoods: Measurement, social inequalities and implications for building healthy communities in urban China. *Journal of Transport and Health, 13*, 259–273. https://doi.org/10.1016/j.jth.2019.05.005
Willberg, E., Fink, C., & Toivonen, T. (2023). The 15-minute city for all?—Measuring individual and temporal variations in walking accessibility. *Journal of Transport Geography, 106*, 103521. https://doi.org/10.1016/j.jtrangeo.2022.103521
Wu, H., Avner, P., Boisjoly, G., Braga, C. K. V., El-Geneidy, A., Huang, J., Kerzhner, T., Murphy, B., Niedzielski, M. A., Pereira, R. H. M., Pritchard, J. P., Stewart, A., Wang, J., & Levinson, D. (2021). Urban access across the globe: An international comparison of different transport modes. *NPJ Urban Sustainability, 1*, 16. https://doi.org/10.1038/s42949-021-00020-2
Zhang, S., Wu, W., Xiao, Z., Wu, S., Zhao, Q., Ding, D., & Wang, L. (2023). Creating livable cities for healthy ageing: Cognitive health in older adults and their 15-minute walkable neighbourhoods. *Cities, 137*, 104312. https://doi.org/10.1016/j.cities.2023.104312

Policy and Public Support

Exploring People's Support of 15-Minute City Policy Implementation and Its Intersection with the Concept's Perceived Benefits and Drawbacks

Jessica Wei-Lin Lam, Ariel Arosemena, and Ehab Diab

1 Introduction

The 15-Minute City (15MC), or X-Minute City planning concept, has gained much recent popularity since its conception in 2016 (Moreno et al., 2021). The concept is a chrono-urbanistic planning approach that can be linked to historical approaches to city planning, in which people can find a wide array of opportunities in closely accessible locations (Allam et al., 2022a; Moreno et al., 2021). This is also known in the literature as proximity planning (Büttner et al., 2024). More importantly, the concept builds upon the overarching accessibility concept, which has been strongly supported by many researchers over the past couple of decades.

Accessibility refers to the ease of reaching specific destinations by certain modes within a certain travel time (Levinson & Krizek, 2018). The concept of a 15MC differs in its ease of communication and simplicity of understanding (Logan et al., 2022), compared to the concept of accessibility, which is often found to be unclearly defined and not operationalized in many city and regional plans (Boisjoly & El-Geneidy, 2017). The concept offered a buzzword easily adopted by policymakers, making it more popular. The concept focuses on improving people's proximity to a wide array of destinations that are needed daily and accessible via active travel modes (e.g., walking and cycling), decreasing the importance of using cars, and reducing the need for longer distance trips (C40 Cities Climate Leadership Group, 2020; Moreno et al., 2021). Ultimately, the 15MC is a proximity enhancement tool that mitigates climate change, one of the

J. W.-L. Lam · A. Arosemena · E. Diab (✉)
Department of Geography & Planning, University of Saskatchewan, Saskatoon, SK, Canada
e-mail: ehab.diab@usask.ca

L. Bouzouina et al. (eds.), *15-Minute Cities*, Footprints of Regional Science,
https://doi.org/10.1007/978-3-032-20840-8_5

greatest problems of the current generation, while also having social sustainability benefits (Marquet et al., 2024b). Thus, a growing number of cities have incorporated the concept into their plans to achieve a wide range of goals (Büttner et al., 2024; Lu & Diab, 2023). For example, recent research has shown that over 15 cities in Canada, the USA, and Australia have already incorporated the concept in their planning documents, while over 10 cities are currently implementing this concept (Lu & Diab, 2023).

A growing number of researchers delved into exploring different aspects of the 15MC concept. Recent works have explored the concept's development and engaged in theoretical discourse (Allam et al., 2022a; Moreno et al., 2021), provided critique (Herbert, 2021; Khavarian-Garmsir et al., 2023), and assessed existing urban conditions through a 15MC lens (Capasso Da Silva et al., 2020; Di Marino et al., 2023; Kamruzzaman, 2022). Other studies have examined how the concept has been implemented by cities across the world (Büttner et al., 2024; Lu & Diab, 2023) and the public acceptance and perception of the concept (Marquet et al., 2024a, 2024b). Nevertheless, very limited efforts focused on modeling the factors affecting people who support the implementation of such a concept and understanding how these factors intersect a person's perception of the benefits and drawbacks of the 15MC concept. Therefore, the purpose of this research is to explore the factors affecting the likelihood of supporting the implementation of the 15MC concept and how these factors are consistent across different groups of populations. More importantly, it aims to explore the association between people's perception of the benefits and drawbacks of the concept and being in favor of implementing such a concept.

To achieve these goals, the research used data from an online survey collected in 2024 in Saskatoon, Canada. Saskatoon is a typical North American medium-scale, car-denominated city. The survey focused on obtaining data from the University of Saskatchewan (USask) community. USask is a major employment, cultural, and educational hub in the city. Through summary statistics, the support of the 15MC concept was examined in relationship to different factors. Afterward, multilevel mixed-effect logistic regression models and a logistic regression model were used to understand the relationship between the previous factors and how likely a person is to favor the implementation of the 15MC concept in the city. This study contributes to the academic literature and practice by offering one of the first quantitative analyses into understanding factors associated with supporting the 15MC planning concept. In addition, it explores the association between people's perception of benefits and drawbacks of implementing 15MC policies and being in favor of such a concept. This is to help cities redefine their communication strategies and understand which groups of society they should work with to better support the implementation of such a policy. Towards a better-supported implementation of 15MC policies, this analysis identifies societal groups that are less likely to support them, hence allowing for the development of specialized and focused communication strategies.

2 Literatue Review

2.1 15MC Concept

Over the history of urban settlements, theoretical and practical theories on the ideal way to organize cities have been developed. Recent twentieth century concepts include Clarence Perry's neighborhood units (Perry, 1929) and Ebenezer Howard's Garden City (Howard, 1965). The garden city model stressed proximity to nature to improve people's enjoyment through access to nature and services. More recently, the emphasis has broadened to incorporate environmental, social, and economic sustainability (United Nations, 2015). In pursuit of this goal, planning concepts like the eco-city and new urbanism movements emerged (Allam et al., 2022b). Researchers have also called for improved accessibility to opportunities by sustainable modes while developing various accessibility measures to guide cities (Levinson & Krizek, 2018).

Within this context, the 15MC, a term and concept conceived by Carlos Moreno in 2016 (Moreno, 2016; Moreno et al., 2021), has gained considerable popularity among policymakers (Marquet et al., 2024b). The 15MC refers to a city where residents access daily needs like "work, housing, food, health, education, culture and leisure" within 15 min (or x min) using active modes such as walking and cycling (TED, 2021). The COVID-19 pandemic conditions provided additional motivation and reasoning for implementing the 15MC, as it created an urgent need for close-to-home access to essential services of food, healthcare, education, and work (C40 Cities Climate Leadership Group, 2020; Moreno et al., 2021). The concept has additionally been promoted by various organizations such as C40 Cities (Khavarian-Garmsir et al., 2023). In its original conception, the 15MC responds to the critique of modern cities being too monofunctional and argues they should focus on "proximity, diversity, density, and ubiquity" (Moreno, 2016), with a recent update replacing ubiquity with digitalization (Moreno et al., 2021).

2.2 15MC Goals and Critique

Existing academic literature on 15MCs includes concept development and theoretical discourse (Allam et al., 2022a; Moreno et al., 2021), critique (Herbert, 2021; Khavarian-Garmsir et al., 2023), assessment of existing urban conditions (Capasso Da Silva et al., 2020; Di Marino et al., 2023; Kamruzzaman, 2022), and policy application (Lu & Diab, 2023). In addition to academic developments and concept modifications, there have been new conceptions of the 15MC through application by cities globally (Lu & Diab, 2023). The concept is commonly introduced to combat climate change by prioritizing active travel modes and reducing the need for longer trips that use motorized transportation (Allam et al., 2022a). Socially, the 15MC concept aims to benefit both individual health and societal cohesion. Usage of active modes allows for more interaction in the local neighborhood (C40 Cities Climate Leadership Group, 2020; Wen et al., 2022), and through collaborative development and implementation of policy (C40 Cities Climate Leadership Group, 2020; Lu & Diab, 2023; Moreno et al., 2021). A recent report found 94 15MC policies globally from five continents (Büttner et al., 2024).

Although not a key pillar of the concept, equity-supporting tools such as community participation in 15MC development have been discussed (Moreno et al., 2021; Pozoukidou & Chatziyiannaki, 2021). Pozoukidou and Chatziyiannaki (2021) argued for more significant equity measures within 15MC policies by including requirements for affordable housing. The concept has also been criticized as a "utopian buzzword" by news media and academic sources (Gower & Grodach, 2022; Herbert, 2021). Others argue that the concept is hardly implementable in North America and Australia (Birkenfeld et al., 2023; Shannon et al., 2019), continents with similar assessed levels of accessibility (Wu et al., 2021). While others argue that the 15MC principles are achievable by any city, including neighborhoods rich and poor (Birkenfeld et al., 2023). A study by Khavarian-Garmsir et al. (2023) critiqued the concept for overlooking the importance of agglomeration economies, where job concentrations in urban areas add value to some industries.

2.3 Acceptance of the 15MC Concept

As it is a new concept, very few academic studies have explored its public acceptance (Marquet et al., 2024a, 2024b). For example, Marquet et al. (2024b) conducted a literature review and comparative analysis of academic and gray literature in response to documented public backlash against the 15MC concept in Canada, Spain, and the United Kingdom. The study reviewed climate change policy acceptability literature, categorizing existing and potential criticisms of the 15MC under the same umbrella. It categorized criticisms into four main areas: (1) social engineering fears (i.e., fear of restrictive climate policies and fear of loss of individual freedoms), (2) distrust of centralized urban planning (i.e., suspicion towards government-led initiatives, distrust of urban lifestyle), (3) automotive freedom concerns (i.e., automobile dependence, fear of restrictive travel policies, and perceived threat to car-oriented lifestyles), and (4) environmental and gentrification risks (i.e., social exclusion, housing & real estate inequalities, and urban homogenization). The main conclusion of the study was that public education must dispel myths and provide accurate information. In another study, Marquet et al. (2024a) argued that successful implementation of the concept requires technical communication and communication regarding individual and societal health benefits, coupled with community engagement in planning and implementing the policies. However, existing studies have not examined the relationship between the perceived criticisms (or drawbacks) and benefits and people's likelihood of supporting the concept.

3 Methodology

3.1 Context and Data Sources

The research used data from an online survey collected in Saskatoon, Saskatchewan, Canada. Saskatoon is a commercial and educational center located in the central area of the province of Saskatchewan (City of Saskatoon, n.d.); it is the largest city in Saskatchewan, with a total population of 266,000 people in 2021 (Statistics Canada, 2024). Commuting in Saskatoon is dominated by personal vehicles, cars account for over 89.2% of commute mode, while only 3.4% and 5.0% of trips are made by using public transit and active transport modes, respectively (Statistics Canada, 2023a). Nevertheless, Saskatoon has a very ambitious vision to double the use of transit and active transportation modes over the coming decade by implementing an infill development strategy to increase land use density and diversity, and building a new bus rapid transit system (City of Saskatoon, 2023). This vision aims to enhance quality of life while supporting more sustainable city growth. The city currently has no 15MC policy; however, there is a growing interest in developing such a policy.

The survey focused on obtaining data from the University of Saskatchewan (USask) community. USask is the main university in the province of Saskatchewan and is a significant employment, cultural, and educational hub in Saskatoon (University of Saskatchewan, 2023b). In 2023, USask registered 26,155 students and 5430 employees, including faculty and staff (University of Saskatchewan, 2023a). The used survey called the "USask Mobility Survey," was developed in collaboration with USask's sustainability office and USask's transportation services office while incorporating feedback from the City of Saskatoon planners. USask Mobility Survey was an online survey collected between February and April 2024, with a target sample of 5% of the USask population. This sample provides a margin of error of less than ±2.5% at a confidence level of 95%. To encourage a higher participation level, incentives for completing the survey were offered.

As recommended by Dillman et al. (2014), recruitment across different channels was performed to have the best possible response rate and representative sample. Channels included direct emails, social media posts and targeted ads, printed posters, and in-person distribution of brochures. To remove unreliable responses, the data cleaning process removed responses that were incomplete, completed too quickly, and/or those that used the same IP address or NSID NSID is USask's Network Service Identifier, a unique ID assigned to each person by USask. It is used to log in to email and other university services. To identify responses that were filled quickly, the average time for different paths to complete the survey was determined. The fastest 5% of responses by the time of completion for each path were identified manually for removal, which excluded participants who responded consciously to the open-ended questions. This led to the exclusion of ~1.8% of responses. After this data-cleaning process, 1655 responses were deemed complete and valid for analysis. Responses were kept for persons located in the main USask campus in the City of Saskatoon. Therefore, people at other campuses and participants who did not provide a valid postal code were removed. 2.2% of responses were removed as a result, yielding 1619 responses for analysis.

The survey was divided into different sections, including open-ended, multiple-choice, and Likert scale questions. Survey sections included general travel behavior information, telecommuting habits, perceived barriers questions, personal information, and the 15MC concept. This 15MC section asked questions about the respondent's familiarity with 15MCs and their opinions about it. Firstly, participants were asked if they were familiar with the 15MC concept. Following this, the concept was introduced briefly to all participants to ensure that they had a shared understanding. Afterward, questions regarding "Do you think you live in a 15-minute neighborhood?" and "Are you in favor of implementing "15-minute city" policies in your neighborhood?" were included in the survey. Finally, participants were asked to rate if they agreed with a few theoretical outcomes of the 15MC policy implementation. These questions included asking participants if they agree that benefits such as "15-minute city" will lead to a stronger sense of community or drawbacks such as "15-minute city" will reduce personal freedoms.

Besides perceptions of the 15MC concept, several variables were prepared for the analysis. Table 1 includes a list of variables that were considered in the study. As there was limited literature about the topic, several demographic factors were extracted from the survey, including age, gender, ethnicity, education, position at USask, and others. Other variables obtained from the survey include those related to transport and commuting, neighborhood, and residential self-selection factors. For example, questions related to commuting mode to USask, the number of cars a respondent had access to, and whether they worked from home were also extracted.

Additionally, several locations and land use factors were developed. For example, the type of dwelling was extracted from the survey, while dwelling density was extracted from the 2021 Canadian census. Neighborhood names were assigned by analyzing surveyed postal codes and geographic boundaries using ArcGIS PRO. Dummy variables according to location in the city were also developed. For example, participants living inside Circle Drive, outside Circle Drive, and outside Saskatoon City were identified. *Circle Drive* is a major highway that was developed as a ring road around the city's most established neighborhoods in the 1960s. Proximity measures, developed by Statistics Canada (2023b), were also extracted and linked to each participant's household. Similarly, accessibility measures by transit during peak and off-peak hours were extracted and linked to each participant.

Proximity measures were generated smallest national spatial unit of analysis managed by Statistics Canada, the Dissemination Block (DB). Each year, the proximity values are generated, and they are normalized across with a minimum value of 0 and maximum value of 1. The values are based on 2021 Statistics Canada Dissemination Blocks, 2021 OpenStreetMap road network, Business Register and Generalized Transit Feeds Specification. The value used is a composite measure of proximity to services at noted distance thresholds: within 1 km: grocery stores, pharmacies, public transportation (based that offers trips between 7:00 a.m. and

Table 1 Variables prepared for the analysis

Variable	Obs.	Mean	Std.	Definition
Socioeconomic factors				
Age	73	35.5	13.6	Calculated age based on the respondent's reported year of birth
Gender				
Women	982	0.61	0.49	Respondent identified as a woman
Men	507	0.31	0.46	Respondent identified as a man
Other genders	130	0.04	0.18	Respondent identified as non-binary, trans, genderqueer/gender non-conforming, two-spirit, or other identity
No answer	57	0.05	0.21	Respondent preferred not to answer
Ethnicity				
Caucasian/White-only	1023	0.63	0.48	Respondent identified as only Caucasian/White
Not Caucasian/White-only	560	0.35	0.48	Respondent identified as not only Caucasian/White
No answer	36	0.02	0.15	Respondent preferred not to answer
Education				
Primary/elementary school	8	0.00	0.07	Respondent's highest education level was a primary/elementary school
Secondary school	286	0.18	0.38	Respondent's highest education level was secondary school
Trade/technical school or diploma	168	0.10	0.31	Respondent's highest education level was trade/technical school or college
Undergraduate degree	531	0.33	0.47	Respondent's highest education level was undergraduate
Graduate degree	571	0.35	0.48	Respondent's highest education level was a graduate
Prefer not to answer	55	0.03	0.18	Respondent did not, or preferred not to answer
Disability status				
Yes	142	0.09	0.28	Respondent identified as having a disability
University position				
Student	831	0.51	0.50	Respondent identified as a student
Faculty/Staff	788	0.49	0.50	Respondent identified as faculty or staff

(continued)

Table 1 (continued)

Variable	Obs.	Mean	Std.	Definition
Annual income group				
Less than 30k/No income	311	0.19	0.39	Annual household income of less than $30,000, and no income
30k to 49k	128	0.08	0.27	Annual household income between $30,000 and $49,000
50k to 99k	311	0.19	0.39	Annual household income between $50,000 and $99,000
100k to 149k	236	0.15	0.35	Annual household income between $100,000 and $149,000
Over 150k	361	0.22	0.42	Annual household income of more than $150,000
Prefer not to answer	272	0.17	0.37	Respondent did not, or preferred not to answer
Dwelling type				
A single-detached house	904	0.56	0.50	Respondent indicated living in a single-detached house
A semi-detached house	69	0.04	0.20	Respondent indicated living in a semi-detached house
A row house	75	0.05	0.21	Respondent indicated living in a row house
An apartment (or condo) in a duplex or triplex	218	0.13	0.34	Respondent indicated living in an apartment (or condo) in a duplex or triplex
An apartment (or condo) in a building that has five or more storeys	185	0.11	0.32	Respondent indicated living in an apartment (or condo) in a building that has five or more storeys
Others	168	0.10	0.31	Respondent indicated living in a seniors' home, mobile home, others, or did not answer
Transport and commuting-related factors				
Main commute mode				
Walk	252	0.16	0.36	Respondent's main commute mode in fall was walking
Ride a bicycle	174	0.11	0.31	Respondent's main commute mode in fall was cycling
Take the bus	385	0.24	0.43	Respondent's main commute mode in fall was the bus
Carpool	126	0.08	0.27	Respondent's main commute mode in fall was carpooling
Drive alone	549	0.34	0.47	Respondent's main commute mode in fall was driving alone
Others	27	0.02	0.13	Respondent's main commute mode in fall was a motorcycle, ride-hailing, or other

Telecommuting all the time	106	0.07	0.25	Respondent's main commute mode in fall was telecommuting
Car access	1413	1.61	0.99	The number of cars the respondent had access to in their household
Neighborhood and location-related factors				
Dwelling density	1586	29,096	128,336	Dwellings per square kilometer
Accessibility measures				
Off-peak public transit accessibility	1619	0.28	0.17	Sum of the normalized value of a DB's (dissemination block) proximity to different facilities by public transit at off peak hour
Peak hour public transit accessibility	1619	0.26	0.16	Sum of the normalized value of a DB's proximity to different facilities by public transit at peak hour
Proximity measures				
Total proximity	1619	0.47	0.27	Sum of the normalized value of a DB's proximity to different facilities
Zone of residence				
Inside city, within Circle Drive	786	0.49	0.50	Respondent resided in Saskatoon, inside Circle Drive
Inside city, outside of Circle Drive	727	0.45	0.50	Respondent resided in Saskatoon, outside Circle Drive
Outside city	106	0.07	0.25	Respondent resided outside Saskatoon
Attitudes and residential self-selection factors				
Future mode preference				
Walk				
Disagree	615	0.38	0.49	Respondent did not desire to walk more in the future
Agree	1004	0.62	0.49	Respondent desired to walk more in the future
Bike				
Disagree	445	0.33	0.47	Respondent did not desire to bike more in the future
Agree	917	0.67	0.47	Respondent desired to bike more in the future
Drive alone				
Disagree	776	0.68	0.47	Respondent did not desire to drive alone more in the future
Agree	371	0.32	0.47	Respondent desired to drive alone more in the future

(continued)

Table 1 (continued)

Variable	Obs.	Mean	Std.	Definition
Residential self-selection, transport preferences				
Prioritize walking				
Not important	679	0.42	0.49	Respondent identified a walking commute as unimportant in the selection of a future residence
Important	895	0.55	0.50	Respondent identified a walking commute as important in the selection of a future residence
N/A	45	0.03	0.16	Respondent did not identify if a walking commute was important in the selection of a future residence
Prioritize biking				
Not important	853	0.53	0.50	Respondent identified a cycling commute as unimportant in the selection of a future residence
Important	714	0.44	0.50	Respondent identified a cycling commute as important in the selection of a future residence
N/A	52	0.03	0.18	Respondent did not identify if a cycling commute was important in the selection of a future residence
Decrease driving cost				
Not important	547	0.34	0.47	Respondent identified driving alone for their commute as unimportant in the selection of a future residence
Important	1025	0.63	0.48	Respondent identified driving alone for their commute as important in the selection of a future residence
N/A	47	0.03	0.17	Respondent did not identify if a driving alone for their commute was important in the selection of a future residence
15MC conceptual perceptions				
Aware of 15MC concept				
Yes	640	0.40	0.49	Respondent heard about the 15MC concept before completing the survey
No	952	0.59	0.49	Respondent had not heard about the 15MC concept before completing the survey
N/A	27	0.02	0.13	Respondent did not answer the question
15MC policy opinion				

Yes	1350	0.83	0.37	Respondent supported the implementation of 15MC city policies in their neighborhood
No	240	0.15	0.36	Respondent supported the implementation of 15MC policies in their neighborhood
N/A	29	0.02	0.13	Respondent did not answer the question
Current neighborhood				
Yes	791	0.49	0.50	Respondent believed they are living in a 15MC neighborhood
No	801	0.49	0.50	Respondent believed they are not living in a 15MC neighborhood
N/A	27	0.02	0.13	Respondent did not answer the question
Benefits of 15MC policy implementation				
Personal automobile reduction				
Disagree	411	0.25	0.44	Respondent disagreed (strongly disagreed or was neutral) to the idea that 15MCs will lead to a reduction in the use of personal automobiles
Agree	1181	0.73	0.44	Respondent agreed or strongly agreed with the idea that 15MCs will lead to a reduction in the use of personal automobiles
N/A	27	0.02	0.13	Respondent did not provide an opinion
Strengthened sense of community				
Disagree	578	0.36	0.48	Respondent disagreed (strongly disagreed or was neutral) to the idea that 15MCs would strengthen the sense of community
Agree	1013	0.63	0.48	Respondent agreed (or strongly agreed) with the idea that 15MCs would strengthen the sense of community
N/A	28	0.02	0.13	Respondent did not provide an opinion
Improved mental health and well-being				
Disagree	522	0.32	0.47	Respondent disagreed (strongly disagreed or was neutral) to the idea that 15MCs would improve the mental health and well-being of residents
Agree	1068	0.66	0.47	Respondent agreed (or strongly agreed) with the idea that 15MCs would improve the mental health and well-being of residents
N/A	29	0.02	0.13	Respondent did not provide an opinion

(continued)

Table 1 (continued)

Variable	Obs.	Mean	Std.	Definition
Drawbacks of 15MC policy implementation				
Increased government surveillance and control				
Disagree	1249	0.77	0.42	Respondent disagreed (strongly disagreed or was neutral) to the idea that 15MCs would increase government surveillance and control
Agree	342	0.21	0.41	Respondent agreed (or strongly agreed) with the idea that 15MCs would increase government surveillance and control
N/A	28	0.02	0.13	Respondent did not provide an opinion
Reduction in personal freedoms				
Disagree	1408	0.87	0.34	Respondent disagreed (strongly disagreed or was neutral) to the idea that 15MCs would reduce personal freedoms
Agree	181	0.11	0.32	Respondent agreed (or strongly agreed) with the idea that 15MCs would reduce personal freedoms
N/A	30	0.02	0.13	Respondent did not provide an opinion
Fines for leaving their "home" districts				
Disagree	1478	0.91	0.28	Respondent disagreed (strongly disagreed or was neutral) to the idea that 15MCs would result in fines if residents left their "home" districts
Agree	112	0.07	0.25	Respondent agreed (or strongly agreed) with the idea that 15MCs would result in fines if residents left their "home" districts
N/A	29	0.02	0.13	Respondent did not provide an opinion

10:00 a.m.), neighborhood parks, within 1.5 km: child care, primary education, and secondary education, and libraries, and finally, within 3 km: healthcare, and within 10 km: employment (Statistics Canada, 2023c).

The accessibility measures, also generated by Statistics Canada at the DB level, are gravity based accessibility measures to seven main categories of amenities: primary and secondary education, postsecondary education, health care, places of employment, grocery stores, cultural and arts facilities, and sports and recreational facilities. There measure is calculated across four modes of transportation—walking, cycling, public transit peak hour, and public transit off-peak hour. The public transit travel times were calculated using the r5r package in R and for the walking and cycling times, Valhalla was used (Statistics Canada, 2023d).

3.2 Data Analysis

This study uses descriptive statistics and statistical models to explore the factors affecting the likelihood of supporting the implementation of the 15MC concept. The descriptive approach explores key socioeconomic, transport and commuting, and perceptions of 15MC benefits and drawbacks variables concerning favoring the implementation of 15MC policies. Descriptive statistics were used to identify patterns, while statistical models were used to isolate and identify the magnitude, direction, and statistical significance of relationships.

The data was modeled using logit and multilevel mixed-effect logistic regression techniques. The use of a multilevel model allows for processing of data that is nested and organized at more than one level (Diab et al., 2014). Whereby differences between entities at a lower level that are systematic are contextualized at a higher level (Luke, 2004). Different nesting approaches were tested, for example, nesting using major city subdivisions versus city neighborhoods. The multilevel mixed-effect logit models were screened using a likelihood ratio (LR) test to compare the multilevel logistic model structure to the logistic model structure. This test was significant for most models, indicating that the multilevel model structure was more appropriate for the analysis using Saskatoon neighborhoods (n=65) at the higher level, while survey participants (n=1579) at the lower level.

The dependent variable for all models was a dummy variable that indicated if a participant was in support of or against the implementation of 15MC policies in the respondent's neighborhood. Initially, different types of variables were explored to understand their statistical significance using T-Test, Chi-square, and Mann-Whitney U-tests. Variables were then tested in the models using a backward stepwise approach, whereby insignificant variables were removed from the models. A Pearson correlation coefficient (greater than 0.7) was another method used to remove variables from the model. The lowest Akaike Information Criterion (AIC) and the highest Log-likelihood (LL) values were used to identify the final model. Accordingly, several factors were removed from the models, including ethnicity, household income, dwelling type, future mode preference, dwelling density, and so on. Awareness of the 15MC concept did not show a significant influence on whether or not someone supported the implementation of the 15MC concept and was therefore removed from the models.

The first model showed differences between genders, leading to the creation of second and third models that were created for the women and men in the sample, respectively. An additional pair of models was generated to compare student versus faculty and staff populations within the data. This is to explore whether the factors associated with being in favor of the 15MC concept are consistent across different groups.

4 Results

4.1 Summary Statistics

4.1.1 Key Socioeconomic Factors

Figure 1 shows support for the 15MC policies implementation by gender identity, ethnicity, and position at USask. By gender identity, support of concept implementation is highest for other gender identities (98%) and women (86%), while men were the least supporting the concept (83%). By ethnicity, those who identified as not white-only showed higher support (87%) than white-only respondents (84%). Among students, support for 15MC policies implementation is considerably higher (90%) than faculty/staff (80%).

4.1.2 Key Transport and Commuting-Related Factors

Figure 2 shows support for 15MC policies among participants by their fall transport mode choices. Support is highest for transit riders (92%), cyclists (91%), and walkers (91%). Among those who drive alone, only 75% are in favor. Participants using other modes (i.e., motorcycle, ride-hailing, or other) have the lowest support, at 70%. Overall, there is a clear pattern of decrease in supporting the concept for people who use cars, either as passengers or drivers, versus those who use other modes. In contrast, people using active transport and transit are the most supportive of the concept implementation.

4.1.3 Neighborhood and Location-Related Factors

Figure 3 shows support for the 15MC policy implementation in relation to proximity and accessibility scores by transit during both peak and off-peak hours. Individuals who support the 15MC implementation tend to have greater proximity

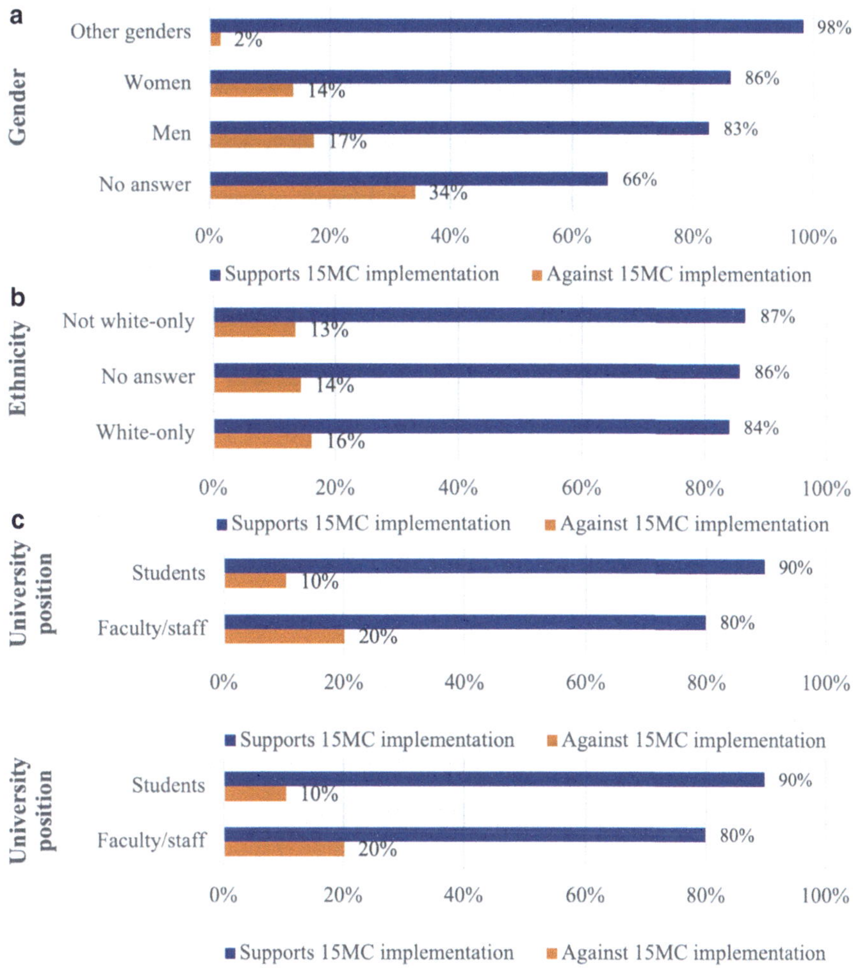

Fig. 1 Proportion of participants who support implementation of the 15MC policies by (**a**) Gender, (**b**) Ethnicity, and (**c**) University position

to facilities (Mean=0.48) compared to those who are against it (Mean=0.40). Where a proximity score of 1 means that a DB has the highest level of relative proximity to ten essential services within Canada and as score of 0, the lowest level of relative proximity to the same ten essential services (i.e., grocery, pharmacy, public transportation, neighborhood parks, childcare, primary and secondary education, libraries, healthcare, and employment) (Statistics Canada, 2023c). Similarly, supporters show higher accessibility via public transport, with the mean of 0.28 and 0.27 for off-peak and peak times, respectively, while participants against have lower accessibility at 0.21 for both periods. Accessibility was calculated as gravity-based accessibility for each DB to seven major destinations,

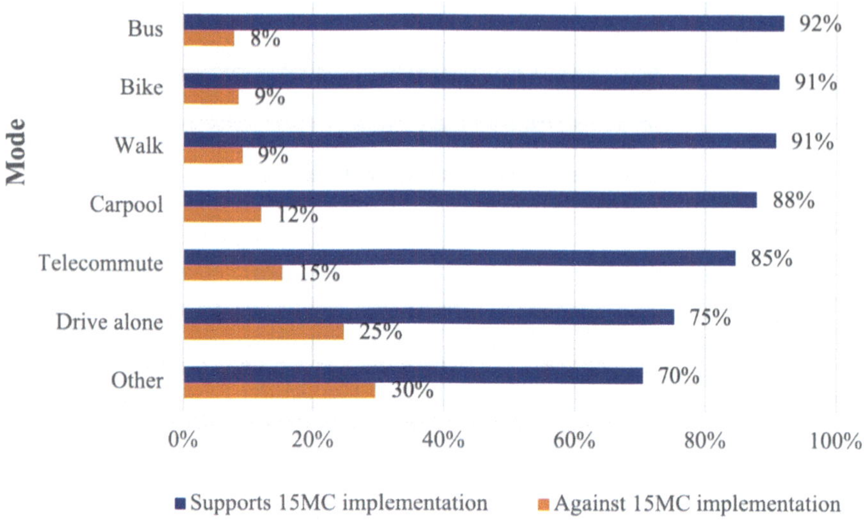

Fig. 2 Proportion of participants who support the implementation of 15MC policies across different commute modes

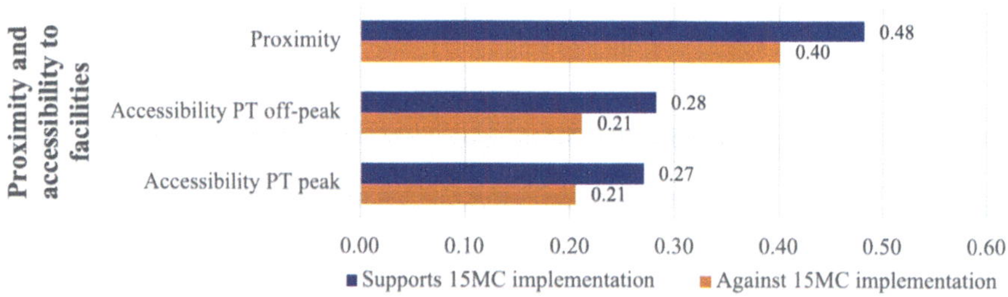

Fig. 3 Proximity and accessibility scores comparing groups in support of and against 15MC policy implementation

including education (primary and secondary, and postsecondary), employment, grocery stores, culture and arts, and sports and recreation facilities (Statistics Canada, 2023d).

Figure 4 shows people's support of the 15MC policy based on the zone of residence, specifically whether individuals live inside Circle Drive and within Saskatoon, outside Circle Drive and within Saskatoon, or outside the city. Among those residing inside Circle Drive, 90% support the 15MC implementation, while 10% are against it. For residents outside Circle Drive, support drops slightly to 84%, with 16% against the policy. Furthermore, individuals living outside the city show lower support with only 54%, for and 46% against the policy implementation.

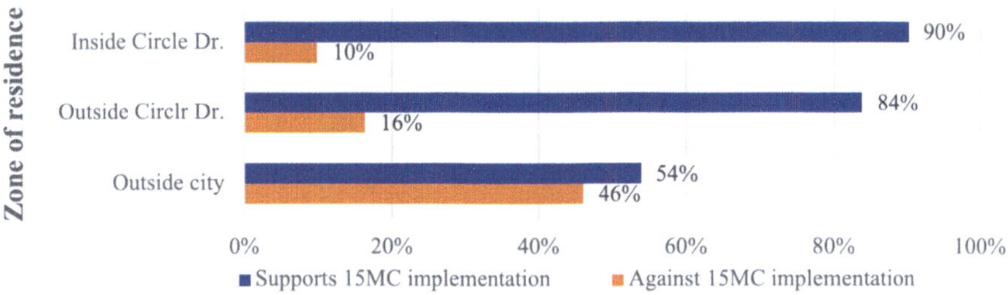

Fig. 4 Proportion of participants who support the implementation of the 15MC policies by zone of residence

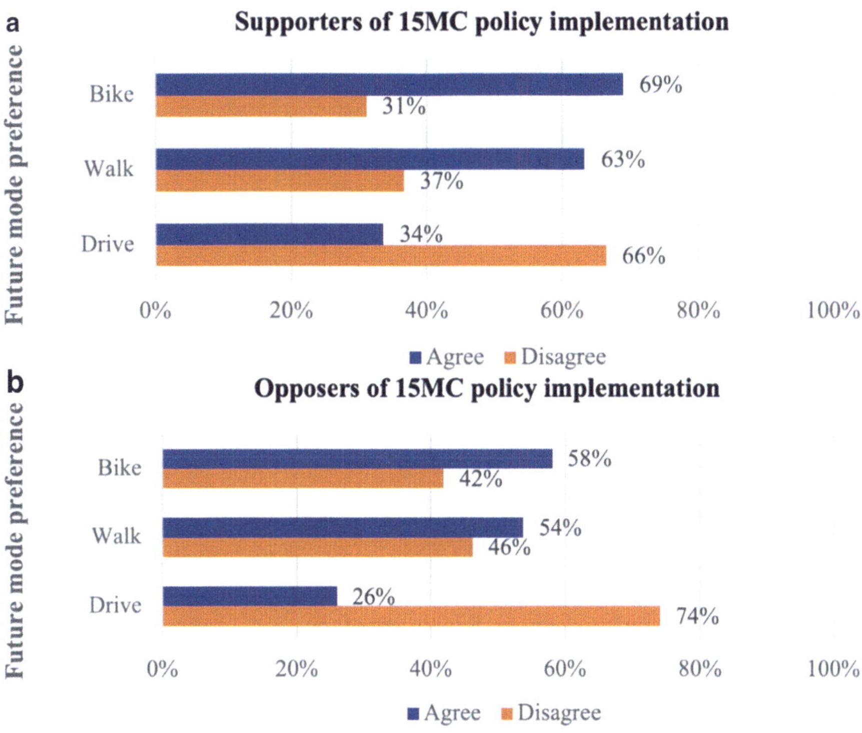

Fig. 5 Mode preference for future trips for (**a**) supporters of 15MC policy implementation, and (**b**) opposers 15MC policy implementation

4.1.4 Attitudes and Residential Self-Selection Factors

Figure 5 presents stated preferences for future mode among groups separated into (a) supporters of 15MC policy implementation and (b) opposers of 15MC policy implementation. Among supporters, most people express a desire to increase cycling (69%) and walking (63%); only 34% wish to drive more. Individuals who opposed

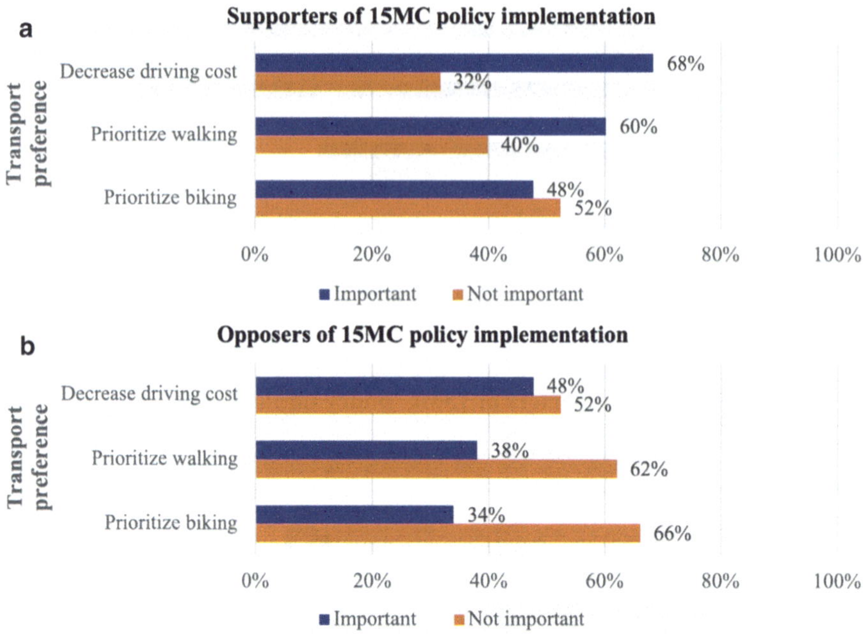

Fig. 6 Transport preferences for residential self-selection of (**a**) supporters of 15MC policy implementation, and (**b**) opposers of 15MC policy implementation

the 15MC policy implementation show a slightly lower inclination toward cycling (58%) and walking (54%). Despite not supporting the implementation of 15MC policies, only a small percentage (26%) wish to drive more. This suggests that future commute mode preferences may not be tied to whether individuals support the 15MC concept or not.

Figure 6 shows transport priorities for future residential location selection among groups separated into (a) supporters of 15MC policy implementation and (b) opposers of 15MC policy implementation. Supporters of the 15MC policy are most interested in decreasing the cost of driving for their commute (68%), followed by prioritizing walking (60%); they do not show a clear preference towards biking to USask (only 52% value its importance). Participants not in favor show less interest in decreasing the cost of driving for their commute (48%). While only 34% and 38% are interested in a new home location that prioritizes walking and cycling to USask, respectively.

4.1.5 Key 15MC Perception Aspects

Figure 7 shows people's support for the implementation of the 15MC concept for those who were aware of the concept before the survey and those who were not. Among individuals who are aware of the concept, 84.6% support its

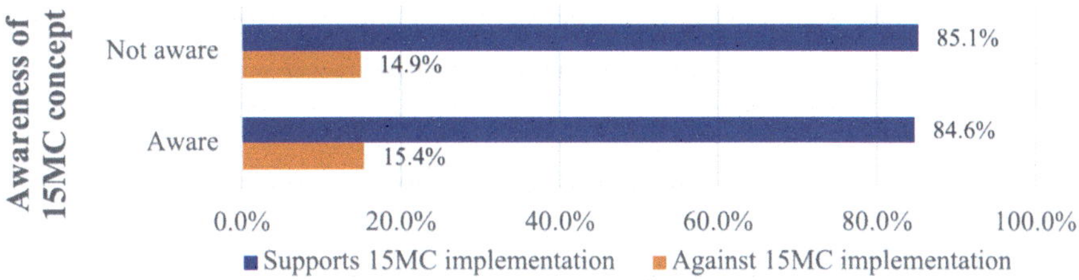

Fig. 7 Proportion of participants who support the implementation of the 15MC policies by previous awareness of the 15MC concept

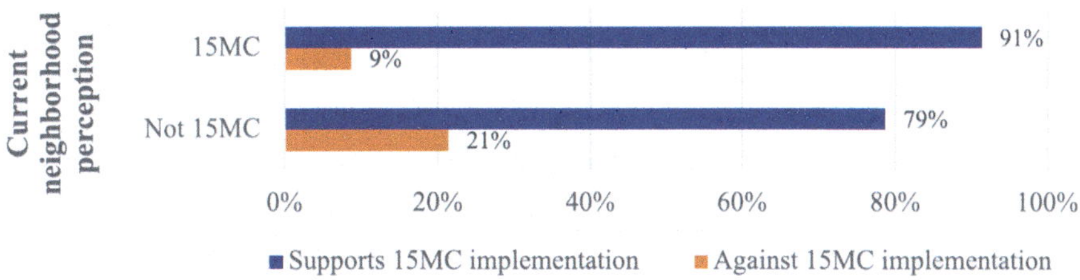

Fig. 8 Proportion of participants who support the implementation of the 15MC policies by perception of current neighborhood

implementation, while 15.4% are against it. Similarly, among those who are not aware of the concept, 85.1% support its implementation, with 14.9% opposing it. These results indicate that previous awareness of the 15MC concept does not significantly influence the level of support, as both groups show similar proportions.

Figure 8 compares support for implementing 15MC policies between those who think they do live in a 15MC neighborhood, and those who do not. Among those who think they already live in a 15MC neighborhood, 91% are in favor of implementing the 15MC policy, and only 9% are not. In contrast, those who think they do not live in such a neighborhood are comparatively less in favor (at 79%) and have a greater proportion who are not in favor (21%). This suggests that those who could possibly already experience the benefits of a 15MC neighborhood are more supportive of the policy.

Figure 9a presents opinions on the benefits and drawbacks of implementing the 15MC concept for those who indicated they support the implementation of the 15MC policy. For positive aspects, 81% agreed that the 15MC policy would reduce the use of personal automobiles, while 19% disagreed. The respondents anticipate a strong sense of community 71%, with 29% disagreeing. Finally, 75% agree it will help increase mental health and well-being, while 25% disagree. On the other hand, for drawbacks, 80% disagree that it will increase government surveillance and control, while 20% agree. Similarly, 8% believe it will reduce personal freedoms, with 92% disagreeing. Additionally, only 6% agree that residents will be fined for leaving their home districts, while 94% disagree.

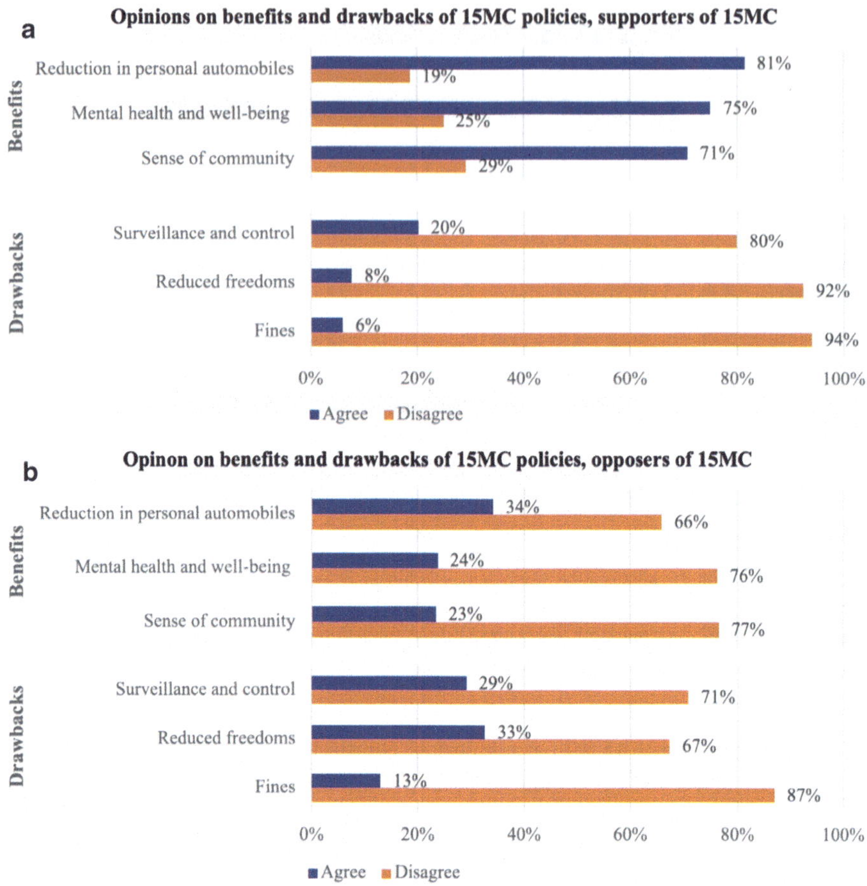

Fig. 9 Opinion on benefits and drawbacks of 15MC policies of (**a**) supporters of 15MC policies; and (**b**) opposers of 15MC policies

Figure 9b shows opinions on the benefits and drawbacks of 15MC policy implementation for those who do not support 15MC policy. Participants do not think that the 15MC policy will help advance goals like the previous group of participants who supported the 15MC policy implementation. Respondents disagree with the benefit categories at rates of 66%, 76%, and 77% with the policy's ability to reduce personal automobile use, improve mental health and well-being, and increase the sense of community, respectively. Despite not supporting the implementation of 15MC policies, the majority (87%) disagree that residents would be fined for leaving their designated districts. Nevertheless, this group is generally more concerned with the idea that 15MC policies could increase government surveillance and control and reduce personal freedoms than the other group of participants who supported 15MC policy implementation.

4.2 Statistical Models

Table 2 shows the results of Model 1, which uses the entire data set (N=1579). Model 1 uses a multilevel mixed-effect logistic regression; it has an LR p-value below 0.001, which indicates a better performance in comparison to an ordinally logistic regression model. Table 3 includes models 3–6, which allow for comparison between subgroups of people who identified as women vs. men, those who held the university position of student vs. faculty and staff, and between those who knew about the 15MC concept before it was explained within the survey. All the models are multilevel mixed-effect logistic regression, with an LR significance

Table 2 Model estimating likelihood of 15MC policy implementation

	Model 1: Entire sample				
	Odds ratio		z	95% CI	
Age	0.99	*	−1.86	0.97	1.00
Gender (ref cat: women)					
Men	0.67	*	−1.92	0.45	1.01
Other genders	1.72		0.50	0.21	14.3
Prefer not to answer	0.37	*	−1.88	0.13	1.04
Transport mode, main (ref cat: drive alone)					
Walk	2.03	**	2.02	1.02	4.05
Bike	3.78	***	3.27	1.70	8.41
Transit	2.79	***	3.45	1.56	5.01
Carpool	3.18	***	2.94	1.47	6.87
Other	0.56		−0.93	0.16	1.93
Telecommute	1.51		1.06	0.71	3.22
Residential self-selection transport preference, decrease driving cost					
Yes	1.49	**	2.03	1.01	2.18
N/A	2.30		0.69	0.21	24.9
15MC concept					
Current neighborhood					
Yes	3.37	***	5.86	2.25	5.06
N/A	1.00	***			
Personal automobile reduction					
Yes	5.09	***	7.64	3.35	7.73
N/A	1.00	***			
Strengthened sense of community					
Yes	2.56	***	4.14	1.64	3.99
N/A	1.00	***			
Improved mental health and well-being					
Yes	3.72	***	5.83	2.39	5.77
N/A	0.46		−0.25	0.00	186

(continued)

Table 2 (continued)

	Model 1: Entire sample				
	Odds ratio		z	95% CI	
Reduction in personal freedoms					
Yes	0.10	***	−8.85	0.06	0.17
N/A	0.16		−0.61	0.00	58.5
Constant	0.73		−0.79	0.34	1.59
Number of observations	1579				
Log likelihood	−393.5				
Wald chi-squared	265.5				
LR test	14.04				
Interclass correlation	0.06				
AIC	829.1				
BIC	941.7				

BIC Bayesian information criterion, *AIC* Akaike information criterion, *LR* likelihood ratio
***Significant at 99.9%, **Significant at 95.0%, *Significant at 90.0%

above 95%, except for the students' model. Therefore, a logistic model is reported for students. Off-peak and peak hour public transit accessibility and total proximity measures were found to be correlated with whether or not the respondent believed they lived in a 15MC neighborhood. Therefore, they were removed from the model.

4.2.1 Model 1: Entire Sample

As can be seen in Table 1, age, gender, main transport mode, residential self-selection priority of reducing the cost of commuting by personal automobile, and 15MC concept factors are significant factors that affect the likelihood of a person supporting the implementation of 15MC policies in their neighborhood. More specifically, age has a negative association with the odds of people supporting the implementation of 15MC policies in their neighborhood. Every 1-year increase in age is associated with a 1.2% reduction in the odds of supporting the 15MC policy implementation. Men, in comparison to women, are 32.5% less likely to support 15MC implementation while keeping all other variables at mean values.

Model 1 shows that people who use walking, biking, transit, and carpooling compared to those whose main mode is the private automobile are more likely to support the 15MC concept. Those who prioritize decreasing the cost of commuting by private automobile when moving to a new residence, a residential self-selection preference, are also more likely to support the 15MC concept implementation than those who indicated that they would not prioritize reducing the cost of commuting by private car when moving to a new residence.

Whether participants believed that they were currently living in a 15MC neighborhood is significantly associated with supporting the 15MC policy implementation, as are aligning beliefs about the positive and lack of agreement with the negative impacts of the policy. People who believe that their current neighborhood is a 15MC neighborhood are 3.4 times more likely to support the implementation of the 15MC concept. For positive impacts of 15MC implementation, those believing that implementation of the 15MC concept would reduce the use of personal automobiles, improve mental health and well-being outcomes, and finally, improve the sense of community were more likely to support the implementation of the 15MC concept by 5.1, 3.7, and 2.6 times, respectively. People who believe implementing 15MC policies will reduce personal freedom were less likely to support implementing the 15MC concept in their neighborhood by 90%.

4.2.2 Models 2 and 3: Women and Men Models

A comparison of those who identified their gender as women vs. men is shown in models in Table 3. Regarding transport modes, men who used cycling or transit as their main commute mode compared to using cars were more likely to support the implementation of the 15MC policies in both genders. In Model 3 (men), the odds ratio for "Transit" (2.49) compares men who are transit users to men who drive alone. Among men, using transit is a positive but not statistically significant (95%) predictor of support compared to driving alone. In contrast, men who cycle are 6.2 times more likely to support the concept than men who drive alone, significant at the 99.9% level. Other main commuting modes were not significant indicators of 15MC concept implementation support for men.

In Model 2 (Women), the odds ratio for "Transit" (4.58) compares women who are transit users to women who drive alone. Among women, using transit is a strong and significant predictor of support compared to driving alone. Dissimilarly, women did not have a significant likelihood of supporting the concept if they cycled versus drove alone. For women, the main transport modes of walking, carpooling, and other modes (i.e., taxi, ride-hailing, e-scoter, and others) were also significant factors in the likelihood of support for implementing the 15MC concept for women. This shows different patterns by gender, regardless of each model's constant value.

Like Model 1, for men and women, the likelihood of supporting the 15MC concept implementation increased if they believed it would provide positive impacts, such as reducing personal automobile use and improving health and well-being. Men and women show a higher likelihood of supporting the implementation of the 15MC policies if they believe that its implementation would improve mental health and well-being or reduce car usage. For women, it was additionally significant that belief of implementing the 15MC concept would improve the sense of community, but this was not the case for men. Women and men also show a higher likelihood of supporting the concept if they believe that their current neighborhood is a 15-minute neighborhood. Both models for men and women show a decrease in the likelihood of supporting the 15MC concept implementation if the respondent believes it reduces personal freedoms.

4.2.3 Models 4 and 5: Students and Faculty/Staff Models

Models generated for the student population (Model 4, logit) and faculty/staff (Model 5, multilevel mixed-effect logit) are shown in Table 3. The students' model includes the zone of residence, where those outside the city are less likely to support 15MC concept implementation by 84%, compared to those who live within the city, within its ring road. Like Model 1, Models 4 and 5 show the significance of variables of mode, whether the respondent identified as living in a 15MC neighborhood, and perceptions of positive and negative aspects of 15MC policies. Students show an increased likelihood of supporting the 15MC concept implementation for modes of cycling, transit, and carpooling. Staff/faculty similarly show support for modes of cycling and carpooling; they are dissimilar because staff/faculty transit users do not show an increase in odds of supporting the 15MC policy. The Staff/faculty group shows an increased odds of supporting the policy for those wanting to decrease the cost of their commute by car, whereas the student model does not. The staff/faculty group model is the only one where those who telecommute show positive odds of supporting the 15MC policy.

4.2.4 Models 6 and 7: Those Were Previously Aware of the 15MC Concept and Those Who Were Not

Table 3 shows models for those who were previously aware of the 15MC concept before the survey (Model 6: Aware) and those who were learning about the 15MC concept for the first time through the survey (Model 7: Unaware), Like Model 1 (Entire Sample), general trends of significance of different variables can be observed. As shown in Model 7 (Unaware), for the subgroup of those introduced to the 15MC concept through the survey, men were 39% less likely to support the concept in comparison to women. Model 1 (Entire sample) and Model 6 (Aware) were the only models to show people who want to decrease the cost of commuting by car in future residential self-selection being more likely to support 15MC policy implementation. Those not previously aware did not show a significant odds of wanting to decrease the cost of their commute by car, nor did they believe that implementing 15MC policies would increase the sense of community. Modes of walking, cycling, and carpooling showed increased odds of supporting 15MC concept implementation for those previously aware. Those previously unaware showed increased odds of supporting the concept for the modes of cycling, transit, carpooling, and other modes. This shows minor differences between the two models.

Overall, the most common factors associated with supporting the 15MC policies across all groups in Table 3 include commuting factors of using bikes and carpooling compared to driving alone. There is a statistically significant association between these factors and the likelihood of supporting the concept. This finding is similar among the designed models. Also, all respondents who felt that they were living in a 15MC neighborhood were more likely to support the 15MC

Table 3 Models estimating likelihood of supporting 15MC policy implementation for different groups

	Model 2: Women				Model 3: Men				Model 4: Student				Model 5: Staff/faculty				Model 6: Aware				Model 7: Unaware			
	Odds		95% CI		Odds		95% CI		Odds		95% CI		Odds		95% CI		Odds		95% CI		Odds		95% CI	
Age	0.99		0.97	1.01	0.98		0.96	1.01	1.01		0.96	1.06	0.98		0.96	1.01	0.98		0.95	1.01	0.99		0.97	1.01
Gender (ref: women)																								
Man									0.59		0.31	1.13	0.71		0.42	1.21	1.26		0.56	2.86	0.61	*	0.37	1.01
Other genders									1.06		0.12	9.38	1.00	***			1.22		0.07	19.8	1.00	***		
No answer									0.50		0.10	2.49	0.28	*	0.06	1.20	3.61		0.41	31.9	0.12	***	0.03	0.41
Transport mode (ref: drive alone)																								
Walk	2.68	**	1.06	6.76	1.79		0.57	5.59	1.82		0.67	4.97	1.45		0.51	4.09	4.75	**	1.10	20.4	1.52		0.66	3.48
Bike	2.49		0.81	7.64	6.16	***	1.61	23.6	4.77	**	1.01	22.5	3.18	**	1.21	8.31	14.5	***	2.41	87.4	2.90	**	1.06	7.93
Transit	4.58	***	1.93	10.9	2.49	*	0.91	6.86	2.87	**	1.28	6.43	1.54		0.53	4.45	1.89		0.61	5.88	3.50	***	1.64	7.49
Carpool	6.34	***	2.01	20.0	1.20		0.32	4.45	6.62	**	1.27	34.6	2.56	**	1.02	6.44	6.86	**	1.14	41.5	2.50	**	1.02	6.17
Other	0.26	*	0.07	1.02	6.56		0.34	126	0.27		0.06	1.34	1.92		0.14	26.1	2.31		0.18	29.3	0.23	*	0.05	1.02
Telecommute	1.24		0.50	3.07	1.25		0.29	5.28	0.88		0.30	2.57	2.75	*	0.83	9.10	3.65		0.76	17.6	0.92		0.36	2.33
Residential self-selection transport preference, decrease driving cost																								
Yes	1.51		0.91	2.51	1.75		0.89	3.45	1.32		0.71	2.44	1.67	**	1.01	2.77	3.78	***	1.66	8.62	1.24		0.77	1.98
N/A	6.55		0.21	208	1.82		0.03	98.8	1.00	***			2.06		0.18	23.6	0.18		0.00	1.647,092	1.00	***		

(continued)

Table 3 (continued)

	Model 2: Women				Model 3: Men				Model 4: Student				Model 5: Staff/faculty				Model 6: Aware				Model 7: Unaware			
	Odds		95% CI		Odds		95% CI		Odds		95% CI		Odds		95% CI		Odds		95% CI		Odds		95% CI	
Zone of residence (ref: Inside city, inside Circle Drive)																								
Inside city, outside Circle Drive									0.81		0.41	1.60												
Outside city									0.16	***	0.05	0.55												
Current neighborhood a 15MC																								
Yes	3.83	***	2.23	6.59	3.30	***	1.60	6.81	4.20	***	2.21	7.98	2.87	***	1.67	4.93	4.55	***	1.95	10.6	3.66	***	2.20	6.09
N/A									1.00	***			1.00											
15MC positive impacts																								
Reduction in personal automobiles																								
Yes	5.45	***	3.11	9.54	5.93	***	2.78	12.7	5.05	***	2.57	9.91	5.56	***	3.20	9.66	4.65	***	2.00	10.8	6.88	***	3.97	11.9
N/A													1.00	***										
Strengthened sense of community																								
Yes	3.24	***	1.81	5.79	1.43		0.62	3.28	2.39	**	1.18	4.85	2.78	***	1.53	5.06	10.1	***	3.91	26.0	1.37		0.79	2.39
N/A					1.00	***			1.00	***			1.00	***							1.00	***		
Improved mental health and well-being																								

Yes	2.43	***	1.39	4.25	10.1	***	4.23	24.4	3.98	***	1.98	8.03	3.41	***	1.88	6.16	5.72	***	2.36	13.9	3.05	***	1.75	5.33
N/A	1.00	***			1.00	***			0.54		0.00	683	1.00	***							1.00	***		
15MC negative impacts																								
Reduction in personal freedoms																								
Yes	0.09	***	0.04	0.18	0.09	***	0.04	0.21	0.11	***	0.05	0.22	0.09	***	0.04	0.19	0.01	***	0.00	0.05	0.24	***	0.12	0.46
N/A	1.00	***			1.00	***			0.11		0.00	130	1.00	***			1.00	***			1.00	***		
Constant	0.50		0.18	1.42	0.55		0.13	2.25	0.49		0.10	2.32	0.79		0.23	2.65	0.24	*	0.05	1.16	0.95		0.36	2.52
Number of observations	979				503				812				748				630				920			
Log likelihood	−234				−131				−161				−225				−107				−259			
Wald chi-squared	152				83.4				–				144				75.8				145			
Interclass correlation	0.08				0.14				–				0.09				0.18				0.07			
LR chi-squared	–				–				222				–				–				–			
Pseudo R-squared	–				–				0.41				–				–				–			
AIC	501				294				364				486				253				553			
BIC	579				362				463				569				337				635			

BIC Bayesian information criterion, *AIC* Akaike information criterion, *LR* likelihood ratio
***Significant at 99.9%, **Significant at 95.0%, *Significant at 90.0%

policies' implementation across all the groups. The perception that 15MC policies can bring environmental, societal, and health factors is also associated with a higher likelihood of supporting the 15MC policies implementation. The idea that 15MC policies will lead to a reduction in personal freedoms consistently impacts all groups by decreasing the likelihood of supporting the 15MC policy implementation.

5 Discussion and Conclusion

The 15MC concept has recently been popularized to achieve various social and environmental benefits. This research explores people's preferences towards supporting the implementation of the 15MC policies and the association between people's perception of the possible benefits and drawbacks of the concept and being in favor of implementing such a policy. To achieve these goals, the study analyzed survey data collected in Saskatoon, Canada, using summary statistics and statistical models to explore how socioeconomic and land use characteristics, travel behavior, and people's beliefs about the positive and negative impacts are associated with the likelihood of supporting 15MC policies.

The results show that older people are less likely to support the implementation of 15MC policies in their neighborhoods. This can reflect a challenge for cities and planning authorities, as older adults are usually more involved in local politics and official elections than younger adults in Canada (Kembhavi, 2013), leading to less support for such policy. Additionally, when analyzing by identified gender, women are more likely than men to support 15MC policy implementation than men while holding all other variables at their means. This highlights an acceptance gap between genders. Therefore, developing strategies that help reach more older adults and men to showcase and communicate the aspects of such a policy during the planning process can be recommended. This is to maximize people's support for such planning policy within these segments.

By commute factors, persons who commute by active modes (i.e., walking and cycling), transit, and carpooling are more likely to support 15MC policies when compared to those who drive alone. Those who value reducing the cost of commuting also show a higher likelihood of supporting 15MC policies. Therefore, shifting people from driving alone to using active transportation modes and transit will reduce car mode share and increase people's likelihood of supporting proximity-based policies such as the 15MC policies. Understanding this link between commuting behavior and support for emerging urban policies such as the 15MC policy can help cities and planning authorities understand the wide range of benefits of their mode share changing efforts, encouraging them to do more.

Participants' previous awareness of the 15MC concept does not significantly impact their support for its implementation. In other words, both those who were aware of the concept and those who were not displayed similar levels of support. This shows that people generally accept the idea of 15MC policy implementations in their neighborhoods, even right after the concept is introduced to them. This also indicates that the concept can be introduced to people without a great risk of them not supporting it. Furthermore, people's perception of their current neighborhood also proves significant in determining their likelihood of supporting the 15MC policy. More specifically, the paper's models indicate that persons who perceive their existing neighborhood as one of the 15MC designs are more likely to support 15MC policy implementation. This finding is consistent across different groups of populations, highlighting the importance of understanding people's current perspectives on their living conditions.

Those who agreed that 15MC implementation would yield positive benefits, including a reduction in the use of personal automobiles and improved health and well-being, show a positive association with being supportive of the concept's implementation across all groups. In contrast, the idea that the 15MC policy would decrease personal freedoms is negatively associated with people supporting the concept. Therefore, it can be important for dedicated information campaigns to communicate (1) the real benefits of such a policy beyond improving people's proximity to different destinations and (2) that the policy will not decrease personal freedoms. These campaigns can include the organization of events and social media campaigns. Encouraging local politicians and city planners to partner with media outlets and local companies to share key policy benefits and dispel myths can also be recommended.

This study explored people's preference towards supporting implementing the 15MC policies in Saskatoon, Canada. Nevertheless, future studies can use similar approaches to understand the factors affecting people's willingness to support such policies in different contexts and locations. Future studies can also focus on collecting data from policymakers, city planners, and other stakeholders to understand their perspectives on the benefits and barriers to implementing proximity-related policies such as 15MC policies. Finally, the paper's results unmask several aspects related to factors affecting people's support of 15MC policies, providing important information for policymakers and transport planners looking forward to developing such a policy. This highlights the importance of communicating the 15MC policies' benefits while combating their misperceptions.

Acknowledgments The authors would like to thank the Natural Sciences and Engineering Research Council of Canada (NSERC) for funding this research. We would like to thank Secretaria Nacional de Ciencia, Tecnología e Innovación (SENACYT) and the Instituto Nacional para la Formación de Recursos Humanos (IFARHU) of Panama for funding Ariel Arosemena to conduct this research at the USask (Scholarship No. 270-2022-040).

Author Contribution The authors confirm contribution to the paper as follows: study conception and design: All; data collection: Arosemena; analysis and interpretation of results: Lam, and Arosemena; draft manuscript preparation: All. All authors reviewed the results and approved the final version of the manuscript.

References

Allam, Z., Bibri, S. E., Chabaud, D., & Moreno, C. (2022b). The theoretical, practical, and technological foundations of the 15-Minute City model: Proximity and its environmental, social and economic benefits for sustainability. *Energies, 15*(16), 6042. https://www.mdpi.com/1996-1073/15/16/6042

Allam, Z., Nieuwenhuijsen, M., Chabaud, D., & Moreno, C. (2022a). The 15-minute city offers a new framework for sustainability, liveability, and health. *The Lancet Planetary Health, 6*(3), e181–e183. https://doi.org/10.1016/S2542-5196(22)00014-6

Birkenfeld, C., Victoriano-Habit, R., Alousi-Jones, M., Soliz, A., & El-Geneidy, A. (2023). Who is living a local lifestyle? Towards a better understanding of the 15-minute-city and 30-minute-city concepts from a behavioural perspective in Montréal, Canada. *Journal of Urban Mobility, 3*, 100048. https://doi.org/10.1016/j.urbmob.2023.100048

Boisjoly, G., & El-Geneidy, A. M. (2017). How to get there? A critical assessment of accessibility objectives and indicators in metropolitan transportation plans. *Transport Policy, 55*, 38–50. https://doi.org/10.1016/j.tranpol.2016.12.011

Büttner, B., Seisenberger, S., McCormick, B., Silva, C., Teixeira, J., Papa, E., & Cao, M. (2024). *Mapping of 15-minute city practices: Report from DUT's 15-minute city transition pathway*. https://dutpartnership.eu/wp-content/uploads/2024/06/DUT_15mC-Mapping_digital_final.pdf

C40 Cities Climate Leadership Group. (2020). *How to build back better with a 15-minute city*. Retrieved July 10, 2024, from https://www.c40knowledgehub.org/s/article/How-to-build-back-better-with-a-15-minute-city?language=en_US

Capasso Da Silva, D., King, D. A., & Lemar, S. (2020). Accessibility in practice: 20-Minute city as a sustainability planning goal. *Sustainability, 12*(1), 129. https://doi.org/10.3390/su12010129

City of Saskatoon. (2023). *Official community plan*. City of Saskatoon. Retrieved June 2, from https://www.saskatoon.ca/business-development/planning/growth-plans/official-community-plan

City of Saskatoon. (n.d.). *Life in Saskatoon*. Retrieved May 22, from https://www.saskatoon.ca/community-culture-heritage/new-saskatoon/life-saskatoon#:~:text=Saskatoon%20is%20located%20the%20central,trails%20along%20the%20river%20valley

Di Marino, M., Tomaz, E., Henriques, C., & Chavoshi, S. H. (2023). The 15-minute city concept and new working spaces: A planning perspective from Oslo and Lisbon. *European Planning Studies, 31*(3), 598–620. https://doi.org/10.1080/09654313.2022.2082837

Diab, E., Wasfi, R., & El-Geneidy, A. (2014). Extraboard team sizing: An analysis of short unscheduled absences among regular transit drivers. *Transportation Research Part A: Policy and Practice, 66*, 27–38. https://doi.org/10.1016/j.tra.2014.04.019

Dillman, D., Smyth, J., & Christian, L. (2014). *Internet, phone, mail, and mixed-mode surveys: The tailored design method*. Wiley.

Gower, A., & Grodach, C. (2022). Planning innovation or city branding? Exploring how cities operationalise the 20-minute neighbourhood concept. *Urban Policy and Research, 40*(1), 36–52. https://doi.org/10.1080/08111146.2021.2019701

Herbert, J. (2021, September 2). Transformation or Gentrification? The hazy politics of the 15-Minute City. *Resilience*. https://www.resilience.org/stories/2021-09-02/transformation-or-gentrification-the-hazy-politics-of-the-15-minute-city/}:~:Text=Originating%20from%20the%20French%2DColombian,or%20cycle%20from%20their%20home

Howard, E. S. (1965). *Garden cities of tomorrow / by Sir Ebenezer Howard; Edited with a preface by F. J. Osborn, with introductory essay by Lewis Mumford*. Faber & Faber.

Kamruzzaman, L. (2022). Typologies of 20-minute neighbourhoods, active transport use, and spatial spillovers. *Findings*. https://doi.org/10.32866/001c.33158

Kembhavi, R. (2013). *Political participation among seniors*. Elections Canada. Retrieved July 15, from https://www.elections.ca/content.aspx?section=res&dir=rec/part/partsen&document=index&lang=e

Khavarian-Garmsir, A. R., Sharifi, A., Hajian Hossein Abadi, M., & Moradi, Z. (2023). From garden city to 15-minute city: A historical perspective and critical assessment. *Land, 12*(2), 512. https://www.mdpi.com/2073-445X/12/2/512
Levinson, D. M., & Krizek, K. J. (2018). *Metropolitan transport and land use: Planning for place and plexus* (2nd ed.). Routledge.
Logan, T. M., et al. (2022). The x-minute city: Measuring the 10, 15, 20-minute city and an evaluation of its use for sustainable urban design. Cities 131: 103924.
Lu, M., & Diab, E. (2023). Understanding the determinants of x-minute city policies: A review of the North American and Australian cities' planning documents. *Journal of Urban Mobility, 3*, 100040. https://doi.org/10.1016/j.urbmob.2022.100040
Luke, D. A. (2004). *Multilevel modeling* (3rd repr. ed.). Sage.
Marquet, O., Anguelovski, I., Nello-Deakin, S., & Honey-Rosés, J. (2024a). Decoding the 15-Minute City debate: Conspiracies, backlash, and dissent in planning for proximity. *Journal of the American Planning Association*, 1–9. https://doi.org/10.1080/01944363.2024.2346596
Marquet, O., Mojica, L., Fernández-Núñez, M.-B., & Maciejewska, M. (2024b). Pathways to 15-Minute City adoption: Can our understanding of climate policies' acceptability explain the backlash towards x-minute city programs? *Cities, 148*, 104878. https://doi.org/10.1016/j.cities.2024.104878
Moreno, C. (2016, October 5). La ville du quart d'heure : pour un nouveau chrono-urbanisme. *La Tribune*. https://www.latribune.fr/regions/smart-cities/la-tribune-de-carlos-moreno/la-ville-du-quart-d-heure-pour-un-nouveau-chrono-urbanisme-604358.html
Moreno, C., Allam, Z., Chabaud, D., Gall, C., & Pratlong, F. (2021). Introducing the "15-minute city": Sustainability, resilience and place identity in future post-pandemic cities. *Smart Cities, 4*(1), 93–111. https://doi.org/10.3390/smartcities4010006
Perry, C. (1929). *The neighborhood unit* (pp. 25–44). Routledge/Thoemmes.
Pozoukidou, G., & Chatziyiannaki, Z. (2021). 15-Minute city: Decomposing the new urban planning eutopia. *Sustainability, 13*(2), 928. https://doi.org/10.3390/su13020928
Shannon, R., Mant, J., Dessewffy, M., & Harrison, L. (2019). 20-Minute neighbourhoods: Creating a more liveable melbourne. *5th International Conference on Transportation and Health.*
Statistics Canada. (2023a). *Census Profile: 2021 Census of Population*. Statistics Canada. Retrieved June 2, from https://www12.statcan.gc.ca/census-recensement/2021/dp-pd/prof/details/page.cfm?Lang=E&SearchText=winnipeg&GENDERlist=1&STATISTIClist=1&DGUIDlist=2021A00054611040&HEADERlist=0
Statistics Canada. (2023b). *Proximity Measures Database, 2021*. Statistics Canada. Retrieved May 3, from https://www150.statcan.gc.ca/n1/pub/17-26-0002/172600022023001-eng.htm
Statistics Canada. (2023c). *Proximity Measures, 2021*. https://www150.statcan.gc.ca/n1/pub/17-26-0002/172600022023001-eng.htm
Statistics Canada. (2023d). *Spatial access measures*. Statistics Canada. Retrieved May 5, from https://www150.statcan.gc.ca/n1/pub/27-26-0001/272600012023001-eng.htm
Statistics Canada. (2024). *Census Profile: 2021 Census of Population*. Statistics Canada. Retrieved June 2, from https://www12.statcan.gc.ca/census-recensement/2021/dp-pd/prof/details/page.cfm?Lang=E&GENDERlist=1,2,3&STATISTIClist=1,4&HEADERlist=0&DGUIDlist=2021A00054711066&SearchText=saskatoon
TED. (2021). *The 15-minute city | Carlos Moreno*. https://www.youtube.com/watch?v=TQ2f4sJVXAI
United Nations. (2015). *Transforming our world: The 2030 agenda for sustainable development*. https://sustainabledevelopment.un.org/content/documents/21252030%20Agenda%20for%20Sustainable%20Development%20web.pdf
University of Saskatchewan. (2023a). *Enrollment reports*. Retrieved May 22, from https://leadership.usask.ca/priorities/reports/enrolment.php
University of Saskatchewan. (2023b). *University of Saskatchewan Act, 1995*. Retrieved May 22, from https://governance.usask.ca/governance/university-of-saskatchewan-act.php#:~:

text=The%20University%20of%20Saskatchewan%20was,provincial%20statute%20created%20the%20university
Wen, L., Marinova, D., Kenworthy, J., & Guo, X. (2022). Street recovery in the age of COVID-19: Simultaneous design for mobility, customer traffic and physical distancing. *Sustainability, 14*(6), 3653.
Wu, H., Avner, P., Boisjoly, G., Braga, C. K., El-Geneidy, A., Huang, J., Kerzhner, T., Murphy, B., Niedzielski, M. A., & Pereira, R. H. (2021). Urban access across the globe: An international comparison of different transport modes. *NPJ Urban Sustainability, 1*(1), 16. https://doi.org/10.1038/s42949-021-00020-2

Urban Factors and Walkability Perception in Medium-Sized Latin American Cities. The Case of Tepic, Mexico

Luis Llamas-Rodriguez and Gabriela Ochoa-Covarrubias

1 Introduction

The systemic crisis facing the planet is a serious warning about the need and urgency to break the current growth model; at the same time this crisis constitutes, as Ban Ki-Moon has stated before the United Nations, an opportunity to introduce deep corrective measures, proposed and proven, still in time to be implemented (Vilches & Gil-Pérez, 2016). The transport sector in Mexico represents the second largest source of greenhouse gas (GHG) emissions, and the country is among the top 10 emitters worldwide (Ruiz et al., 2022). Despite these data, walking predominates as a mean of transport for school trips (54.1%) and ranks third (27.4%) for commuting to work (INEGI, 2022). This context has marked changes in the evolution of the expansive urban development model towards a people-centred one that seeks to increase walking in the modal split of mobility, democratising public space (Gehl, 2016) and improving quality of life (Baobeid et al., 2021). Nevertheless, 74% of public spending on transport infrastructure goes to motorised transport, and only 9.68% is dedicated to pedestrian infrastructure, creating a shortage of walkable environments (Méndez et al., 2018).

In this context, there is a lack of tools to assess and monitor walkability in its multidimensionality, especially in complex urban contexts such as in Latin America (Arellana et al., 2020). In this work, a Walkability Assessment Model (WAM) is developed to identify priority areas for walkable, accessible and sustainable urban projects through a participatory and multicriteria approach. The study area is the city of Tepic, with 368,364 inhabitants, 75% of the inhabitants of the Metropolitan

L. Llamas-Rodriguez · G. Ochoa-Covarrubias (✉)
Department of Habitat and Urban Development, ITESO, Universidad Jesuita de Guadalajara, Guadalajara, Mexico
e-mail: ochoagabriela@iteso.mx

L. Bouzouina et al. (eds.), *15-Minute Cities*, Footprints of Regional Science,
https://doi.org/10.1007/978-3-032-20840-8_6

Zone of Tepic, the most populated region of the state of Nayarit (INEGI, 2020). The mobility in Tepic is dominated by walking, with 38.9% of trips, and collective public transport (35.8%), while 21% of trips are made by car (IMPLAN Tepic, 2017).

This research offers a methodology that is replicable in other urban contexts, integrating open data, citizen perceptions and technical analysis. Specifically, two research questions are answered: What are the non-observable factors and observable components of walkability (FCW) that influence the decision to walk in Tepic? To what extent are these characteristics are present in different areas of the city?

This paper is structured in five sections. The first section presents the literature review. The methodology section details the creation of the WAM, based on a mixed analysis (qualitative and quantitative) that integrates tools such as geospatial analysis in Geographic Information Systems (GIS), collaborative mapping and focus groups, as well as the Analytical Hierarchical Process (AHP). The results section presents the non-observable factors and observable components that influence people's walking decisions in Tepic and the extent to which these characteristics are present in different areas of the city. The discussion analyses the implications of the FCW, contrasts the findings with other related studies and proposes lines of action to promote pedestrian mobility in Mexico. The paper ends with a conclusion section and policy recommendations.

2 Literature Review

The concept of walkability started to be used in transport and planning studies in the early 2000s, but quickly became a widely used concept in sustainable mobility and other disciplines such as health and environmental sciences (Fonseca et al., 2022). Walkability is increasingly valued and has become an important basis for street construction practice and related research (Balsas, 2017). Although its definition is still ambiguous, some authors define it as the degree to which the built environment is pedestrian-friendly (Habibian & Hosseinzadeh, 2018), based on supporting and encouraging safe, comfortable and interesting urban walking trips (Moura et al., 2017; Bartzokas & Yorgos, 2016).

Walkability has become a critical research topic in developed countries. From 2009 to 2018, most research was conducted in North America (44.8%), followed by Europe (22.4%), Asia (14.9%) and Oceania (10.4%). Meanwhile, Africa and Latin America each accounted for only 1.5% of the research (Arellana et al., 2020). The above percentages suggest that walkability has not received sufficient attention in the Global South, especially in Latin America. In Mexico, the studies found on walkability are limited to cases located in Baja California, Guanajuato and Mexico City (Tena, 2018; IMPLAN Valle de Santiago, 2021; Martínez et al., 2021; Berrones-Sanz, 2022).

There is no consensus on how to measure walkability and how to analyse the various attributes of the built environment related to this practice (Shashank &

Schuurman, 2019), as there are multiple important criteria. For the most part, walkability has been assessed quantitatively in metrics that measure several attributes of the urban built environment. One of the initial frameworks for its measurement was the '3Ds' (Cervero & Kockelman, 1997), which consists of three elements: density, diversity and design; later, destination and distance were added to become the '5Ds' (Ewing & Cervero, 2010).

However, these objective walkability metrics do not include important qualitative characteristics such as features of the urban environment, the feeling of safety or the pleasure of walking (Fonseca et al., 2022). They do not consider components such as the physical conditions of pavements, which are closely related to perceived walkability (Jensen et al., 2017). This offers new possibilities for assessing perceived walkability using additional, more precise scales.

Studies of walkability are relatively recent; therefore, it is important to recognise the integral structure of the analysis, the selection of factors and components that impact it, as well as the conditions that discourage it (Talavera et al., 2012), since the set of properties present in the environment enables the occurrence of this or any other behaviour (Gibson, 1977). To understand behaviours, ecological models have been used as a more productive hierarchy for the promotion of physical activity, with several variables and levels: intrapersonal (biological, psychological), interpersonal/cultural, organisational, physical (built and natural environment) and political (laws, rules, regulations, and codes) (Sallis et al., 2006).

In that sense, other researchers have categorised the evaluation of walkability based on feasibility, accessibility, safety, comfort and pleasure (Alfonzo, 2005); image, openness, human scale, transparency and complexity (Ewing & Handy, 2009) or sidewalk condition, traffic flow, comfort and attractiveness (Arellana et al., 2020). However, Li et al. (2023) mentions that Alfonzo's hierarchy of walking needs can function as a skeletal structure to incorporate features of the built environment both at the meso-scale, at the zone level, at the micro-scale and at the street level (Basu et al., 2022; Erturan & Van der Spek, 2022; Guzman et al., 2022; Koo et al., 2022); therefore, this hierarchy is taken as a reference for this study.

The hierarchy of walking needs, according to Alfonzo (2005), is organised into five levels. The first, feasibility, addresses the practicality of the walking journey considering mobility, time and responsibilities. The second, accessibility, refers to the quantity and quality of activities in the environment, as well as street conditions. The third, safety, focuses on crime and urban design. The fourth, ease and satisfaction, analyses elements that influence the walking experience, such as the condition of pavements. And the fifth, pleasure, relates to the attractiveness and aesthetic experience of the environment (Alfonzo, 2005).

From this hierarchy, it is necessary to define non-observable factors and observable components to evaluate objective and perceived walkability. For this purpose, Guzmán et al. (2022) present a selection at the meso-scale and micro-scale from a Latin American perspective to generate an index for the city of Bogotá, Colombia. Since no approximations were found in Mexico at both scales, meso and micro-scales are considered as a starting point, especially in urban contexts in countries of the Global South, where the state of pedestrian infrastructure is often a barrier to walking (Larrañaga et al., 2016; 2019).

In this case study, it is necessary to consider the urban design factors that influence pedestrian walking behaviour (Lotfi & Koohsari, 2009), and to recognise that walking behaviour is influenced by how the environment is perceived (Liao et al., 2022) and how physical features influence the quality of the walking environment both directly and indirectly (Ewing & Handy, 2009).

3 Methodology

3.1 Study Area

The city of Tepic, Nayarit, Mexico (Fig. 1) was chosen to apply the walkability assessment model (WAM) for three reasons: first, the large share of walking in the modal split, particularly for school and work trips; second, the proximity to opportunities, given the limited urban sprawl as a medium-sized city; third, the availability of official data on the physical, economic and social context at the street scale.

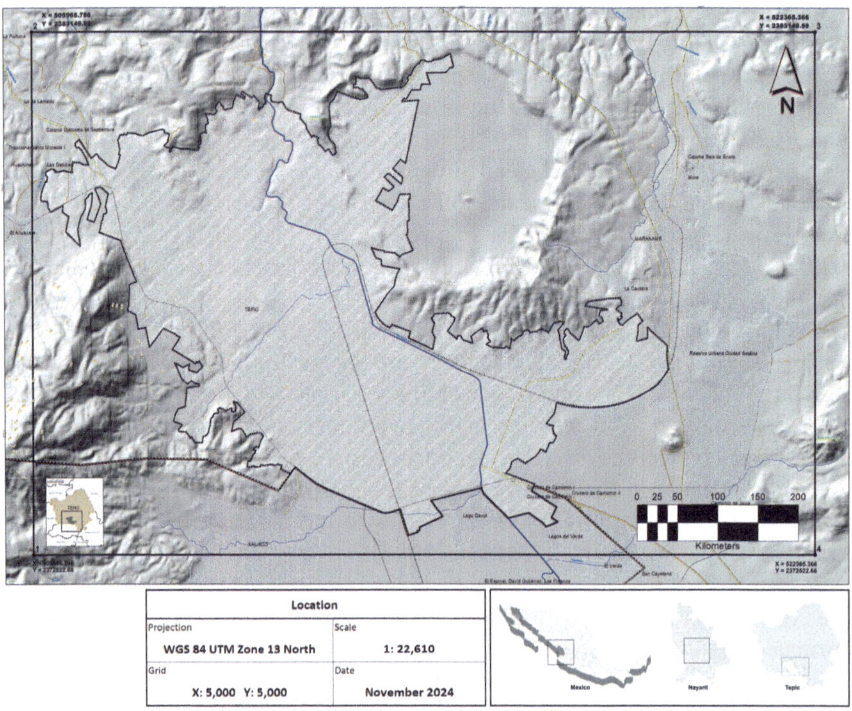

Fig. 1 Location of the city of Tepic, Nayarit, Mexico

Table 1 General methodological outline for the creation of a walkability assessment model for Tepic

Stage	Method	Tool	Result
FCW matrix definition	Focus groups	Atlasti 24	FCW matrix in Tepic
	Collaborative maps		
	Multicriteria analysis (AHP)	Excel	
FCW modelling in Tepic	Map calculator	QGIS 3.40	WAM in Tepic

FCW non-observable factors and observable components of walkability, *WAM* walkability assessment model

The assessment of walkability in Tepic, Mexico, is elaborated based on two methodological steps: the definition of an FCW matrix and the mapping of these elements. Table 1 describes the methods, tools and expected results in each stage.

3.2 *Definition of Non-observable Factors and Observable Components of Walkability*

During this stage, the FCW matrix for Tepic was defined and weighted, adapting the hierarchy of walking needs of Alfonzo (2005) and the FCW of Guzmán et al. (2022). The definition of the matrix was done through four focus groups and a collaborative mapping exercise whose discourses were analysed to identify non-observable factors and observable components, while their weighting results from a multicriteria analysis.

First, four focus groups (FGs) were formed with public officials, professionals and relevant stakeholders who had an impact on the issue of urban mobility and walkability, and their discourses were analysed with Atlasti 24. First, their socio-psychological profile was characterised through questions such as Who walked today? What motivated you to walk? Why do you usually walk? When do you prefer not to walk and why? Second, walking behaviours were assessed with respect to elements such as streets' physical conditions (What are the physical characteristics of the streets where you usually walk?), street safety (From your experience, what gives you a perception of street safety perception when you walk?), personal safety (What elements provide you with personal safety when you walk?), urban accessibility (What are your usual destinations when you walk?) and comfort (What are the elements that make a street more comfortable?). In addition, a general question was added (What would you like the streets where you usually walk to be like?).

A collaborative mapping was carried out to consider the perspective of children and care mobility. The objective was to identify the characteristics, perceptions or elements that facilitate or impede the walking of children, adolescents and their caregivers on the streets. Information was sought on the impact of the streets' physical conditions, street safety, personal safety (day and night), comfort and urban accessibility (on the way to school, to public transport, to shops, etc.). On a map given to the participants, the children identified routes with drawings, stickers, shapes or texts (perceptions or experiences) and expressed how they feel when walking through the streets near their school, both during the day and at night, guided by the above-mentioned elements to be evaluated.

Teachers as well as parents reported the perspective of care mobility and described what safe environments for children should look like. Their perspective was incorporated into the flap of the map that the students had previously worked on. Using paths, shapes or drawings, they also identified areas on the map that, from their point of view as adults, were suitable or should be improved. This methodological stage, which includes both focus groups and collaborative mapping, results in an FCW matrix adapted to Tepic.

Subsequently, the weighting of the components of the FCW matrix was obtained with a multicriteria analysis, the Analytical Hierarchical Process (AHP), which prioritises through pairwise comparisons (Saaty, 1980) defined by urban public policy experts, organised in a focus group. First, each participant considered and ranked the components of FCW individually. Then, they briefly explained the results and the mode was calculated to assign the weighting to each element of the matrix for the group. Finally, for each factor and component, the quality of the process was checked through the consistency ratio as a quality indicator. If this ratio was less than 0.10, the hierarchical analysis process was considered reasonable. Several rounds were applied to evaluate the FCW successively. For each round, the quality of the process was checked with the consistency ratio as a quality indicator. This resulted in normalised values for the six factors and 24 components, in sum, the weighted FCW matrix.

3.3 Modelling Walkability Characteristics with Mapping of Non-observable Factors and Observable Components

The walkability characteristics in the city were modelled by mapping the weighted FCW matrix adapted to the city, with a map calculator in a free Geographic Information System (GIS). Cartographic data available in open access and with national coverage were used to make the process transferable to the cities of the country. Where available, data from local and state institutions were integrated to provide greater detail than the national agency's data. First, 24 maps representing the distribution of observable components were produced. Depending on the type of variable, categories were assigned using binary methods, areas of influence or inverse distance weight (IDW). The categorical values were assigned with criteria derived from federal, state and municipal regulations (see Table 2).

The 24 component maps with categories were then converted to raster format to calculate the raster maps based on the weighting defined in the FCW matrix. This results in six maps of non-observable factors. Finally, the walkability assessment model was calculated with the sum of the six maps. To analyse the micro-scale,

Table 2 Reference standards for defining values in component categories

Name	Alignment
Political Constitution of the United Mexican States	Arts. 6, 11, 16, 17, 21 and 60.
General Law of Mobility and Street Safety	Arts. 2, 70 and 74.
General Law of Human Settlements, Territorial Planning and Urban Development	Principle: Don't leave anyone behind, don't leave anyone out; Governing axis: Social policy; Strategic axis: Sustainable development; Programme: Urban development and housing.
National Development Plan 2019–2024	Priority objective 3; Strategy 3.5; Priority action: 3.5.5.
National Programme of Territorial Planning and Urban Development 2021–2024	Principle 5; Goal 15: Sustainable mobility; National axes 1 and 2; Priority objectives 1.1, 2.4; General guidelines 1.1.4 and 2.4.1.
National Territorial Planning Strategy 2020–2040	Strategy 3; Strategy Objectives 3.1, 3.3. and 3.4; Action lines 3.1.2, 3.1.3, 3.3.2, 3.3.3 and 3.4.1.
National Mobility and Street Safety Strategy 2023–2042	Axis 3; Strategic objectives 3.1, 3.3 and 3.4; Action lines 3.1.2, 3.1.3, 3.3.2, 3.3.3 and 3.4.1.
NOM-SEDATU-004-2023/Mexican Street Manual	Urban design standards.
NOM-034-SCT2/SEDATU-2022	Signalling standards and street devices.
Human Settlement, Territorial Planning and Urban Development Law for the State of Nayarit	Arts. 2 and 3.
Nayarit State Mobility Law	Arts. 1, 2, 3, 4, 6, 7, 9, 10, 11, 12 and 16.
Sectorial Programme for Urban Development and Inclusive Mobility 2021–2027	Strategic objective 2; Action lines 2.2 and 2.3; Goal 2.2.1.
Regulation of the Nayarit State Mobility Law	Arts. 5, 28, 36, 38 and 40.
Sustainable Urban Mobility Programme for Tepic	General objective 1; Strategic objective 3 y 5; Territorial urban objective; Project 14.

areas with contrasts of social deprivation were analysed in terms of the degree of walkability resulting from the model.

4 Results

4.1 *Non-observable Factors and Observable Components of Walkability in Tepic*

The identification of FCW was based on the results of the four focus groups (FGs) and the collaborative mapping. The FGs were composed of relevant professionals and stakeholders with influence over urban mobility and walkability in

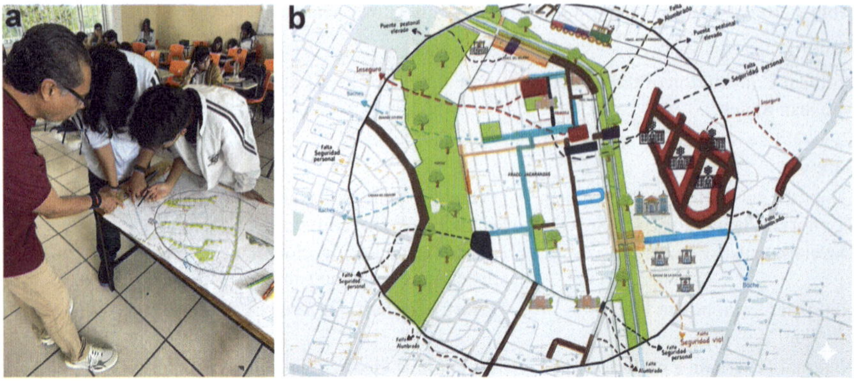

Fig. 2 Evidence from collaborative mapping with children and caregivers. (**a**) Children and caregivers in collaborative mapping; (**b**) Example of resulting maps

Tepic. FC_01 included seven technicians and public officers of the Municipal Institute of Planning of Tepic; FC_02, consisted of eight teachers and students from higher education programmes related to the development of mobility projects in the city of Tepic; FC_03, included seven technicians, public officers and professionals from the Government of the State of Nayarit and the City Council of Tepic; and FC_04, assembled ten women professionals and activists in Tepic. In the mapping with children and caregivers, two primary and three secondary school groups (five schools) representing a variety of social conditions participated (Fig. 2).

The discourse analysis of the four focus groups and the collaborative mapping conducted in Atlasti 24 resulted in six non-observable factors and 24 observable components affecting walkability in Tepic. Subsequently, the AHP of the expert group was applied and resulted in the weighting of the elements of the FCW matrix as shown in Table 3.

Table 3 shows the different proportions within the six factors:

1. Streets' physical conditions: The presence of pavement was the most important, followed by the presence of physical obstacles, ramps and type of pavement with 53%, 26%, 15% and 5%, respectively.
2. Street safety: The observable components that had the greatest impact on Tepic residents' walking were the presence of crosswalks and vertical and horizontal signage with 61% and 20%, respectively, while traffic speed and the width of the vehicle lane section were rated lower.
3. Security: The presence of conflictive groups and the existence of street lighting accounted for over 84% of the importance, while active shops and services and the lack of visual permeability due to urban trees at night only accounted for 18% of the importance of this factor.
4. Urban accessibility: The importance of the components of this factor was more homogeneous, with three out of four having similar weights, i.e., distance to

Table 3 Weights of non-observable factors and observable components of walkability in Tepic, Nayarit

Non-observable factor	Factor weight	Observable component	Component weight	Total	Position by factor	General position
Streets' physical condition	0.1074	Presence of sidewalk	0.5286	0.05677164	1	6
		Obstacles on the sidewalk	0.2596	0.02788104	2	10
		Presence of ramps	0.1509	0.01620666	3	16
		Pavement	0.0609	0.00654066	4	22
Street safety	0.1202	Presence of signage	0.2003	0.02407606	2	12
		Traffic speed	0.1374	0.01651548	3	15
		Presence of crosswalks	0.6137	0.07376674	1	5
		Section width of the vehicular stream	0.0487	0.00585374	4	24
Security	0.3190	Presence of public lightning (positive)	0.3956	0.1261964	2	2
		Presence of trees at night (negative)	0.0742	0.0236698	4	13
		Crime records (negative)	0.4432	0.1413808	1	1
		Presence of open shops and services (positive)	0.0870	0.027753	3	11
Urban accessibility	0.0466	Distance to public transport	0.3373	0.01571818	1	17
		Distance to shops and services	0.2401	0.01118866	2	19
		Distance to urban equipment	0.2817	0.01312722	3	18
		Density of inhabited dwellings	0.1409	0.00656594	4	21
Comfort	0.0877	Presence of urban furniture	0.1196	0.01048892	3	20
		Length of the block	0.2149	0.01884673	2	14
		Presence of historical heritage	0.0683	0.00598991	4	23
		Presence of urban trees (positive)	0.5972	0.05237444	1	8
Natural environment	0.3190	Temperature	0.1232	0.0393008	4	9
		Topographic slopes	0.1736	0.0553784	3	7
		Swamp or flood zones	0.3865	0.1232935	1	3
		Landslide areas	0.3167	0.1010273	2	4
	1.00			1.00		

public transport, urban facilities and shops with 34%, 28% and 24%, respectively, while the density of inhabited dwellings only contributed 14%.
5. Comfort: The presence of street trees and block length accounted for more than 80% of the weighting, while street furniture and historical heritage were marginal.
6. Natural physical environment: Flooding and landslides had high ratings (39% and 32%, respectively), while topographic slopes and temperature were rated lower (17% and 12%, respectively).

4.2 *Modelling Walkability Characteristics in Tepic by Mapping Non-observable Factors and Observable Components*

To understand the extent to which walkability characteristics are present in different areas of the city, spatial analysis and modelling were conducted in the GIS by modelling the weighted FCW matrix. Table 4 describes the variables, value ranges, spatial scale and data source for each of the 24 components of the FCW matrix used in the mapping.

The WAM for Tepic resulted from the application of the parameters of Table 3 to raster maps with the GIS map calculator. Figure 3 shows that the central and north-western areas have more favourable elements for walkability, while the opposite is reflected in the eastern and peripheral areas of the city, where the topography is hilly.

To analyse the micro-scale, areas with contrasts of social deprivation are illustrated with respect to the degree of walkability resulting from the model (Fig. 4). The WAM provides insights into how individuals adapt their behaviour in response to the built urban environment and the social contexts in which they reside. The findings are divided into five key observations, represented in Figures (a) to (e).

1. Figure (a): Walkability in Contexts with High Levels of Social Deprivation

In areas with high levels of social deprivation, pedestrian activity is highly concentrated, yet many streets exhibit very low walkability. This paradox is attributed to factors such as the streets' poor conditions, security, road safety and conditions of the natural environment. Despite these barriers, the volume of pedestrian movements does not decrease, suggesting that socioeconomic dynamics may override infrastructural limitations. This context underscores the need to address walkability barriers in less advantaged neighbourhoods, where there are more pedestrians.

2. Figure (b): Heterogeneity in Contexts of Medium-to-High Levels of Social Deprivation

Streets in areas of medium-to-high levels of social deprivation display improvements in urban design but continue to face challenges related to urban accessibility, security and road safety. The WAM indicates a fragmented heterogeneity in urban development, manifesting as structural barriers that constrain pedestrian movements in socially disadvantaged areas.

Table 4 Parameters assigned to walkability non-observable factors and observable components in the mapping process

Non-observable factors	Observable component	Measurement elements	Categories	Range of values	Spatial scale	Source of information
Streets' physical condition	Presence of sidewalk	Presence of sidewalks per block in both sides of the street	Yes	1	Segment/Raster in georeferenced units at 5 m	Characteristics of the urban environment 2020 (INEGI)
			No	0		
	Obstacles on the sidewalk	Presence of public telephones, semi-fixed posts or public lighting poles per block in both sides of the street	Yes	0	Segment/Raster in georeferenced units at 5 m	Characteristics of the urban environment 2020 (INEGI)
			No	1		
	Presence of ramps	Presence of ramps per block on both sides of the street	Yes	1	Segment/Raster in georeferenced units at 5 m	Characteristics of the urban environment 2020 (INEGI)
			No	0		
	Pavement	Type of surface in street section	Asphalt or hydraulic pavement	1	Segment/Raster in georeferenced units at 5 m	Municipal geographic information (IMPLAN Tepic)
			Asphalt pavement or stone	0.75		
			Surface within a private zone	0.5		
			Unpaved	0.25		

(continued)

Table 4 (continued)

Non-observable factors	Observable component	Measurement elements	Categories	Range of values	Spatial scale	Source of information
Street safety	Presence of road signs	Presence of pedestrian or traffic lights and street signs per block on both sides of the street	Yes	1	Segment/Raster in georeferenced units at 5 m	Characteristics of the urban environment 2020 (INEGI)
			No	0		
	Traffic speed	Official speed limit of public transport plus the proposed speed according to the street category of the State of Nayarit	Up to 30 km/h	1	Line/Raster in georeferenced units at 5 m	Government geographical information (Government of the State of Nayarit/City of Tepic)
			31–60 km/h	2/3		
			61 km/h or more	1/3		
	Presence of crosswalks	Presence of crosswalks per block in both sides of the street	Yes	1	Segment/Raster in georeferenced units at 5 m	Characteristics of the urban environment 2020 (INEGI)
			No	0		
	Section width of the vehicular stream	Section width according to the street category of the State of Nayarit	Local 12 m	1	Line/Raster in georeferenced units at 5 m	Municipal geographic information (IMPLAN Tepic)
			Collector 15 m	2/3		
			Main 30 m	1/3		
Security	Presence of public lightning (positive)	Presence of public lighting in the streets	Yes	1	Point/Raster in georeferenced units at 5 m	Municipal geographic information (IMPLAN Tepic)
			No	0		
	Presence of trees at night (negative)	Amount of street cover	Yes	0	Polygon/Raster in georeferenced units at 5 m	Municipal geographic information (IMPLAN Tepic)
			No	1		
	Crime records (negative)	Presence of crime records	Yes	1	Point/Raster in georeferenced units at 5 m	Municipal geographic information (Public Security)
			No	0		
	Presence of open shops and services (positive)	Presence of shops and services	Yes	1	Point/Raster in georeferenced units at 5 m	National Statistical Directory of Economic Unit 2024 (INEGI)
			No	0		

Urban accessibility	Distance to public transport	Distances to public transport stops	0–333 m	1	Polygon/Raster in georeferenced units at 5 m	Government geographical information (Secretary of Mobility)
			334–666 m	2/3		
			667–1000 m	1/3		
	Distance to shops and services	Distance in ranks to shops and services	0–333 m	1	Polygon/Raster in georeferenced units at 5 m	National Statistical Directory of Economic Unit 2024 (INEGI)
			334–666 m	2/3		
			667–1000 m	1/3		
	Distance to urban equipment	Distance in ranks to urban equipment	0–333 m	1	Polygon/Raster in georeferenced units at 5 m	Municipal geographic information (IMPLAN Tepic)
			334–666 m	2/3		
			667–1000 m	1/3		
	Density of inhabited dwellings	Concentration of inhabited dwellings per block	81–100% inhabited	1	Polygon/Raster in georeferenced units at 5 m	Population and housing census 2020 (INEGI)
			61–80% inhabited	0.80		
			41–60% inhabited	0.60		
			21–40% inhabited	0.40		
			0–20% inhabited	0.20		

Table 4 (continued)

Non-observable factors	Observable component	Measurement elements	Categories	Range of values	Spatial scale	Source of information
Comfort	Presence of urban furniture	Presence of public transport stops, bicycle ports or public telephones per block on both sides of the street	Yes	1	Segment/Raster in georeferenced units at 5 m	Characteristics of the urban environment 2020 (INEGI)
			No	0		
	Length of the block	Block length	0–100 m	1	Line/Raster in georeferenced units at 5 m	Characteristics of the urban environment 2020 (INEGI)
			101–300 m	0.75		
			301–500 m	0.50		
			501 m or more	0.25		
	Presence of historical heritage	Presence of historical heritage and its adjacent constructions protected by the INAH	Yes	1	Point/Raster in georeferenced units at 5 m	Government geographic information (INAH/ Government of the State of Nayarit)
			No	0		
	Presence of urban trees (positive)	Amount of tree cover	Yes	1	Polygon/Raster in georeferenced units at 5 m	Municipal geographic information (IMPLAN Tepic)
			No	0		

Natural environment	Temperature	Surface temperatures in degrees Celsius	14.1–15.7	1	Polygon/Raster in georeferenced units at 5 m	Surface temperature images of MODIS, Terra satellite. February 2000-December 2023. Average product of 8 days at a resolution of 1 km
			15.8–17.4	2/3		
			17.5–19.1	1/3		
	Topographic slopes	Percentage of topographic slopes	0–6%	1	Polygon/Raster in georeferenced units at 5 m	Level curve (INEGI)
			6.1–15%	0.50		
			15.1% or more	0		
	Swamp or flood zones	Flood areas in ranks	Very low	1	Polygon/Raster in georeferenced units at 5 m	Municipal geographic information (Risks atlas)
			Low	0.80		
			Moderate	0.60		
			High	0.20		
			Very High	0		
	Landslide areas	Landslide areas in ranks	Very low	1	Polygon/Raster in georeferenced units at 5 m	Municipal geographic information (Risks atlas)
			Low	0.80		
			Moderate	0.60		
			High	0.20		
			Very High	0		

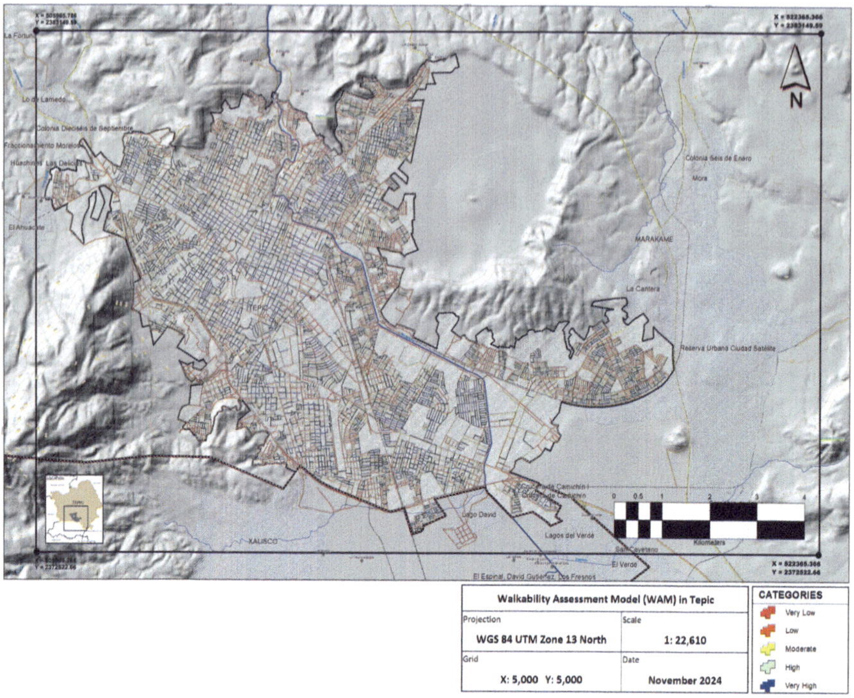

Fig. 3 Walkability assessment model in Tepic

3. Figure (c): Improvements in Areas of Medium-to-Low Levels of Social Deprivation

Neighbourhoods with medium-to-low levels of social deprivation show a predominance of streets with moderate, high and very high walkability levels. Urban design improvements are particularly evident near urban equipment, commercial zones and essential services. These findings highlight the influence of urban planning and the vitalisation of public space in enhancing walkability in semi-privileged contexts.

4. Figure (d): Paradox of High Walkability and Reduced Pedestrian Activity

High and very high walkability levels are prevalent in certain areas, yet pedestrian activity decreases. This phenomenon suggests a complex interplay between improved urban design and social dynamics. This context highlights the need to explore additional non-observable factors and observable components that could promote walking in these areas, such as improving comfort, urban accessibility or the conditions of the natural environment.

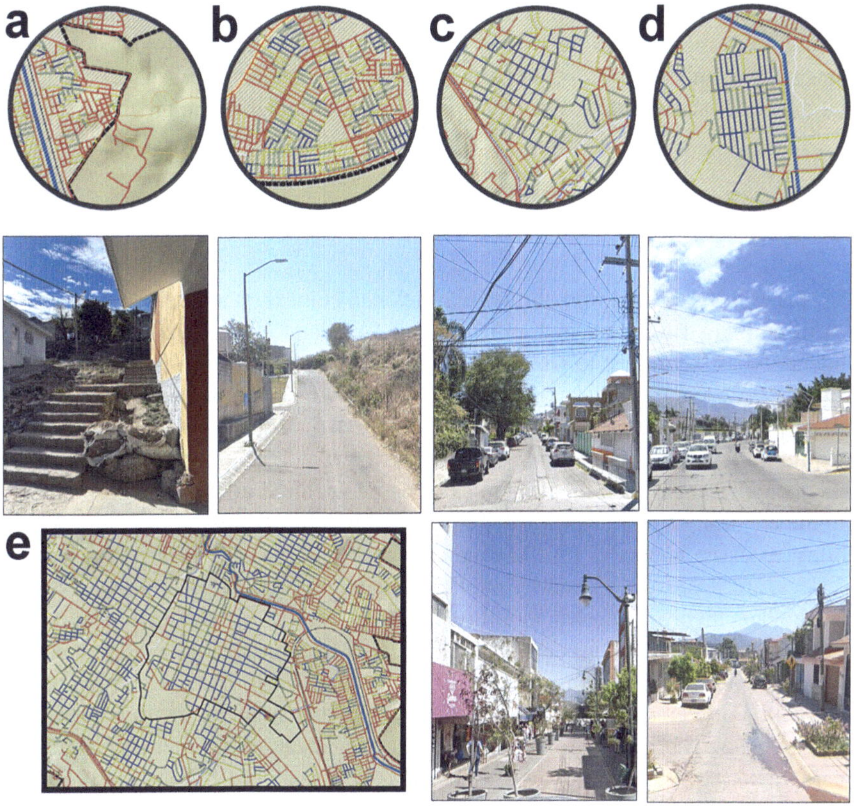

Fig. 4 Degree of walkability at micro-scale with varying degrees of pedestrian movement and social deprivation in Tepic. (**a**) Zitakua neighbourhood and proximity areas with a high degree of pedestrian trips and with a high and very high degree of deprivation; (**b**) Vistas de la Cantera neighbourhood with a high concentration of pedestrian traffic and high and medium social deprivation; (**c**) Colonia Morelos and proximity areas with a high degree of pedestrian displacement and with a medium and low degree of deprivation; (**d**) The Colonia Jacarandas and proximity areas with a low degree of pedestrian displacement and a low degree of social deprivation; (**e**) Colonia Centro and surroundings

5. Figure (e): High Walkability in Central and Western Zones

The central and western parts of the city exhibit the highest walkability levels. Streets rated as very high have undergone interventions under the 'complete street' model, incorporating semi-pedestrianised and fully pedestrianised designs that prioritise pedestrians over vehicles. These improvements reflect the potential of targeted urban interventions to create walkable environments.

The findings underscore the critical role of urban design in fostering walkable communities, particularly in addressing social disparities. The WAM provides valuable insights for urban planners and policymakers, emphasising the need to enhance accessibility and pedestrian-friendly initiatives across all socioeconomic contexts. Prioritising such efforts can contribute to equitable urban development that integrates social equity, ultimately promoting sustainable and inclusive cities.

5 Discussion

This paper has analysed walkability in a Mexican context, at a time when there is a dearth of related research in Latin America (Arellana et al., 2020). Walkability, understood as the degree to which the built environment is pedestrian-friendly and promotes safe, comfortable and interesting walking trips (Habibian & Hosseinzadeh, 2018; Moura et al., 2017; Bartzokas & Yorgos, 2016). The observable components identified in collaboration with citizens to foster walkable environments in Tepic reflect an alignment with the principles of people-centred urban design, emphasising the creation of more sustainable cities. The results obtained through the walkability assessment model (WAM) indicate that spaces, which integrate specific features, are positively valued by public officials, professionals and other relevant stakeholders involved in urban mobility. However, walkability should not be limited to physical or functional aspects. It is also crucial to consider how to improve people's perceptions of their environment. This multidimensional approach is consistent with Fonseca et al.'s (2022) observations, which highlight the impact of walkability across multiple disciplines, including health and environmental sciences. In this context, and in line with the ideas of Balsas (2017), walkability should increasingly be considered as a fundamental factor in the design and management of 15-minute cities. Furthermore, its potential as a strategic tool to address challenges related to pedestrian mobility and to improve the quality of the urban built environment is underlined, contributing to the development of more inclusive, safe and sustainable communities.

Theoretical frameworks proposed in the late twentieth and early twenty-first centuries, such as Cervero and Kockelman's (1997) '3D' model, have been fundamental to understanding walkability, although their ability to capture all observable street-level components that influence pedestrian travel is limited. These models require adaptations that consider the particularities of cities, especially in the context of the Global South, where factors such as the streets' physical condition, perceptions of safety and human-scale urban design are crucial to the pedestrian experience. Although there is no definitive consensus on the non-observable factors and observable components to assess walkability (Shashank & Schuurman, 2019), in Tepic the direct relationship between safety, both personal and road safety and the conditions of the physical-natural environment is evident.

A multiscale approach addresses both the meso-scale, which focuses on the dynamics of urban areas, and the micro-scale, which examines specific street conditions. Recent research (Basu et al., 2022; Erturan & Van der Spek, 2022; Guzman et al., 2022; Koo et al., 2022) demonstrates that this holistic perspective can identify the key factors that affect the walking experience, from physical infrastructure to social perceptions and dynamics. The study reaffirms the importance of assessing qualitative street-level characteristics, such as a sense of safety and walking enjoyment, which are essential to understanding and improving urban walkability (Fonseca et al., 2022). The physical conditions of pavements, for example, emerge as a fundamental component in any walkability analysis, as pointed out by Jensen et al. (2017). In this sense, it is vital to consider specific aspects of the Mexican context, such as the existence of obstacles on pavements, the quality of signage, the speed of vehicular traffic, street lighting, urban trees and accessibility to transport and services.

It is important to note that, as Talavera et al. (2012) point out, it is vital to identify not only the factors that encourage walking but also those that discourage it. Therefore, walkability assessment models should consider both the positive and negative aspects of their components. In addition, it is crucial to recognise that some of these factors have time-specific impacts, as their influence may vary according to the time of year or the time of day. For example, trees can increase comfort and improve environmental conditions for walking, but in certain locations, they can affect visibility and increase risks associated with personal safety, especially in Latin American contexts.

This study reinforces the importance of understanding walkability as a complex phenomenon that encompasses multiple hierarchical levels and dimensions. As Gibson (1977) points out, it is essential to analyse the intrinsic properties of urban environments that condition human behaviour, in order to identify both barriers and opportunities to promote pedestrian mobility. Within this framework, walkability should be conceived not only as a matter of physical design but also as a process that integrates intrapersonal, cultural, organisational, political and contextual factors, in line with Alfonzo (2005) and Sallis et al. (2006).

6 Conclusion

This paper addressed two research questions: (1) What are the non-observable factors and observable components influencing the decision to walk in Tepic? and (2) To what extent are these factors present in different areas of the city? In response, a walkability assessment model (WAM) was developed for Tepic, Nayarit, Mexico, utilising a method that analyses non-observable factors and observable components through the perspectives of public officials, professionals and other relevant stakeholders involved in urban mobility and walkability.

The results offer a valuable tool for urban planning focused on pedestrian mobility. The WAM enables the evaluation of both tangible elements of the built environment—such as street conditions—and intangible factors, such as safety and walking comfort, based on citizens' perceptions. Pedestrian mobility cannot be solely understood through physical infrastructure, as, particularly in the Latin American context, people adapt to the limitations of the built environment driven by their need to move. This adaptability is essential to understanding the broader dynamics of pedestrian movement.

Walkability is influenced by a hierarchical, socio-ecological framework, beginning with individual experiences and cultural factors and extending to political contexts shaped by laws and regulations. This highlights the fact that pedestrian mobility is a dynamic and complex outcome, influenced by the integration of these various factors. Effective planning must consider this multidimensionality to foster accessible and walkable environments that promote increased pedestrian activity.

The study further reveals that while there are shared patterns in the factors influencing walkability from a Latin American perspective, it is essential to tailor solutions to the geographical, environmental and socio-cultural specifics of each city. The findings emphasise that 15-minute city solutions should not be uniform but should consider local dynamics to ensure that interventions align with community perceptions. Additionally, the qualitative and quantitative methods used in this study demonstrated their complementary effectiveness.

The interplay between non-observable factors, such as perceptions of the physical environment and walking comfort, with socio-cultural dynamics and citizens' perceptions, highlights the need to adopt a transdisciplinary approach to walkability. This approach recognises its role not only in improving urban mobility but also in enhancing sustainable human development and advancing the 15-minute city paradigm.

This case study marks a significant step forward in assessing walkability in tropical cities of western Mexico and serves as a foundation for similar research in other regions. It complements ongoing efforts to promote 15-minute cities, providing a tool that can enhance the effectiveness of public policies and strengthen Mexican cities' capacity to address urban sustainability challenges. The model offers a replicable framework for future studies and public policies focused on pedestrian mobility and the 15-minute city concept in Mexico and across Latin America.

Moreover, the physical-natural environment, shaped by the city's morphology, plays a key role in pedestrian experiences. In tropical cities, environmental factors such as precipitation and extreme temperatures must be considered in the design of walkable spaces. These findings stress the need for adaptive approaches that incorporate both universal and local characteristics when assessing walkability.

Ultimately, this study underscores the value of walkability assessment models in the Global South as innovative tools for crafting solutions tailored to local contexts. Such models help identify specific urban needs and contribute to designing equitable, sustainable environments. Walkability, therefore, should be regarded not only as an indicator of urban quality but as a fundamental tool for realising the vision of 15-minute cities, where access to services and public spaces is universally available, improving the quality of life for all residents.

References

Alfonzo, M. A. (2005). To walk or not to walk? The hierarchy of walking needs. *Environment and Behavior, 37*, 808–836.

Arellana, J., Saltarín, M., Larrañaga, A., Álvarez, M., & Henao, C. (2020). Urban walkability considering pedestrians' perceptions of the built environment: A 10-year review and a case study in a medium-sized city in Latin America. *Transport Reviews, 40*(2), 183–203.

Balsas, C. (2017). The right to walk in cities, a comparative review of Macau, Lisbon and Las Vegas. *International Journal of Law in the Built Environment, 9*(2), 123–142.

Basu, N., Oviedo, O., King, M., Kamruzzaman, M., & Haque, M. (2022). The influence of the built environment on pedestrians' perceptions of attractiveness, safety and security. *Transportation Research Part F: Traffic Psychology Behavior, 87*, 203–218.

Baobeid, A., Koç, M., & Al-Ghamdi, S. (2021). Walkability and its relationships with health, sustainability, and livability: Elements of physical environment and evaluation frameworks. *Frontiers in Built Environment, 7*(7), 6.

Bartzokas, A., & Yorgos, N. (2016). What matters when it comes to "Walk and the city"? Defining a weighted GIS-based walkability index. *Transportation Research Procedia, 24*, 523–530.

Berrones-Sanz, L. (2022). Walking in Mexico City: Sociodemographic characteristics of the pedestrian. *Theorical and Empirical Researches in Urban Management, 17*(2).

Cervero, R., & Kockelman, K. (1997). Travel demand and the 3Ds: Density, diversity, and design. *Transportation Research Part D: Transport and Environment, 2*(3), 199–219.

Ewing, R., & Cervero, R. (2010). Travel and the built environment: A meta-analysis. *American Planning Association, 76*(3), 265–294.

Ewing, R., & Handy, S. (2009). Measuring the unmeasurable: Urban design qualities related to walkability. *Journal of Urban Design, 14*(1), 65–84.

Erturan, A., & Van der Spek, S. (2022). Walkability analyses of Delft city centre by Go-Along walks and testing of different design scenarios for a more walkable environment. *Journal of Urban Design, 27*(3), 287–309.

Fonseca, F., Papageorgiou, G., Tondelli, S., Ribeiro, P., Conticelli, E., Jabbari, M., & Ramos, R. (2022). Perceived walkability and respective urban determinants: insights from Bologna and Porto. *Sustainability, 14*, 1.

Gehl, J. (2016). *La ciudad vital, segura, sana y sostenible en Ciudades para la gente. Páginas 105,114*. Ediciones El Equilibrista y Kaluz.

Gibson, J. (1977). The theory of affordances. In R. Shaw & J. Bransford (Eds.), *Perceiving, acting, and knowing: Toward an ecological psychology* (pp. 67–82). Lawrence Erlbaum.

Guzmán, L., Arellana, J., & Castro, W. (2022). Desirable streets for pedestrians: Using a street-level index to assess walkability. *Transportation Research Part D: Transport and Environment, 111*.

Habibian, M., & Hosseinzadeh, A. (2018). Walkability index across trip purposes. *Sustainable Cities and Society, 42*, 216–225.

Instituto Nacional de Estadística y Geografía. (2020). *Censo Nacional de Población y Vivienda 2020*.

Instituto Nacional de Estadística y Geografía. (2022). *Principales resultados del Censo Nacional de Población y Vivienda 2020* (pp. 111–123).

Instituto Municipal de Planeación de Tepic. (2017). *Plan Integral de Movilidad Urbana Sustentable*. Ayuntamiento de Tepic.

Instituto Municipal de Planeación del Valle de Santiago. (2021). *Índice de caminabilidad de Valle de Santiago 2020-2021*. IMPLAN.

Jensen, W., Stump, T., Brown, B., Werner, C., & Smith, K. (2017). Walkability, complete streets, and gender: Who benefits most? *Health & Place, 48*, 80–89. https://doi.org/10.1016/j.healthplace.2017.09.007

Koo, B., Guhathakurta, S., & Botchwey, N. (2022). How are neighborhood and street-level walkability factors associated with walking behaviors? A big data approach using street view images. *Environment Behavior, 54*(1), 211–241.

Larrañaga, A. M., Arellana, J., Rizzi, L. I. et al. (2019) Using best–worst scaling to identify barriers to walkability: a study of Porto Alegre, Brazil. *Transportation 46*, 2347–2379. https://doi.org/10.1007/s11116-018-9944-x

Larrañaga, A., Rizzi, L., Arellana, J., Strambi, O., & Cybis, H. (2016). The influence of built environment and travel attitudes on walking: A case study of Porto Alegre, Brazil. *International Journal of Sustainable Transport, 10*, 332–342.

Li, Y., Yabuki, N., & Fukuda, T. (2023). Integrating GIS, deep learning, and environmental sensors for multicriteria evaluation of urban street walkability. *Landscape and Urban Planning, 230*.

Liao, B., Van den Berg, P., Van Wesemael, P., & Arentze, T. (2022). Individuals' perception of walkability: Results of a cojoint experiment using videos of virtual environments. *Cities, 125*(103650).

Lotfi, S., & Koohsari, M. (2009). Analyzing accessibility dimension of urban quality of life: Where urban designers face duality between subjective and objective reading of place. *Social Indicators Research, 94*, 417–435.

Martínez, M., Gortari, J., & Gaona, L. (2021). Marginación y percepción del ambiente construido. *Estudio cualitativo en zonas urbanas. Espacialidades, 11*(01), 63–82.

Méndez, G., Velazco, V., Menéndez, K., Alduena, I., Escalante, D., Tapia, J., Adaya, E., & Hernández, J. (2018). *Anatomía de la Movilidad en México, Hacia Dónde Vamos* (pp. 25–26).

Moura, F., Cambra, P., & Gonçalves, A. (2017). Measuring walkability for distinct pedestrian groups with a participatory assessment method: A case study in Lisbon. *Landscape and Urban Planning, 157*, 282–296. https://doi.org/10.1016/j.landurbplan.2016.07.002

Ruiz, A., Hernández, J., & García, M. (02 de febrero de 2022). *Tres líneas de acción para descarbonizar el sector transporte en México*. World Resources Institute.

Sallis, J., Cervero, R., Ascher, W., Henderson, K., Kraft, K., & Kerr, J. (2006). An ecological approach to creating active living communities. *Annual Reviews Public Health, 27*, 297–322.

Saaty, T. L. (1980). The analytic hierarchy process, New York: McGraw Hill. International, Translated to Russian, Portuguese, and Chinese, Revised editions, Paperback (1996, 2000), Pittsburgh: RWS Publications.

Shashank, A., & Schuurman, N. (2019). Unpacking walkability indices and their inherent assumptions. *Health & Place, 55*, 145–154.

Talavera, R., Valenzuela-Montes, L., & Soria-Lara, J. (2012). La calidad peatonal como método para evaluar entornos de movilidad urbana. *Documents d'Anàlisi Geogràfica 2014, 60*(1), 161–187.

Tena, F. (2018). *Relación entre caminabilidad física y percibida en Tijuana, México*. El Colegio de la Frontera.

Vilches, A., & Gil-Pérez, D. (2016). La transición a la Sostenibilidad como objetivo urgente para la superación de la crisis sistémica actual. *Revista Eureka sobre Enseñanza y Divulgación de las Ciencias, 13*(2), 395–407. https://rodin.uca.es/handle/10498/18296.

Transport Planning and the 15-Minute City: Opportunities and Challenges

Clara Loubet, Francesco Bruzzone, Louafi Bouzouina, Cristina Bircu, and Silvio Nocera

1 Introduction

In an era where the goal of carbon neutrality by 2050 is imperative, the rapid growth of urban populations poses substantial challenges to achieving sustainable development. Currently, over 55% of the world's population resides in cities, and this figure is expected to rise to nearly 68% by 2050 (United Nations, 2020). Cities, while driving economic growth and innovation, are responsible for more than 60% of global greenhouse gas emissions. As the world works toward limiting global warming to a maximum of 1.5 degrees Celsius, as outlined in the Paris Agreement, significant transformations in various sectors, particularly urban planning and transport, are urgently needed (United Nations, 2020).

The transport sector alone accounts for approximately 23% of global greenhouse gas emissions, making it a focal point for sustainable innovation (UNECE, 2024). Addressing these challenges requires rethinking how cities function and how people move within them. One of the most promising frameworks for such innovation is the 15-minute city, a concept developed and popularized by urban planner Carlos Moreno since 2016. The 15-minute city envisions a decentralized urban model where residents can meet most of their daily needs-whether living, working, shopping, healthcare, education, or entertainment-within a 15-minute walk or bike ride

C. Loubet · L. Bouzouina
LAET, ENTPE, University of Lyon, Vaulx-en-Velin, France
e-mail: louafi.bouzouina@entpe.fr

F. Bruzzone · C. Bircu · S. Nocera (✉)
Department of Architecture and Arts, Università Iuav di Venezia, Venice, Italy
e-mail: nocera@iuav.it

L. Bouzouina et al. (eds.), *15-Minute Cities*, Footprints of Regional Science,
https://doi.org/10.1007/978-3-032-20840-8_7

Fig. 1 The 15-minute city framework (based on Khavarian-Garmsir et al., 2023)

from their homes (Moreno et al., 2021). This approach not only reduces the reliance on cars but also fosters healthier, more resilient communities.

To concretize the vision of a 15-minute city, several urban planning principles must be upheld (Fig. 1). First, proximity is a central aspect, ensuring that key services and amenities are accessible within a short distance. Second, achieving an optimal density is critical: a more compact urban environment enables accessibility to essential functions on foot or by bicycle: higher population density and well-distributed services can significantly reduce travel times and dependence on private motorised vehicles. Third, flexibility in the design and use of public space is vital. Multifunctional spaces can accommodate a variety of uses, maximizing land efficiency, enhancing social interaction, and fostering more vibrant urban areas. Thus, planners must focus on creating multi-use areas where residential, commercial, and recreational functions are integrated within the same neighborhood. Finally, an efficient, sustainable transportation network is indispensable. All neighborhoods should be well-connected through infrastructure to support active modes of transport (walking and cycling) and public transit, thus avoiding the isolation of peripheral areas and reducing reliance on car travel (Khavarian-Garmsir et al., 2023).

This chapter delves into the advantages and limitations of the 15-minute city, enhancing the perspective of transport planning. First, we provide a thorough literature review on opportunities and challenges for the 15-minute city. Then, using Milan, Italy, as a case study, we examine how the city's urban and transport systems have evolved to align with the principles of the 15-minute city. By understanding both the potential and the challenges inherent in this model, we can better assess its applicability in other urban contexts around the world.

2 Opportunities for the 15-Minute City

The 15-minute city offers, according to its supporting arguments, a compelling response to contemporary urban challenges, particularly the need to reduce greenhouse gas emissions. By minimizing car dependency, this urban model directly addresses traffic congestion and promotes active mobility, such as walking and cycling, which in turn reduces CO_2 emissions and improves air quality (Allam et al., 2022; Bassolas et al., 2019). The integration of multi-use districts, where residential, commercial, recreational, healthcare, education, and entertainment functions coexist, shortens the distance and frequency of trips, resulting in a notable reduction in the daily use of private cars (Khavarian-Garmsir et al., 2023). Research indicates that adopting the 15-minute city framework, alongside the use of multiple transport modes such as walking, cycling, and public transport, could potentially reduce CO_2 emissions by 50% to 88% (Jin et al., 2024). In some cities where active modes of transportation are already well-developed, the totality of the population can reach essential services within 15 minutes by active modes (Knap et al., 2023). To extend this condition to a greater number of cities, significant investment is required in pedestrian infrastructure and walkability, as well as the promotion of walking as a viable transport option (Abdelfattah et al., 2022; Giuffrida et al., 2024).

In addition to addressing greenhouse gas (GHG) emissions, the 15-minute city mitigates air and noise pollution at the local level. Outdoor air pollution is responsible for approximately 37% of premature deaths globally, translating into 4.2 million early deaths annually, with motor transport contributing 70–80% of nitrogen dioxide (NO_2) emissions (Allam et al., 2022; WHO, 2022a). Without substantial improvements in air quality, many countries will face increasing mortality rates and economic losses. By concentrating services and populations in denser, more walkable areas, the 15-minute city model reduces motorized traffic in favor of active transportation, resulting in cleaner air. In Barcelona, for example, the introduction of superblocks has redirected 230,000 motorized vehicles towards public transport or soft mobility, leading to a 24% reduction in NO_2 levels and a 5% decrease in noise pollution (Mueller et al., 2020). On a broader scale, a 20–25% reduction in car traffic within dense cities could lead to a 27% drop in $PM_{2.5}$ and a 38% reduction in NO_x levels (Shoina et al., 2024).

Beyond environmental benefits, the 15-minute city promotes physical health by encouraging active mobility. At a time when nearly a quarter of the global population is insufficiently active and faces a higher risk of premature death by 20 to 30%, city designs that prioritize walkable and cyclable environments can significantly improve public health outcomes (Ferretto et al., 2021; WHO, 2022b). In addition, environments designed for active modes of transport improve road safety, reducing the incidence of injuries, falls, and traffic accidents (EIT, 2022).

For the 15-minute city to succeed in tackling pollution and emissions equitably, social attributes like inclusion and equality must be prioritized. Equal access to transportation is vital to ensure that all residents, regardless of their socioeconomic status, can benefit from the mobility supply, particularly public and shared modes

(Bruzzone et al., 2023a). This requires thoughtful integration of infrastructure, such as parking and charging stations, near residential areas and public transport hubs to facilitate inter-modality. Strategic urban planning and transport policies can also allocate more space for active transport modes, improving accessibility and attractiveness (Pozoukidou & Angelidou, 2022).

However, improving access to public transport is insufficient on its own. In a pedestrian-focused 15-minute city, the capacity of the transport network must also be expanded to ensure neighborhoods remain well-connected to the rest of the city (Wolański, 2023). A shift in public habits, such as diversifying work, school, and service schedules, could help spread the demand for transport over a longer period, alleviating rush-hour congestion and improving the overall quality of life for city residents (Abdelfattah et al., 2022).

Ultimately, the well-being of residents is central to the 15-minute city's success. By reducing car congestion and travel times, residents can reclaim time for activities that enhance their mental and physical health (Moreno et al., 2021). Furthermore, the tangible nature of the changes brought about by 15-minute city policies, such as improved walkability and air quality, can make these benefits more immediately perceptible to residents. In contrast to larger-scale, city-wide emissions reduction initiatives, the localized effects of a 15-minute city, such as pedestrian-friendly spaces and accessible services, are often easier for residents to experience directly. Engaging the community through participatory planning and highlighting the benefits of alternative transport modes can strengthen local support and contribute to the model's long-term success (Marquet et al., 2024).

3 Challenges for the 15-Minute City and Forward-Looking Solutions

Despite its promise, the vision of the 15-minute city -a dense, connected, accessible, and digitized urban environment- faces several significant challenges. One of the most prominent criticisms is the risk of *gentrification* that often accompanies its implementation. Improving accessibility and enhancing certain neighborhoods typically leads to a rise in property values, especially in urban centers that already experience intense real estate pressure. This can push low-income households to the urban periphery, turning these upgraded neighborhoods into exclusive enclaves for wealthier residents (Marquet et al., 2024; Pozoukidou & Chatziyiannaki, 2021). As a result, gentrification can exacerbate social inequality, undermining one of the key goals of the 15-minute city: creating equitable access to urban services for all residents. In contrast, suburban and peripheral areas, which are often less densely populated, find it far more difficult to adopt the 15-minute city concept. In many of these areas, residents remain heavily reliant on private cars, whereas people in urban centers have gradually shifted toward cycling and walking over the past decade and a half (Bruzzone et al., 2021; Poorthuis & Zook, 2023). This growing disparity between the urban core and the periphery highlights the uneven accessibility to essential services (Bertocchi et al., 2025). While city centers may successfully embody the 15-minute city ideals, suburban areas frequently remain monofunctional, with low walkability and limited public transport options (Akrami et al., 2024; Yang et al., 2023). This raises fundamental questions about whether and how suburban environments can adapt to the 15-minute city model.

One key barrier is population density, which is often insufficient in suburban areas to justify the expansion of public services and transport infrastructure. For the 15-minute city to succeed, public transport must be frequent and extensive enough that car use becomes not only unnecessary but inconvenient for daily needs (EIT, 2022). Without sufficient density, the financial and operational feasibility of delivering essential services, whether public transport or local amenities, becomes highly problematic (Bruzzone et al., 2023b). The challenge of uneven accessibility extends beyond spatial considerations to demographic disparities. Accessibility is inherently tied to personal mobility, and reduced walking speeds can severely limit the ability of certain groups or individuals to access services within 15 minutes (Dunning et al., 2023). Seasonal variations further complicate accessibility, particularly for vulnerable groups like the elderly, who may find it difficult to navigate outdoor spaces during the colder months or even at night (Willberg et al., 2023). The 15-minute city, therefore, cannot adopt a one-size-fits-all approach but must be adapted to account for the needs of different demographic and socio-economic groups.

Another issue is that many cities have not adequately integrated employment into their 15-minute city strategies. While the model works well for accessing local services, it is far more difficult to bring workplaces within a 15-minute reach of all residents (Lu & Diab, 2023). Studies show that the likelihood of finding suitable employment increases with the distance a person is willing to travel (Cremaschi, 2022). In cities that aim to increase local accessibility, commuting to work remains a significant challenge. Even in cities that are considered close to achieving the 15-minute city ideal, 35% of home-to-work journeys are still made by car (EMEF, 2022). This is especially true for jobs located in industrial zones or central business districts, which are often far from residential neighborhoods. As such, connecting districts via public transport is crucial because expecting all journeys -particularly work-related ones- to fall within the 15-minute threshold remains unrealistic (Wolański, 2023).

There are also conceptual limitations in the relationship between proximity and travel time. Although the 15-minute city focuses on reducing travel for essential services, research shows that people spend only 12 minutes of their daily 75-minute travel budget on services and consumption. This suggests that daily journeys are often more complex, encompassing a wide range of social, cultural, and political activities (Poorthuis & Zook, 2023). Consequently, the assumption that reducing proximity to essential services will drastically cut overall travel time may be oversimplified. For example, Abbiasov et al. (2022) report that only 12% of daily trips are made for basic amenities within a 15-minute walk, though this varies widely by city and socio-economic context. Furthermore, the 15-minute city model may conflict with the very nature of urban life, which thrives on the concentration of key institutions and services on a city-wide scale. Replicating certain amenities, such as universities, museums, or theaters, at a neighborhood level is neither feasible nor desirable, as these institutions benefit from the large markets and cultural diversity that cities provide (Ratti & Florida, 2021). Therefore, while the 15-minute city can improve access to day-to-day services, it may struggle to fully address the broader needs and functions of a vibrant urban environment.

Despite the challenges associated with the 15-minute city, several innovative solutions can help mitigate these issues through adaptation and technological integration. One potential solution for less densely populated areas, particularly those further from city centers, lies in the deployment of micro-mobility options. In addition to bicycles, micro-mobility vehicles such as electric scooters and e-bikes can play a critical role in bridging distances that are too far to walk yet too short to justify car usage (Moreno et al., 2021; Papas et al., 2023). Studies indicate that the average distance traveled by electric scooter users ranges between 0.7 km and 2.4 km, with travel times between 8 and 12 minutes, while e-bike users typically cover 3 to 4.5 km in 15 to 20 minutes (Şengül & Mostofi, 2021). These vehicles effectively extend the range of the 15-minute city by making it feasible for residents to travel farther than they could by walking alone, thus enhancing local mobility options. Micro-mobility also aligns well with multiple use cases within the 15-minute city framework. For example, 42% of micro-mobility users rely on these vehicles for leisure, 41% for commuting, and 36% for running errands (Lang et al., 2022). Given its versatility, micro-mobility could play an important role in enabling suburban and less dense areas to overcome the spatial limitations that make walkability more challenging. A study conducted in two car-dependent areas of Auckland, New Zealand, revealed that while these areas had structural limitations, micro-mobility significantly improved residents' accessibility to key services. This indicates the feasibility of micro-mobility as a strategy to enhance connectivity in car-dependent neighborhoods (Jiang et al., 2024). To further encourage cycling and micro-mobility, a range of supportive infrastructure must be developed. This includes promoting inter-modality, ensuring that bicycles and scooters are easily integrated with public transport systems, providing secure and supervised bicycle storage facilities at key transport hubs, offering e-bike rentals, and improving the overall cycling infrastructure. By enhancing connectivity between cycle paths and intermodal centers, cities and regions can foster greater mobility, aligning with the objectives of the 15-minute city (Mishra et al., 2022).

In addition to micro-mobility, technology plays a significant role in advancing the 15-minute city concept. Technological innovations offer numerous solutions to optimize local mobility, connectivity, and efficiency (Bruzzone et al., 2025). For instance, the introduction of autonomous vehicles holds great potential for reshaping urban mobility. Autonomous vehicles could significantly reduce the demand for parking in city centers, as these vehicles could park themselves after dropping off passengers, potentially cutting parking space requirements by up to 67%. This reduction could be even greater if shared autonomous vehicles are incorporated into the system (Fayyaz et al., 2022). Autonomous mobility could help alleviate some of the parking and congestion issues faced by many cities, thereby supporting the dense and car-light environment envisioned by the 15-minute city. Beyond autonomous vehicles, smart city technologies provide further opportunities to address mobility challenges such as congestion and parking management. The use of big data and advanced technological solutions can help optimize urban functions by managing micro-mobility systems, monitoring environmental conditions, and controlling traffic flows (Khavarian-Garmsir et al., 2023). Smart technologies also have the potential to enhance accessibility by ensuring that micro-mobility and public transport services are better coordinated and responsive to residents' needs. However, these digital solutions must be implemented in a way that fosters inclusive access to mobility, ensuring that all residents, including the elderly and other vulnerable groups, benefit from the technological advancements in the 15-minute city (Pozoukidou & Angelidou, 2022).

Despite the advantages of smart technology, there are concerns that the adoption of autonomous vehicles may paradoxically lead to an increase in car journeys and a reduction in urban density (Bircu et al., 2025; Cremaschi, 2022). If autonomous vehicles become too convenient, they might encourage longer commutes or increase travel distances, which would undermine the principles of proximity and density that are central to the 15-minute city. Thus, while technology offers numerous solutions, it is essential to carefully manage its implementation to ensure that it aligns with the sustainable and equitable goals of the 15-minute city model.

Despite several doubts, many of the challenges facing the 15-minute city can be addressed through the strategic deployment of micro-mobility solutions and smart technologies. However, these solutions must be thoughtfully integrated to prevent unintended consequences, such as increased vehicle usage or exacerbated inequality, ensuring that the 15-minute city remains a viable and equitable model for future urban development.

4 15-Minute City Initiatives Worldwide: The Case of Milan

Numerous 15-minute city initiatives have been launched globally, with prominent examples including Paris, where Mayor Hidalgo championed the concept as part of her political campaign, Barcelona's super-blocks, and the "high streets" project in the United Kingdom. In the United States and Australia, cities like Portland and Melbourne have embraced the concept with the introduction of 20-minute neighborhoods (Fig. 2; Plan Melbourne, 2017; Pozoukidou & Chatziyiannaki, 2021). Some cities used the COVID-19 pandemic as an opportunity to redefine urban development policies, and Milan, one of Italy's hardest-hit cities during the crisis, serves as a prime example. The municipality of Milan decided to "redefine the use of roads and public spaces to increase non-polluting mobility" and to "rediscover the neighborhood dimension" by implementing a 15-minute city model (Milan Municipality, 2020; Pinto & Akhavan, 2021). This section will explore how Milan's approach, particularly in terms of transport planning, has sought to achieve the goals of the 15-minute city.

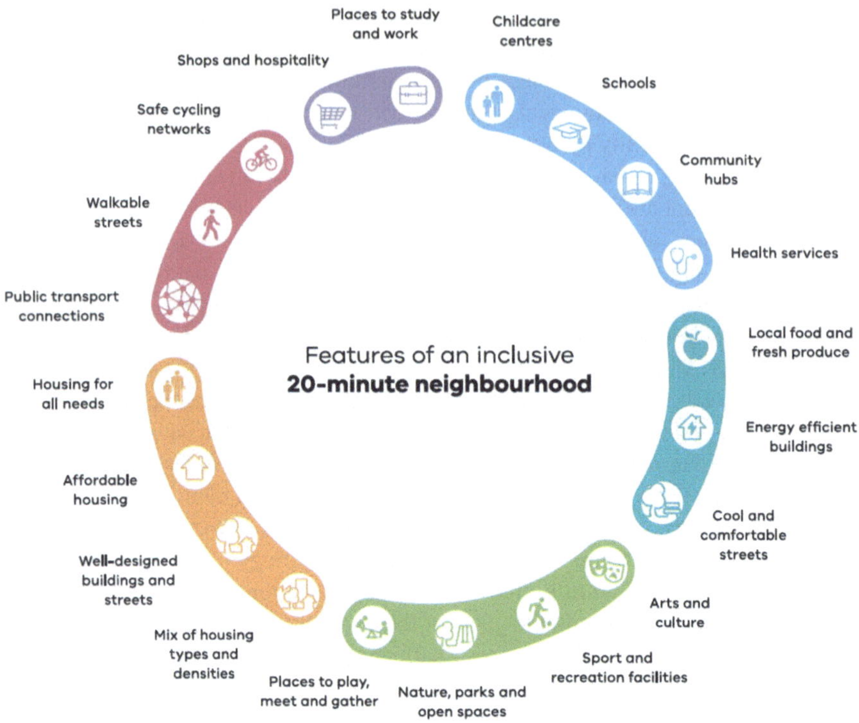

Fig. 2 Features of a 20-minute neighborhood (Plan Melbourne, 2017)

Focusing on local services and amenities, Milan aimed to improve access to essential services within a 15-minute walk. The overarching goal was to reclaim the urban environment for residents by extending public spaces, such as parks, and creating more pedestrian zones (Milan Municipality, 2020). Even before the pandemic, similar projects had been launched, with efforts to replace roads or intersections with public squares. Redesigning public spaces to prioritize pedestrian and cycling pathways is crucial for mitigating risks posed by extreme weather events and improving overall well-being, particularly in response to crises like COVID-19. High rush-hour traffic on public transportation networks exacerbates health risks, which the city addressed by advocating for lifestyle changes and altering residents' mobility patterns (Pinto & Akhavan, 2021). The city envisions a future where residents can live, work, and access most of their daily needs within the same neighborhood.

However, the challenge of reaching work within a 15-minute radius remains significant. To address this, Milan has proposed solutions such as adapting schedules for public services and businesses to avoid peak traffic times and regulate mobility demand. The city also encourages companies to implement telework policies and provide coworking spaces, allowing residents to work closer to home (Milan Municipality, 2020). The ultimate goal is to create neighborhoods where work, housing, green spaces, leisure, and public services coexist, reducing the need for long commutes (Papas et al., 2023).

Despite these efforts, the 15-minute city model cannot fully eliminate the need for commuting. Although initially planned as temporary, many new public squares, such as Piazza Dergano, Piazza San Luigi, and Piazza Angilberto II, remain in place in 2023, with some transformed into permanent fixtures. Milan's initiatives have not been limited to the city center; 29 of 40 projects have focused on the outskirts. These projects include new cycle paths and the addition of 30 km/h limited speed zones (Milan Municipality, 2023; Report della Mobilità, 2022). It is estimated that one out of every two Milanese citizens now has access to a public open space within a 15-minute walk from their home (Fabris et al., 2023).

Milan has also prioritized the development of cycling infrastructure. Over the past 5 years, the city has created 91 km of bike lanes, reaching 11% urban coverage. Much of this development occurred during the COVID-19 crisis in 2020 but continued in the following years. As a result, cycling's modal share has increased dramatically, from less than 10% in 2019 to nearly 22% in 2022 along key corridors like Corso Buenos Aires (Fig. 3). In tandem with this expansion, Milan has increased its supply of shared bikes, offering conventional bikes, electric bikes, and bikes equipped with child seats. Residents can rent these bikes using the same card as for public transport, in line with the city's aim to integrate mobility services (Mobility as a Service model). The municipality has also invested in shared electric scooters, mopeds, and cars, with shared car rentals rising by 6% between 2021 and 2022.

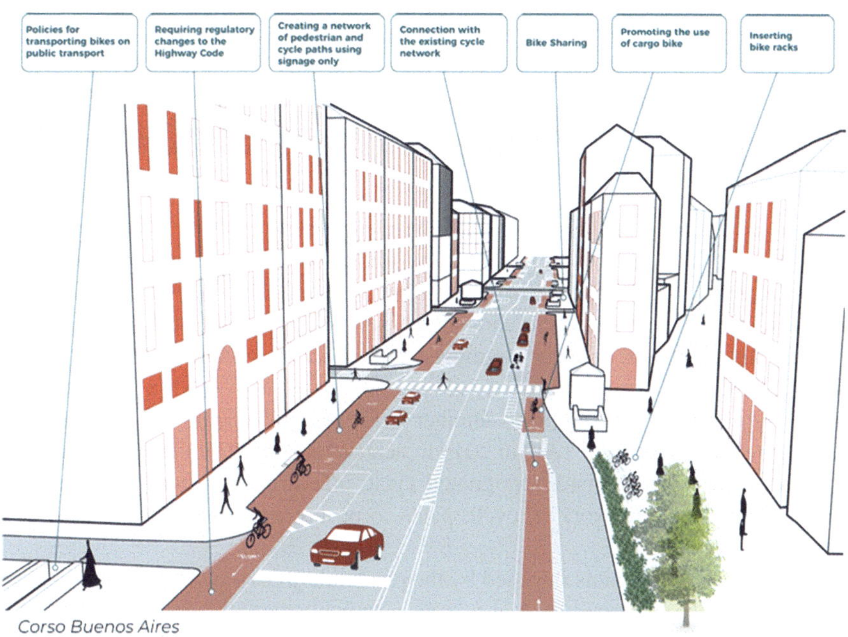

Fig. 3 Corso Buenos Aires (Milan)

Electric scooter rentals saw a 54% increase, while moped rentals increased by 23% during the same period. Although the overall impact on road congestion has been modest, the city's congestion index has slightly decreased from 31% in 2019 to 29% in 2022 (European Commission, 2023).

Accessibility and inclusivity are also major components of Milan's 15-minute city concept. The city has made 100% of its buses accessible to people with reduced mobility, while 85% of trams and trolleybuses are similarly equipped. Bus lines, such as the 63 and 54, and tram lines, such as the 1 and 16, have been extended to reach peripheral areas, ensuring that much of the city is accessible within 30 minutes by public transport (Report della Mobilità, 2022).

The regeneration of several railway stations has provided further opportunities for sustainable urban planning. For example, the Porta Romana station is being redeveloped to incorporate parks, local shops, services, and residential buildings, alongside workspaces. The mobility plan for this area includes a fully pedestrianized zone and extensive cycling facilities, reinforcing the 15-minute city's focus on easy access to services. The neighborhood is also well-connected to the broader city via public transport, including train, metro, bus, and tram services. The reduced role of cars in this plan underscores the city's commitment to more sustainable forms of transportation (Abdelfattah et al., 2021; COIMA, 2021).

Despite significant progress, several challenges remain. Milan still has areas, particularly in its outer suburbs, that cannot yet be classified as part of a 15-minute city. However, the issue is less severe compared to other European cities such as Madrid and Amsterdam, partly due to Milan being surrounded by multiple small centers on the outskirts, which have high population densities and localized services (EIT, 2022). Additionally, pedestrian areas remain concentrated in the city center, limiting safe pedestrian access to essential services in outlying areas. Traffic congestion, while somewhat reduced, persists, indicating that many residents are still unable to travel to work within a 15-minute window (Report della Mobilità, 2022).

One risk Milan faces is gentrification. The city's numerous investments to improve neighborhood viability, public spaces, and services have led to increased property prices. Some districts that have undergone redevelopment have seen price increases of up to 20% over the past 5 years (Pinto & Akhavan, 2021). Furthermore, while it's still too early to assess the full impact of these measures on air pollution and greenhouse gas emissions, the results remain uncertain (Fabris et al., 2023).

The rapid rise in cycling poses another challenge: safety. Bicycle-related accidents have increased, from 1200 in 2019 to nearly 1600 in 2021. Given that 76% of these accidents occur outside designated cycle infrastructure, there is a pressing need to expand the network of cycle paths to ensure cyclists' safety (Transform Transport, 2024).

In Milan, the constraints imposed by the pandemic were transformed into opportunities. The city embraced tactical urbanism, quickly and efficiently redeveloping public spaces, particularly in peripheral neighborhoods (Fig. 4; Fabris et al., 2023). This strategy, combined with substantial investments in soft mobility, has led to

Fig. 4 Via Spoleto-Venini before and after the "open piazza" project (Report della Mobilità, 2022)

better connections between pedestrian areas and major traffic arteries, making the 15-minute city model more accessible to its residents. Today, 88% of Milan's population can access essential services (public transport, leisure, education, health, shops) within a 15-minute walk, and 36% within just 5 minutes (EIT, 2022).

While there are still areas outside the 15-minute threshold, especially on the city's outskirts, the introduction of e-powered micro-mobility vehicles and mobility-on-demand services could help bridge these gaps, improving accessibility and connecting residents to public transport networks (Abdelfattah et al., 2021). Although much work remains to be done, Milan's efforts, such as the open squares project, represent a solid foundation for further urban renewal and development (Fabris et al., 2023).

5 Conclusion

The concept of the 15-minute city revolves around creating neighborhoods where all essential services, such as shops, schools, healthcare, and leisure facilities, are within a 15-minute walk or cycle ride. Similar to other, more famous examples such as Paris or Barcelona, Milan leveraged the pandemic confinement period to initiate a transformation of its urban landscape, starting with temporary open-space initiatives that later became permanent fixtures. The city's broader initiatives aim to renew certain districts entirely, making them more self-sufficient by offering residents all the services they need nearby. This urban renewal aligns with Milan's active mobility policy, which seeks to reduce car dependency, alleviate traffic congestion, and significantly enhance residents' quality of life. Key investments include expanding cycling and pedestrian infrastructure, as well as strengthening public transport to better connect neighborhoods.

However, challenges remain, particularly in the outskirts where service density is lower-Milan should take advantage of it being somewhat unique compared to other cities. As its outer areas are home to smaller, densely populated centers, Milan should work at metropolitan level, making the 15-minute city model more feasible (Vitulano et al., 2025). This highlights a key aspect of the concept: it must be adapted to the unique urban environment of each city and cannot be applied uniformly across all regions.

The 15-minute city carries profound implications for the future. By reducing the need for long car-based commutes it contributes to lowering air pollution and greenhouse gas emissions, crucial factors in combating climate change. Cleaner air, improved public health, and a reduced environmental footprint are direct benefits of this model. Compact cities with greener public spaces and less reliance on motorized vehicles also foster safer, more vibrant neighborhoods, enhancing residents' overall well-being.

Additionally, by encouraging mixed-use neighborhoods, where housing, shops, schools, and leisure spaces coexist, the 15-minute city model promotes more efficient resource use and reduces the need for costly infrastructure. From a transportation perspective, realizing this vision requires a redesign of existing networks, focusing on creating safe cycle paths, improving pedestrian walkways, expanding micro-mobility options, and developing more flexible, interconnected public transport systems that can facilitate smooth travel between neighborhoods without relying on private cars.

In conclusion, despite several challenges, the 15-minute city represents a forward-thinking approach for cities to adopt when planning their development. Even when implemented partially or in subsequent steps, it offers a viable solution to many of the pressing challenges of the twenty-first century: environmental sustainability, public health, and quality of life. Achieving this ideal can set cities on a path toward a more sustainable, livable, and equitable future.

References

Abbiasov, T., Heine, C., Edward L. Glaeser, Ratti, C., Sabouri, S., Salazar Miranda, A., & Santi, P. (2022). *The 15-minute city quantified using mobility data*. https://doi.org/10.3386/w30752

Abdelfattah, L., Deponte, D., & Fossa, G. (2021). The 15-minute city as a hybrid model for Milan. *Tema, Journal of Land Use, Mobility and Environment, 14*, 71–86. https://doi.org/10.6092/1970-9870/8653

Abdelfattah, L., Deponte, D., & Fossa, G. (2022). The 15-minute city: Interpreting the model to bring out urban resiliencies. *Transportation Research Procedia, 60*, 330. https://doi.org/10.1016/j.trpro.2021.12.043

Akrami, M., Sliwa, M., & Rynning, M. K. (2024). Walk further and access more! Exploring the 15-minute city concept in Oslo, Norway. *Journal of Urban Mobility, 5*, 100077. https://doi.org/10.1016/j.urbmob.2024.100077

Allam, Z., Bibri, S. E., Chabaud, D., et al. (2022). The '15-minute city' concept can shape a net-zero urban future. *Humanit Soc Sci Commun, 9*, 126. https://doi.org/10.1057/s41599-022-01145-0

Allam, Z., Nieuwenhuijsen, M., Chabaud, D., & Moreno, C. (2022). The 15-minute city offers a new framework for sustainability, liveability, and health. *The Lancet Planetary Health, 6*, E181. https://doi.org/10.1016/S2542-5196(22)00014-6

Bassolas, A., Barbosa-Filho, H., Dickinson, B., et al. (2019). Hierarchical organization of urban mobility and its connection with city livability. *Nature Communications, 10*, 4817. https://doi.org/10.1038/s41467-019-12809-y

Bertocchi, D., Bruzzone, F., & Nocera, S. (2025). A big data analysis of the gender gap in mobility at the regional scale: Insights from northern Italy. *Journal of Transport Geography, 128*, 104390. https://doi.org/10.1016/j.jtrangeo.2025.104390

Bircu, C., Bruzzone, F., Cavallaro, F., Indelicato, A., & Nocera, S. (2025). Reducing road congestion and air pollution with ICT-based platooning. *Transportation Research Procedia, VSI: TIS ROMA 2024, 90*, 35–42. https://doi.org/10.1016/j.trpro.2025.06.030

Bruzzone, F., Scorrano, M., & Nocera, S. (2021). The combination of e-bike-sharing and demand-responsive transport systems in rural areas: A case study of Velenje. *Research in Transportation Business & Management, 40*, 100570.

Bruzzone, F., Cavallaro, F., & Nocera, S. (2023a). The definition of equity in transport. *Transportation Research Procedia, 69*, 440–447.

Bruzzone, F., Nocera, S., & Pesenti, R. (2023b). Feasibility and optimization of freight-on-transit schemes for the sustainable operation of passengers and logistics. *Research in Transportation Economics, 101*, 101336. https://doi.org/10.1016/j.retrec.2023.101336

Bruzzone, F., Cavallaro, F., & Nocera, S. (2025). Accessibility potential of long-distance Mobility-as-a-Service. Transportation Research Part A: Policy and Practice, 195, 104466. https://doi.org/10.1016/j.tra.2025.104466

COIMA. (2021). *Porta Romana railway yard: the team led by OUTCOMIST, Diller Scofidio + Renfro, PLP Architecture and Carlo Ratti associati wins the tender for the preliminary urban regeneration masterplan with the Parco Romana project*. Retrieved July 18, 2024, from https://coima.com/en/media/press-releases/porta-romana-railway-yard

Cremaschi, M. (2022, 28 avril). *Ville du quart d'heure, ville des GAFA?* Métropolitiques. https://www.urbanisme-francophonie.org/wp-content/uploads/2022/12/M.-Cremaschi-Metropolitiques.pdf

Dunning, R. J., Dolega, L., Nasuto, A., Nurse, A., & Calafiore, A. (2023). Age and the 20-min city: Accounting for variation in mobility. *Applied Geography, 156*, 103005. https://doi.org/10.1016/j.apgeog.2023.103005

EIT Urban Mobility. (2022). *15-minute city: Human-centred planning in action*. https://www.eiturbanmobility.eu/wp-content/uploads/2022/11/EIT-UrbanMobilityNext9_15-min-City_144dpi.pdf

Enquesta de Mobilitat en Dia Feiner 2022 (EMEF). (2022). *Fulletó Principals Resultats*. Retrieved June 28, 2024, from https://www.omc.cat/documents/662112/1182871/EMEF+2022_Fullet%C3%B3.pdf

European Commission, Joint Research Centre. (2023). Christidis, P., Vega Gonzalo, M., Ulpiani, G. et al., *Post-pandemic trends in urban mobility*, Publications Office of the European Union. https://data.europa.eu/doi/10.2760/627403

Fabris, L. M. F., Camerin, F., Semprebon, G., & Balzarotti, R. M. (2023). How 15-min city, tactical urbanism, and superblock concepts are affecting major cities in the postCovid-19 era? In Z. Allam (Ed.), *Sustainable urban transitions* (Urban sustainability). Springer. https://doi.org/10.1007/978-981-99-2695-4_10

Ferretto, L., Bruzzone, F., & Nocera, S. (2021). Pathways to active mobility planning. *Research in Transportation Economics, 86*, 101027. https://doi.org/10.1016/j.retrec.2020.101027

Fayyaz, M., González-González, E., & Nogués, S. (2022). Autonomous mobility: A potential opportunity to reclaim public spaces for people. *Sustainability, 14*(3), 1568. https://doi.org/10.3390/su14031568

Giuffrida, N., Mölter, A., Pilla, F., Carroll, P., & Ottomanelli, M. (2024). On the equity of the x-minute city from the perspective of walkability. *Transportation Engineering, 16*, 100244. https://doi.org/10.1016/j.treng.2024.100244

Jiang, J., Qiao, W., Chuang, I.-T., Li, Y., Wang, T., & Beattie, L. (2024). Mapping liveability: The "15-min city" concept for car-dependent districts in Auckland, New Zealand. *Applied Geography, 163*, 103197. https://doi.org/10.1016/j.apgeog.2024.103197

Jin, T., Wang, K., Xin, Y., Shi, J., Hong, Y., & Witlox, F. (2024). Is a 15-minute city within reach? Measuring multimodal accessibility and carbon footprint in 12 major American cities. *Land Use Policy, 142*, 107180. https://doi.org/10.1016/j.landusepol.2024.107180

Khavarian-Garmsir, A. R., Sharifi, A., Hajian Hossein Abadi, M., & Moradi, Z. (2023). From garden city to 15-minute city: A historical perspective and critical assessment. *Land, 12*(2), 512. https://doi.org/10.3390/land12020512

Khavarian-Garmsir, A. R., Sharifi, A., & Sadeghi, A. (2023). The 15-minute city: Urban planning and design efforts toward creating sustainable neighborhoods. *Cities, 132*, 104101. https://doi.org/10.1016/j.cities.2022.104101

Knap, E., Ulak, M., Geurs, K. T., Mulders, A., & Van der Drift, S. (2023). A composite X-minute city cycling accessibility metric and its role in assessing spatial and socioeconomic inequalities—A case study in Utrecht, the Netherlands. *Journal of Urban Mobility, 3*, 100043. https://doi.org/10.1016/j.urbmob.2022.100043

Lang, N., Schellong, D., Hagenmaier, M., Herrmann, A., & Hohenreuther, M. (2022). *Putting micromobility at the center of urban mobility*. https://imo.unisg.ch/wp-content/uploads/2022/05/220520_Joint-Article-BCG-HSG_Putting-Micromobility-at-the-Center-of-Urban-Mobility.pdf

Lu, M., & Diab, A. (2023). Understanding the determinants of x-minute city policies: A review of the North American and Australian cities' planning documents. *Journal of Urban Mobility, 3*, 100040. https://doi.org/10.1016/j.urbmob.2022.100040

Marquet, O., Mojica, L., Fernández-Núñez, M. B., & Maciejewska, M. (2024). Pathways to 15-minute city adoption: Can our understanding of climate policies acceptability explain the backlash towards x-minute city programs? *Cities, 148*, 104878. https://doi.org/10.1016/j.cities.2024.104878

Milan Municipality. (2020). *Milan 2020: Adaptation strategy*. https://www.comune.milano.it/documents/20126/7117896/Milano+2020.+Adaptation+strategy.pdf/d11a0983-6ce5-5385-d173-efcc28b45413?t=1589366192908

Milan Municipality. (2023). *Open squares*. Retrieved July 17, 2024, from https://www.comune.milano.it/aree-tematiche/quartieri/piano-quartieri/piazze-aperte

Mishra, S., Bhattacharya, D., Chaturvedi, A., & Singh, N. (2022). Assessing micro-mobility services in pandemics for studying the 10-minutes cities concept in India using geospatial data analysis: An application. In *Proceedings of the 15th ACM SIGSPATIAL International Workshop on Computational Transportation Science (IWCTS '22)*. Association for Computing Machinery, Article 3, 1–10. https://doi.org/10.1145/3557991.3567788

Moreno, C., Allam, Z., Chabaud, D., Gall, C., & Pratlong, F. (2021). Introducing the "15-minute city": Sustainability, resilience and place identity in future post-pandemic cities. *Smart Cities, 4*(1), 93–111. https://doi.org/10.3390/smartcities4010006

Mueller, N., Rojas-Rueda, D., Khreis, H., Cirach, M., Andrés, D., Ballester, J., Bartoll, X., Daher, C., Deluca, A., Echave, C., Milà, C., Márquez, S., Palou, J., Pérez, K., Tonne, C., Stevenson, M., Rueda, S., & Nieuwenhuijsen, M. (2020). Changing the urban design of cities for health: The superblock model. *Environment International, 134*, 105132. https://doi.org/10.1016/j.envint.2019.105132

Papas, T., Basbas, S., & Campisi, T. (2023). Urban mobility evolution and the 15-minute city model: From holistic to bottom-up approach. *Transportation Research Procedia, 69*, 544. https://doi.org/10.1016/j.trpro.2023.02.206

Pinto, F., & Akhavan, M. (2021). Scenarios for a post-pandemic city: Urban planning strategies and challenges of making "Milan 15-minutes city". *Transportation Research Procedia, 60*, 370. https://doi.org/10.48295/ET.2021.85.12

Plan Melbourne. (2017). *20-minute neighbourhoods*. Retrieved July 1, 2024, from https://www.planning.vic.gov.au/guides-and-resources/strategies-and-initiatives/20-minute-neighbourhoods

Poorthuis, A., & Zook, M. (2023). Moving the 15-minute city beyond the urban core: The role of accessibility and public transport in the Netherlands. *Journal of Transport Geography, 110*, 103629. https://doi.org/10.1016/j.jtrangeo.2023.103629

Pozoukidou, G., & Angelidou, M. (2022). Urban planning in the 15-minute city: Revisited under sustainable and smart city developments until 2030. *Smart Cities, 5*(4), 1356–1375. https://doi.org/10.3390/smartcities5040069

Pozoukidou, G., & Chatziyiannaki, Z. (2021). 15-minute city: Decomposing the new urban planning Eutopia. *Sustainability, 13*(2), 928. https://doi.org/10.3390/su13020928

Ratti, C., & Florida, R. (2021). World Economic Forum. *The 15-minute city meets human needs but leaves desires wanting. Here's why*. https://www.weforum.org/agenda/2021/11/15minute-city-falls-short/

Report della Mobilità. (2022). https://www.amat-mi.it/it/progetti/report-mobilita/

Şengül, B., & Mostofi, H. (2021). Impacts of e-micromobility on the sustainability of urban transportation—A systematic review. *Applied Sciences, 11*(13), 5851. https://doi.org/10.3390/app11135851

Shoina, M., Voukkali, I., Anagnostopoulos, A., Papamichael, I., Stylianou, M., & Zorpas, A. A. (2024). The 15-minute city concept: The case study within a neighbourhood of Thessaloniki. *Waste Management & Research, 42*(8), 694. https://doi.org/10.1177/0734242X241259926

Transform Transport. (2024). *The right to cycle: Investigating cycle safety in Milan.* https://transformtransport.org/research/livable-streets/the-right-to-cycle-investigating-cycle-safety-in-milan/

UNECE. (2024). *United Nations adopts landmark global decarbonization strategy in transport by road, rail and inland waterway.* https://unece.org/climate-change/press/united-nations-adopts-landmark-global-decarbonization-strategy-transport-road#:~:text=Transport%20accounts%20for%20some%2023%25%20of%20annual%20GHG,79%25%20and%20freight%20transport%20by%20100%25%20by%202050

United Nations. (2020). *Cities and pollution.* https://www.un.org/en/climatechange/climate-solutions/cities-pollution

Vitulano, V., Pantaloni, G. G., Bocca, A., & Bruzzone, F. (2025). Accessibility and Spatial Conditions in Northern Italian Metropolitan Areas: Considerations for Governance After Ten Years of Metropolitan Cities. https://doi.org/10.3390/urbansci9120526

Willberg, E., Fink, C., & Toivonen, T. (2023). The 15-minute city for all?—Measuring individual and temporal variations in walking accessibility. *Journal of Transport Geography, 106*, 103521. https://doi.org/10.1016/j.jtrangeo.2022.103521

Wolański, M. (2023). The potential role of railway stations and public transport nodes in the development of "15-minute cities". *Infrastructures, 8*(10), 141. https://doi.org/10.3390/infrastructures8100141

World Health Organization. (2022a). *Ambient (outdoor) air pollution.* https://www.who.int/news-room/fact-sheets/detail/ambient-(outdoor)-air-quality-and-health

World Health Organization. (2022b). *Physical activity.* https://www.who.int/news-room/fact-sheets/detail/physical-activity

Yang, Y., Qian, Y., Zeng, J., Wei, X., & Yang, M. (2023). Walkability measurement of 15-minute community life circle in Shanghai. *Land, 12*(1), 153. https://doi.org/10.3390/land12010153

Social and Spatial Equity

Rethinking the Walkability and Accessibility of the Ageing Society: Milan 15-Minute City for Whom?

Mina Akhavan, Fulvia Pinto, and Michael Rodrigues

1 Introduction: Motivation, Research, Aims and Methods

In a rapidly ageing society, it is widely accepted that mobility and accessibility are essential in the perception of well-being and quality of life in older age (Banister & Bowling, 2004; Pinto & Sufineyestani, 2018). Older adults often face mobility limitations due to age-related physical and cognitive declines. Such limitations can lead to increased social isolation, as older adults who are less mobile may struggle to engage in social activities, which can further exacerbate feelings of loneliness and depression (Wong et al., 2017). Adapting mobility patterns to the needs of senior citizens also implies understanding individual capabilities, environmental factors, and economic and social context (Pinto & Akhavan, 2018). Scholars argue that designing age-friendly environments, which prioritise walkability and proximity, significantly contributes to the overall quality of life for older adults, as these settings facilitate social connections (Nieboer & Cramm, 2017). The design of urban spaces, including the availability of pedestrian-friendly infrastructure, plays a crucial role in enabling senior citizens to navigate their neighbourhoods safely and independently (Buffel et al., 2012; Rosso et al., 2011).

In the contemporary urban society with rapid urbanisation and ageing, the design of cities is increasingly directed towards sustainable and inclusive development models, such as the '15-minute city' (hereinafter 15mC). This concept was

M. Akhavan (✉)
Department of Human Geography, Planning and International Development, Group of Urban Planning, University of Amsterdam (UvA), Amsterdam, The Netherlands
e-mail: m.akhavan@uva.nl

M. Rodrigues
TU Delft, Delft, The Netherlands

F. Pinto
Politecnico di Milano, Milan, Italy

L. Bouzouina et al. (eds.), *15-Minute Cities*, Footprints of Regional Science,
https://doi.org/10.1007/978-3-032-20840-8_8

introduced as an urban planning pandemic response, aiming to create walkable communities, where residents can access essential services—such as work, healthcare, education, and leisure—within a 15-minute walk or bike ride from their homes, emphasising the importance of 'proximity' (Moreno et al., 2021). This approach not only promotes sustainable modes of transport (primarily walkability and cycling) but also aims to reduce dependency on cars, thereby contributing to environmental sustainability efforts (e.g., lowering carbon emissions) and enhancing public health and quality of life (Allam et al., 2022).

Although the origins of the principles are not fully new, the term 15mC (or X-minute city/neighbourhood) appeared in the literature very recently. Since 2019, growing attention has been reserved for exploring its theoretical principles, practicality and innovative research methodologies. While the 15mC is the most recognised term, variations such as the '20-minute city' or 'X-minute city' have also been investigated, reflecting different urban contexts and planning philosophies (Logan et al., 2022). Several European cities, such as Paris, Barcelona and Milan, are leading the way in adopting the 15mC principles, each tailoring their planning and design strategies to enhance proximity and accessibility, promote sustainable mobility, and improve the overall quality of urban life (Alberti & Radicchi, 2022; Pinto & Akhavan, 2022a; Pinto & Akhavan, 2022b).

Despite its promising framework, the 15mC concept has been questioned for its credibility in ensuring social equity in certain neighbourhoods and accessibility for all—see, for instance, the seven pitfalls discussed by Mouratidis (2024). Casarin et al. (2023) argue that this concept may reinforce existing social inequalities and lead to gentrification, where wealthier residents move into regenerated neighbourhoods. Other scholars underline the limitations of adopting a one-size-fits-all urban planning approach, highlighting the importance of inclusivity in designing walkable environments and considering the needs of older adults and children, as traditional walkability metrics often overlook factors such as safety, comfort, and accessibility for different age groups (Zysk & Zalewska, 2024).

So, here the question is whether the 15mC model is really inclusive for all population segments, including older adults. Within the context of this question, this chapter aims to contribute to the rather fast-growing literature on this topic from the lens of the ageing society. We focus on the case of Milan, with its 88 defined neighbourhoods (NIL[1]). By means of GIS data, isochrones are utilised to determine walkable areas in the neighbourhoods: choropleth maps are created to show walkability distribution. A ranking index based on walkability metrics is introduced to classify neighbourhoods, aiding in clustering analysis for a better understanding of their walkability profiles, focusing on senior citizens. Furthermore, the main urban

[1] In Milan, "Nuclei di Identità Locale" (NIL) refers to distinct neighbourhood units within the city that are part of an urban planning initiative to better manage and understand each area's diverse socio-demographic and territorial characteristics. These neighbourhoods reflect Milan's historical and social identity, aiming to strengthen community ties, improve service provision, and promote sustainable development (see Comune di Milano: https://www.pgt.comune.milano.it/psschede-dei-nil-nuclei-di-identita-locale/nuclei-di-identita-locale-nil)

policies and actions promoted by the Municipality of Milan for making a 15mC are presented to discuss their potential advantages and disadvantages for senior citizens (aged 65 years and over).

The remainder of this chapter is structured as follows: Section 2 systematically reviews the 15mC model and its relevance for senior citizens. Section 3 presents the empirical analysis, focusing on the case of Milan, where GIS and isochrone mapping are used to assess service accessibility for older adults in the city's 88 neighbourhoods; it also presents the urban policies and planning actions promoted by the Municipality of Milan to make a 15mC. Finally, Sect. 4 discusses the findings, their implications for urban policies, and suggestions for improving inclusivity in the application of the 15mC model.

2 The 15-Minute (or X-Minute) City Model and Neighbourhood Walkability for the Senior Citizens: A Review

2.1 Defining the Concept and State-of-the-Art Studies

Although the origins of human-oriented planning paradigms can be traced back to the early 1900s and Clarence Perry's 'Neighbourhood Unit' (Perry, 1929), Carlos Moreno's '15-minute' city lies in the novel perspective of the chrono-urbanism, which advocates for time-based urban planning that prioritises accessibility and the efficient use of urban spaces (Moreno et al., 2021). His idea got particular attention during and after the COVID-19 pandemic. This was due to the major impact on cities considering the changes in ways of working (hybrid working), shopping (e-commerce), and the need to adapt the built environment, deisgn and streetscapes for future health and climate emergencies (Florida et al., 2023). In fact, the 2020 pandemic has further highlighted the need for urban resilience and sustainability, prompting a reevaluation of existing urban frameworks. The pandemic underscored the disparities in access to essential services, making the 15mC model particularly relevant as it promotes proximity and accessibility, thereby addressing social inequalities (Foster et al., 2023; Khavarian-Garmsir et al., 2023). In this context, the model is seen as a response to the challenges posed by car-centric urban planning, advocating for a shift towards more pedestrian-friendly environments that prioritise active transportation (Ferrer-Ortiz et al., 2022).

There is a fast-growing body of literature on the topic of the 15mC in general. To understand the state-of-the-art studies and find the gaps, we conducted a systematic review of the literature from the Scopus database using the following strings: '15-minute city', '20-minute city', 'X-minute city', 'neighbourhood planning', 'proximity-based planning', and 'walkable city', which resulted in more than 500 articles—including only the papers in English, and limited to the subject areas of 'Social Sciences, Engineering, Environmental Science, Computer Sciences, Energy,

Environmental Science, Business and Economics. After the initial search, duplicate and irrelevant records were removed through a three-step screening process. First, titles and keywords were reviewed to exclude papers not directly related to proximity-based or walkability concepts. Second, abstracts were screened to verify relevance to the 15-minute (or X-minute) city framework and to exclude purely technical or non-urban studies. Finally, full texts were read when abstracts were ambiguous. This process reduced the initial 500 records to 252 relevant publications focusing on the social, spatial, and policy dimensions of the 15-minute city and walkability. The results confirm that the term '15mC' appeared in the literature only since 2019 (see Fig. 1).

Based on our database of relevant publications and using the VOSviewer open-source software, we analysed the co-occurrence of the keywords to discover the central themes in the research. Figure 2 shows the network visualisation, where six main thematic clusters have been identified: (i) sustainable development, travel time and neighbourhood (in red with 21 items); (ii) climate change, quality of life and urban area (in green with 18 items); (iii) urban planning and walkability (in dark blue with 14 items); (iv) 15mC, proximity and sustainability (in yellow with 10

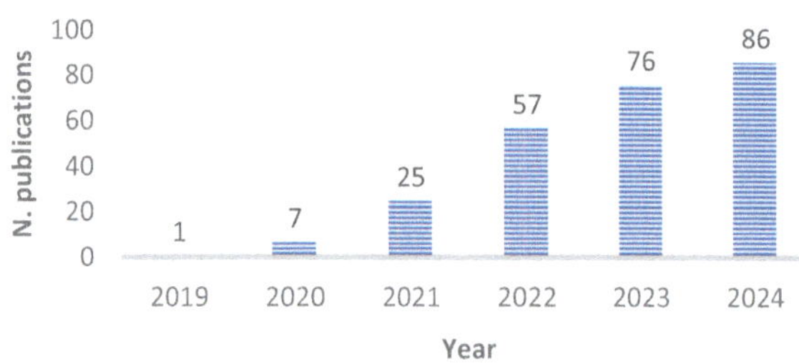

Fig. 1 The number of publications for each year, from the systematic literature review. Source: Own elaboration, based on the Scopus database

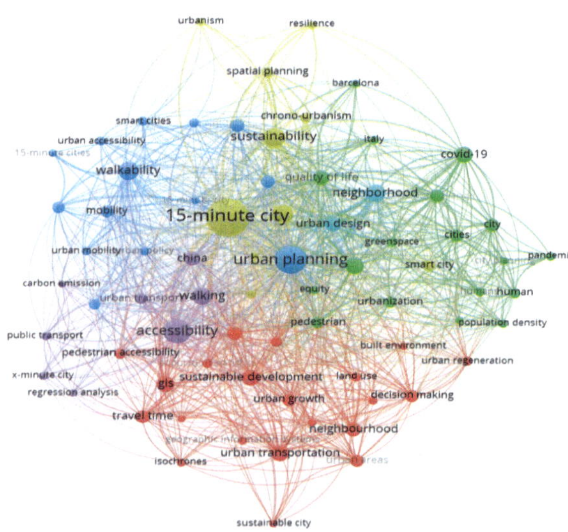

Fig. 2 Co-word analysis of keywords from Scopus publications (2019–2024). Source. Own elaboration, based on the Scopus database and using the VOSviewer (open source)

items); (v) accessibility (in purple with 8 items); (vi) neighbourhood and urban design (in light blue with 2 items).

The majority of studies advocate the 15mC concept, emphasising that its framework can significantly contribute to urban sustainability, liveability, and health by creating environments that prioritise proximity to essential services (Allam et al., 2022). The concept has also been linked to improved public health outcomes by ensuring that health services, recreational areas, and other essential amenities are within easy reach, thereby fostering a healthier urban populace (Song et al., 2022). Other lines of research associate the concept with neighbourhood walkability, including accessibility, design, and implications for urban living, improving the citizens' quality of life. Gorrini et al. (2023) examine walkability specifically for children in Bologna, Italy, within the context of the 15mC framework. It highlights the importance of designing urban spaces that are not only walkable but also safe and engaging for children. The findings suggest that enhancing walkability can lead to increased outdoor activity and social interaction among children, thereby contributing to healthier lifestyles.

Despite its advantages, the 15mC concept has faced criticism, particularly concerning its potential to amplify social inequalities and gentrification. Elldér (2024) discusses the dilemma of balancing the risk of gentrification with local accessibility, suggesting that the implementation of the 15mC could lead to socio-economic stratification within urban areas. Others question the practicality of the model in diverse urban contexts. Mouratidis (2024) points out that the 15mC may not adequately address the needs of all demographics, particularly those who rely on public transport for longer commutes or those living in areas with limited service availability. Furthermore, the concept's reliance on a holistic approach to urban planning raises questions about the adequacy of current systems of urban governance to implement such comprehensive changes effectively. Critics argue that without robust policy frameworks and community engagement, the 15mC could remain an ideal rather than a practical reality for all (Casarin et al., 2023).

2.2 *Mobility and Accessibility of the Ageing Society: The 15-Minute City for Whom?*

As populations age, understanding the factors that influence mobility and accessibility becomes essential for promoting independence, social participation, and overall well-being among seniors. Mobility for older adults encompasses not only the ability to move physically but also the capacity to access essential services and engage in social activities. Pantelaki et al. (2021) highlighted that older adults with restricted mobility face significant challenges in accessing services such as healthcare and community centres, which can adversely affect their quality of life. Others also discuss that proximity to resources and social support positively influences mobility and social participation among older adults, emphasising the importance of accessible environments (Levasseur et al., 2015). The built environment also plays a crucial role in facilitating or hindering mobility for older adults. Features such as walkable neighbourhoods, safe pedestrian pathways, and proximity to essential services are associated with increased daily walking among older adults and allow them to navigate their communities effectively (Rosso et al., 2011).

One of the main challenges in designing the 15Mc is to consider the different mobility patterns of age groups, as they have different walking speeds and needs in terms of accessibility and comfort during urban travel. Many studies have shown that the average walking speed differs between the young and the elderly and depends, in addition to age and gender, on the morphology of the terrain and weather conditions (Arango & Montufar, 2008; Duim et al., 2017). In fact, one of the critiques of the 15Mc model is that walkability assessments often fail to consider the varying speeds at which different demographics move, particularly children and the elderly (Gorrini et al., 2023). The authors suggest that urban planners and designers should incorporate speed variability into their models to ensure that all community members can effectively access the services in their neighbourhoods. Another study by Willberg et al. (2023) measures individual and temporal variations in walking accessibility within the 15mC framework. The findings show how different demographic groups experience walkability and the implications for urban planning, underlining the need for inclusive designs that cater to diverse citizens.

In the already rich and fast-growing literature, only a limited number of studies have specifically addressed the ageing society and incorporated the (mobility and accessibility) needs of older adults in the context of the 15mC concept. Ulloa-León et al. (2023) discuss the implications of the 15mC for elderly populations, emphasising the need for urban planning and design that prioritises accessibility to services and recreational spaces for seniors, thus promoting active and healthy lifestyles. Zhang et al. (2023) argue that 15-minute walkable neighbourhoods and creating urban spaces that facilitate social interaction and physical activity among older adults are essential for maintaining cognitive function and overall health. A recent publication by Buliung (2024), focuses on the 'last millimetre' issues that seniors face when accessing services, underlining the need for infrastructure that supports mobility for older populations. The findings highlight that improving walkability and accessibility is crucial for enabling seniors to engage with their communities and access necessary services.

3 Empirical Analysis: Ageing Society and the 15-Minute City Adaptability in the CASE of Milan

3.1 Introducing the Ageing Milan

Milan is divided into 9 Municipalities and 88 neighbourhoods—Nuclei di Identità Locale—NIL—which corresponds to the basic planning units of the Territorial Government Plan (in Italian *Piano di Governo del Territorio-PGT)*. The city is considered demographically old, as almost a quarter of the total population—above 1.3 million—is 65 years or older (The Italian National Institute of Statistics—ISTATA,

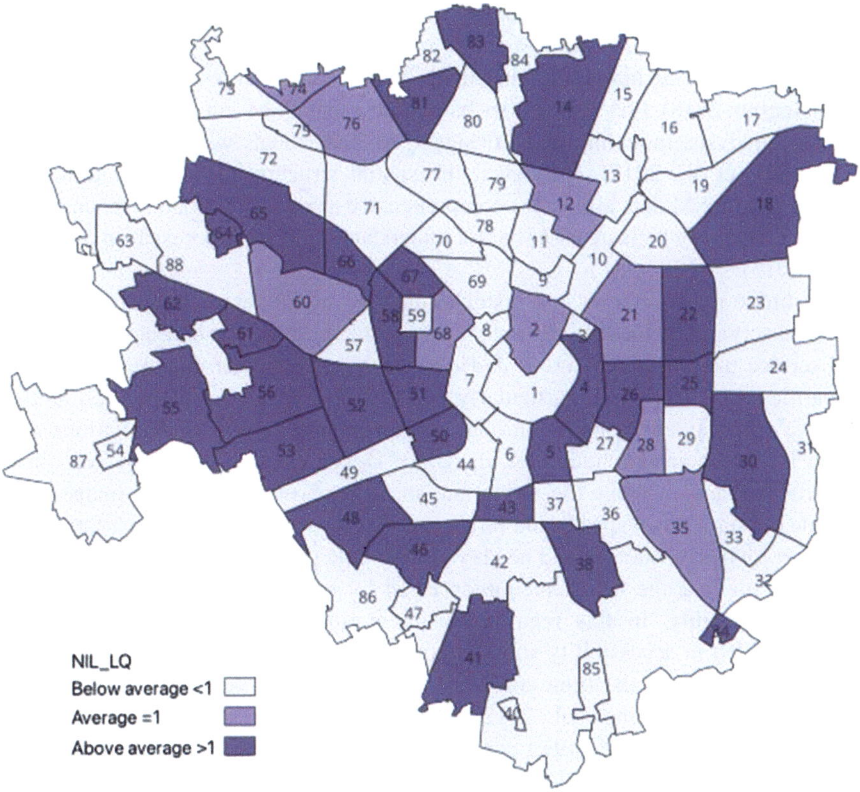

Fig. 3 Percentage of the senior citizens in the neighbourhoods (NILs) over the total in Milan (2023). Source: Elaborated by authors, data from ISTAT—The Italian National Institute of Statistics

n.d). Figure 3 shows the spatial distribution of the elderly in the neighbourhoods of Milan: 36 neighbourhoods have a location quotient (LQ) greater than 1 (dark grey), so the concentration of the older adults in these neighbourhoods is higher than the city average. Many peripheral neighbourhoods—namely, Bande Nere, Baggio, Barona and Quarto Cagnino—are among the ones that inhibit a greater percentage of seniors over the total population of the city. On the map, each NIL polygon is labelled with its corresponding identification number (ID_NIL); for the full NIL names, see Appendix 1.

3.2 Data, Methods and Analysis of Walkability Distribution for the Elderly in the 88 Neighbourhoods of Milan

This research has used OpenStreetMap (OSM) data for network modelling and service location analysis. OSM, a community-driven, open-source platform, provides detailed geographical data on urban features, such as roads, making it highly valuable for these purposes. Its granular data enhances the assessment of service accessibility, making it particularly useful for urban researchers and planners (Boeing, 2018). Previous studies have shown that OSM can effectively support network analysis, including those designed for individuals with limited mobility (Mobasheri et al., 2017). Moreover, the spatial structure of mobility networks derived from OSM data allows for comprehensive analyses of urban form and connectivity, both of which are essential for understanding service accessibility (Louail et al., 2015).

Isochrone maps offer crucial insights into how urban design impacts access to essential services by identifying areas where residents may face barriers to mobility and service use. Research shows that isochrone mapping can uncover significant disparities in accessibility, particularly for socially excluded groups, such as the elderly, who may encounter additional challenges due to physical limitations and urban infrastructure (Achuthan et al., 2010; Dovey et al., 2017). However, it is important to acknowledge the limitations of this approach: isochrone maps often assume constant travel speeds and rarely account for terrain difficulty, variations in transport modes, or the specific needs of vulnerable populations. These simplifications can obscure the actual inequities faced by senior citizens and others with restricted mobility. In this regard, the integration of Geographic Information Systems (GIS) in accessibility studies provides a more nuanced understanding of travel behaviour and the time constraints that affect access to services (Bonotti et al., 2015; Delafontaine et al., 2012).

In Milan's Nuclei di Identità Locale (NIL), service accessibility can be analysed using isochrones, which depict the spatial distribution of services within specific walking time intervals. Walking speed among senior citizens is an important indicator of health and mobility. Research has shown that walking speed is strongly linked to balance and physical activity levels in older adults (Yamamoto et al., 2014). While there is no consensus in the literature on a specific walking speed for senior citizens, we used an indicative speed of 1.1 m/s (3.96 km/h) for this study. To simplify the modelling process, we assumed this speed to always remain constant over a 15-minute walking period.

The analysis was carried out for the seven groups of services in Milan's Nuclei di Identità Locale (NIL):

(i) **Cultural Activities**: Access to cultural amenities (museums, theatres, etc.).
(ii) **Healthcare Services**: Access to healthcare services, including clinics and hospitals, excluding pharmacies.
(iii) **Places of Worship**: Proximity to religious institutions (churches, mosques, etc.).
(iv) **Social Services**: Availability of social support services within walking distance.
(v) **Public Green Spaces**: Access to parks and open green areas.
(vi) **Sports Facilities**: Proximity to sports and recreational facilities.
(vii) **Supermarkets**: Access to food retail outlets like supermarkets.

The analysis provides valuable insights into the spatial distribution and accessibility of services within Milan's NILs, taking into account the walking speed of senior citizens (Fig. 4). These maps illustrate the accessibility of various services in Milan based on walking times for senior citizens, measured in minutes. They are particularly useful for understanding how senior citizens can access essential services across the city. The maps categorise services into different types and display three accessibility zones: 0–5 minutes (green), 5–10 minutes (yellow), and 10–15 minutes (red). Generally, those living in the city centre have better access to a broader range of services within shorter walking distances. While peripheral NILs do offer access to some services, there are potential gaps in accessibility to certain amenities in these areas.

Figure 5 illustrates accessibility levels to different categories of services across NILs. Each cell represents the relative accessibility (0–100%), with green

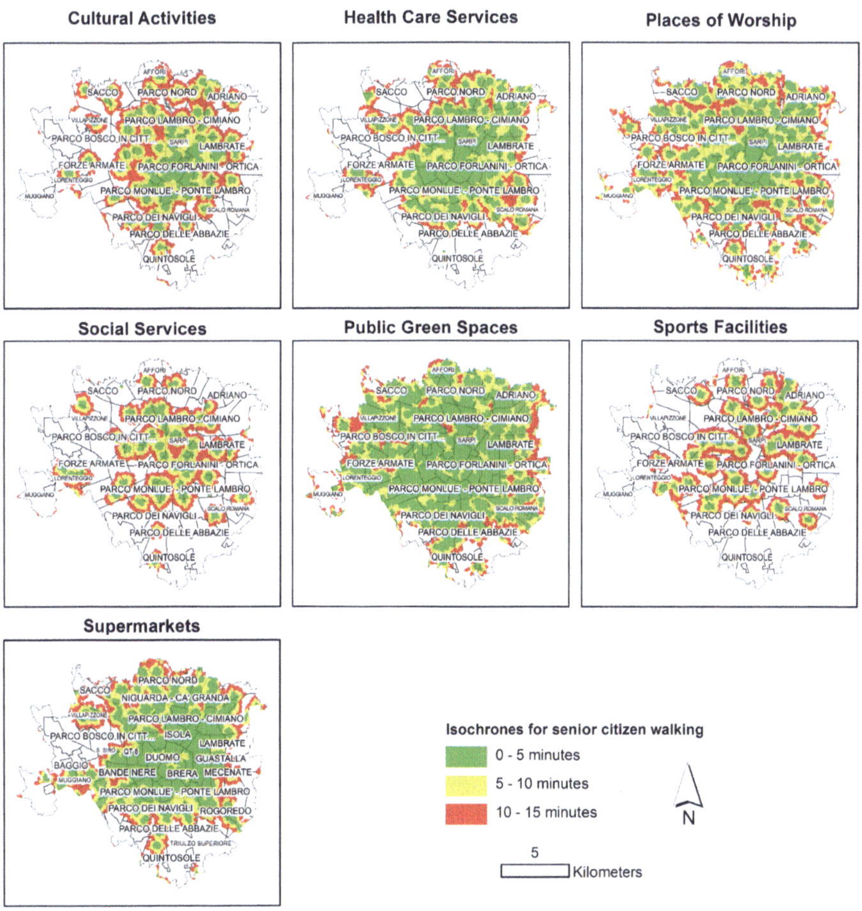

Fig. 4 Accessibility of various services in Milan's Nuclei di Identità Locale (NIL). Source: Elaborated by the authors

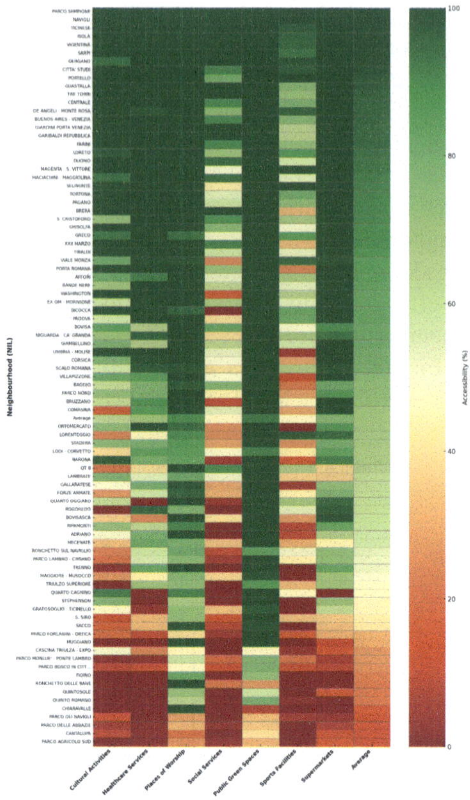

Fig. 5 Accessibility to services by neighbourhood (sorted by average). Source: Elaborated by the authors

indicating higher access and red lower access. NILs are sorted by their overall average accessibility, making it possible to quickly identify areas with systemic disadvantages. The heatmap reveals a clear spatial divide in accessibility across Milan's NILs. Central and historically consolidated neighbourhoods such as Parco Sempione, Navigli, Ticinese, Isola, and Guastalla exhibit consistently high accessibility levels across all essential services, often exceeding 80–90%, reflecting dense service provision and overlapping urban functions. In contrast, peripheral and more suburban NILs including Parco Agricolo Sud, Parco delle Abbazie, Chiaravalle, Figino, and Quinto Romano, rank lowest, with averages below 40–50%, indicating significant gaps in access despite their larger territorial extent. Overall, the results underscore a strong centre–periphery gradient, where residents in central NILs benefit from service density, while those in peripheral areas face longer travel distances and limited accessibility to essential services.

From the spatial analysis, the following interpretations are drawn for each category of essential services—see Figs. 4 and 5 and the detailed percentage of NIL area within 15-minute walking distance of each service in Appendix 1:

(i) *Cultural Activities*

Most cultural activities are concentrated in the city centre, especially within the 0–5 minute isochrone. While the highest concentration of cultural offerings is found here, activities are also distributed throughout the city, providing accessibility for residents in peripheral areas within a 10- to 15-minute walking distance. Many neighbourhoods, particularly in central Milan, enable access to cultural facilities within a 5-minute walk. However, significant portions of the periphery still require 10–15 minutes of walking to reach these facilities. This indicates that cultural activities are less accessible to senior citizens living on the outskirts of the city.

The city-wide average percentage of NIL area within a 15-minute walking distance to cultural activities is 62.65%. NILs with the highest accessibility (100%, +37.35%) include Parco Sempione, Tre Torri, Magenta—S. Vittore, Sarpi, and Isola. In contrast, NILs such as Chiaravalle, Parco Agricolo Sud, Quinto Romano, Quintosole, and Rogoredo record 0% coverage (−62.65%), highlighting major accessibility gaps.

(ii) *Health Care Services*

Healthcare services are relatively well-distributed throughout the city, with facilities located in various NILs, indicating a well-developed healthcare network in Milan. Most NILs offer high accessibility to healthcare services. However, there is a notable concentration of facilities in the city centre, particularly within the 0–5 minute isochrone. The concentration of healthcare services is evident in the city centre and some larger neighbourhoods, as shown by the large patches of green on the accessibility maps. Generally, most of Milan is within a 10-minute walk of healthcare services. However, accessibility declines the further one moves from the city centre, with some peripheral areas falling into the 10- to 15-minute range. This distribution is particularly important for elderly residents, who may require more frequent access to healthcare services.

Almost all areas are well-served by healthcare services, with many approaching or achieving 100% coverage. This indicates that healthcare is generally accessible throughout the city. The city-wide average percentage of NIL area within a 15-minute walking distance to healthcare services is 69.48%. NILs such as Bande Nere (100%, +30.52%) and Affori (91.62%, +22.14%) show excellent access. On the other hand, Quarto Cagnino (47.77%, −21.71%) and Ronchetto delle Rane (59.95%, −9.53%) fall below the average.

(iii) *Places of Worship*

Places of worship are distributed throughout the city, reflecting Milan's diverse religious landscape. Like cultural activities, there is a significant concentration of places of worship in the city centre. Overall, the accessibility of these locations is fairly well-balanced across Milan, with most of the city situated within a 5 to 10-minute walking distance from a place of worship.

The city-wide average percentage of NIL area within a 15-minute walking distance to places of worship is 89.27%. Several NILs, including Affori and

Bande Nere, achieve 100% coverage (+10.73%). Barona, however, records 82.89% (−6.38%), falling below the city-wide average.

(iv) *Social Services*

The accessibility of social services varies across the NILs, with some areas enjoying excellent coverage while others have limited options. Although the city centre provides a good range of social services, there are notable accessibility gaps in some peripheral NILs. Social services are concentrated in specific areas of Milan, leading to large zones where walking times exceed 15 minutes, particularly in the outskirts. However, there are also clusters with better access (green), primarily located in the city centre. This distribution indicates that senior citizens living farther from these central locations may face challenges in accessing social services.

The city-wide average percentage of NIL area within a 15-minute walking distance to social services is 45.66%. Vigentina (100%, +54.34%) and Affori (53.03%, +7.38%) perform well. In contrast, Adriano (0%, −45.66%) and Barona (4.63%, −41.02%) are severely underserved.

(v) *Public Green Spaces*

Most areas of Milan enjoy full or near-full access to public green spaces, such as parks and recreational areas, indicating that the city's green infrastructure is well-distributed. Public green spaces have the highest level of accessibility compared to all other services. A significant portion of the city is within a 5-minute walk, and only a few areas fall into the 10- to 15-minute range. This widespread access is particularly beneficial for senior citizens, as proximity to parks and green spaces is essential for their well-being.

Access to public green spaces in Milan is notably high, with a city-wide average percentage of NIL area within a 15-minute walking distance of 92.42%. Most areas are close to this average, showing only minor deviations. This indicates that the city generally offers excellent access to green spaces across its territorial units.

(vi) *Sports Facilities*

Sports facilities tend to cluster in specific areas, particularly in the city centre and certain neighbourhoods. While many areas fall within the 5–10 minute walking range, there are also several peripheral zones where access requires 10–15 minutes of walking. This distribution suggests that senior citizens in some parts of the city may need to travel farther to reach fitness facilities.

The city-wide average percentage of NIL area within a 15-minute walking distance to sports facilities is 46.40%. Vigentina (95.27%, +48.87%) and Affori (68.37%, +21.97%) demonstrate strong coverage. By contrast, Adriano (8.82%, −37.58%) and Villapizzone (12.02%, −34.38%) fall far below the average.

(vii) *Supermarkets*

Supermarkets are relatively well-distributed throughout the city, with many areas within the 0–5 minute and 5–10 minute walking ranges. While some peripheral areas have slightly longer walking times of 10–15 minutes, overall access to supermarkets for groceries and daily needs appears to be adequate for most residents, including seniors. Like other services, there is a slight concentration of supermarkets in the city centre.

The city-wide average percentage of NIL area within a 15-minute walking distance to supermarkets is 75.92%. Affori (100%, +24.08%) and Bande Nere (100%, +24.08%) show excellent access. Adriano (66%, −9.92%) falls slightly below the average.

3.3 Policies and Actions for Making the Milan 15-Minute City

In his recent book, Carlos Moreno (2024), who modernised the concept of the 15mC, dedicates a chapter to Milan as an example where many efforts have been made to improve social relations and the quality of life of citizens. In fact, Milan's sustainability approaches and path towards a 15mC started before the Covid-19 pandemic and was then accelerated by the 'Milan 2020 Adaptation Strategy' launched by Municipality.[2] The plan outlines a range of comprehensive measures aimed at improving and diversifying mobility options (walking, cycling and other soft mobility solutions) and supporting social innovation while creating community cohesion. One of the objectives of the plan concerns the recovery of public spaces for leisure and sport. As acknowledged by the C40 Knowledge Hub, Milan's vision for making a 15-minute city is centred on two storylines (see Table 1), with the aim of changing land use to improve accessibility to the city's functions and expand opportunities for social interaction in public spaces. However, in these initiatives, there are no specific references to the most vulnerable age groups, such as the over-65 s, who represent almost a quarter of the total population in Milan.

Here, it is worth introducing in more detail the main actions and policies that led to the creation of the Milan 15mC. In 2018, the Municipality of Milan launched the

Table 1 Milan's path towards the 15-minute city concept

Storylines	Supporting initiatives
1. Becoming a 15mC, Milan becomes sustainability connected.	– Investing on public transport to connect the city with sustainable mobility options. – Introducing digital public services to reduce travel needs and make public services more accessible to citizens.
2. Giving back urban public spaces to the citizens.	– *Open Squares* programme, using tactical urbanism to transform and revitalise parking lots and underutilised square into places for local communities and social gathering. – Milan 2020 adaptation plan—*Open Streets* programme focuses on improving cycling and walking infrastrcutures.

Source. Adapted by the authors, from C40 (2022). The 15-minute city: International experiences

[2]The open document of Milan 2020 Adaptation Strategy is published on the website of the Municipality of Milan: https://www.comune.milano.it/documents/20126/71178 96/Milano+2020.+Adaptation+strategy.pdf

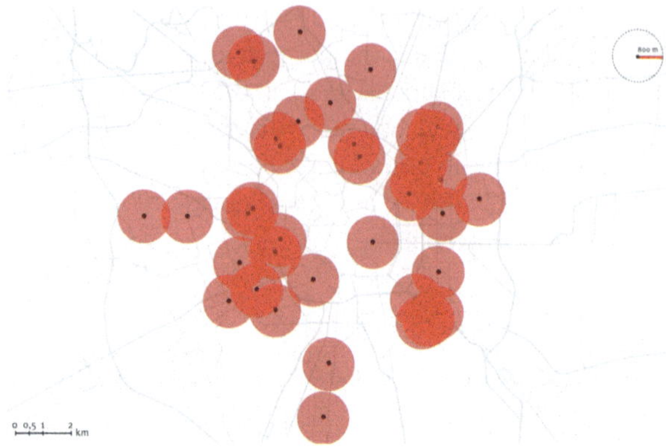

Fig. 6 Proposals for 'Open Squares' in Milan. Source: The report 'Piazze Aperte: A Public Space Program for Milan. Retrieved from: https://www.comune.milano.it/aree-tematiche/quartieri/piano-quartieri/piazze-aperte

'Piazze Aperte' programme, developed together with Agenzia Mobilità Ambiente Territorio (AMAT), in collaboration with Bloomberg Associates and the Global Designing Cities Initiative. The programme aims to regenerate public spaces by transforming them into community gathering places, expanding pedestrian zones, and promoting sustainable modes of transport for the sake of the environment and the citizens' quality of life. It uses bottom-up approaches and tactical urbanism to bring public spaces back to the centre of the neighbourhood and the life of its inhabitants. From 2018 to 2021, 38 suggestions were submitted in response to the call 'Piazza Aperte in Ogni Quartiere' (see Fig. 6); this number was more than doubled by 2023. Thus, one in two residents in Milan will have a square within 15 minutes (800 mt) from their home (Comune di Milano, 2023).

Following efforts to revitalise and rethink the liveability of public spaces, the Municipality launched a city-wide initiative in the wake of the COVID-19 pandemic, in 2020: 'Open Streets Program' (in Italian, Strarde Aperte)—aimed to improve the quality of life in Milan by promoting sustainable mobility, expanding public spaces, and fostering a healthier environment.[3] The programme proposes a network of pedestrian and cycling infrastructures. Once again, tactical urbanism interventions are used as light, fast, and economical solutions on an experimental basis. Through temporary pedestrianisation interventions on some streets, widening of sidewalks, establishment of 30 km/h zones as shared streets, connections with existing cycle paths, and installation of outdoor seating, public space is freed up to serve citizens' (see Fig. 7).

In 2022, the Municipality proposed 'Mi15 Spazi e Servizi per Milano a 15 minuti' to support social impact investments by profit and non-profit businesses with non-repayable grants. The goal is to activate or enhance neighbourhood spaces and services to promote the 15mC strategies. With this call for proposals, the Municipality of Milan invited businesses to submit investment projects capable of activating or enhancing spaces and services that are currently absent or

[3] https://www.comune.milano.it/-/mobilita-sostenibile.-strade-aperte-granelli-vicini-all-obiettivo-dei-35-km-di-ciclabili-realizzate-

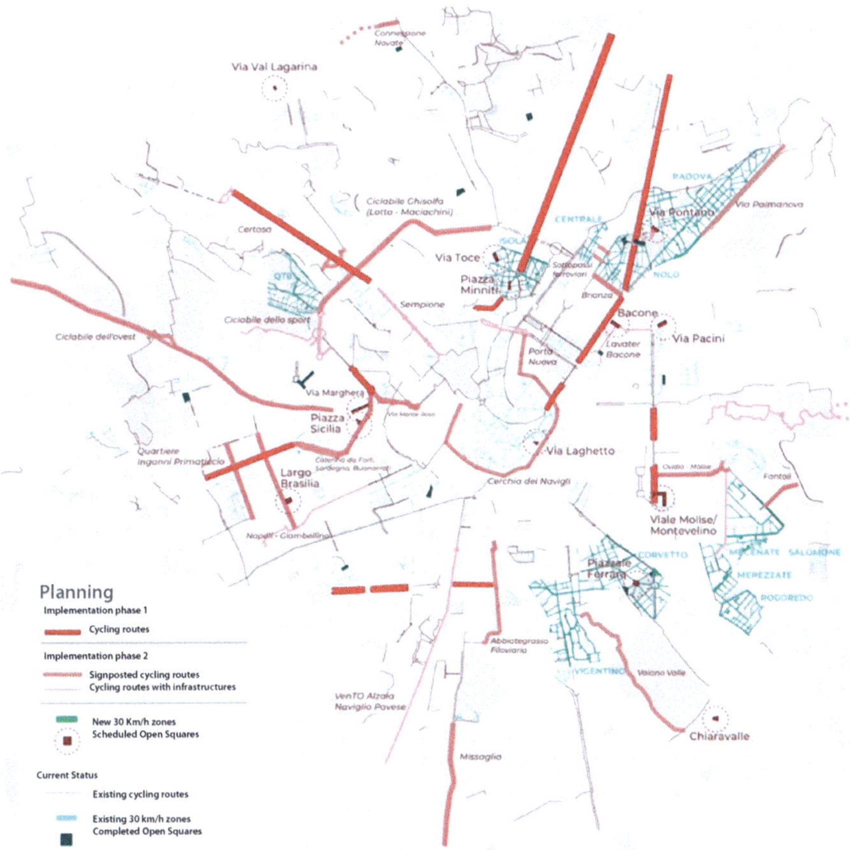

Fig. 7 Planned intervention for walkability and cyclability in the city of Milan. Source: Comune di Milano (2020, p. 25). Retrieved from https://www.sporteimpianti.it/wp-content/uploads/2021/10/Milano-Strade-Aperte.pdf

insufficient, generating a positive social impact at the neighbourhood scale. The eligible project areas covered all the dimensions of the 15mC: learning and playing, working and business, leisure and cultural activities, manufacturing, reusing and repairing, healthily food and shopping, taking care of oneself and feeling good, sports and social activities. After the evaluation, 22 projects were selected in different neighbourhoods. Reviewing all the projects, although there are cases addressing children and youth—e.g. the project '15 Minuti: Arte e servizi vicini a bimbi/e donne e famiglie', for the neighbourhoods Stadera—Chiesa Rossa, Gratosoglio, Vigentino—Fatima—there is no specific focus on older adults (Comune di Milano, 2022).

The Municipality of Milan continues to prioritise local commerce, crafts, and neighbourhood services. In May 2024, the first three funding calls for the Three-Year Plan for the Local Economy 2024–2027 were launched, offering nearly 5 million euros to support both profit and non-profit economic activities in the city's neighbourhoods. The 2024 and 2025 calls include *Scuola dei Quartieri*, *Mi-15*, and *Negozi e Botteghe di Quartiere*. The *Mi-15* programme, titled 'Spaces and services for Milan in 15 minutes', is aimed at companies capable of generating a positive social or environmental impact on the local community. Through this initiative, the Municipality encourages businesses to contribute to the 15mC concept by co-financing the necessary investments to establish and enhance spaces, services, or socially beneficial activities within local neighbourhoods (Comune di Milano, 2024).

4 Discussion and Conclusion: Towards a 15-Minute City for All

With the increase in life expectancy and the decline in birth rates, Milan is facing a growing elderly population. This demographic shift requires rethinking urban planning and mobility policies to ensure that older people can easily access essential services and actively participate in community life. A neighbourhood suitable for senior citizens includes health services, easy access to public transportation, and the availability of spaces for socialisation. Whether Milan is an inclusive city for older people requires an in-depth analysis of the policies, infrastructures, and services available at the neighbourhood scale. This article explores the strengths and areas for improvement to make Milan truly inclusive for its older citizens. In light of the 15mC strategies, Milan is promoting proximity and accessibility to services and public transport. Although innovative and ambitious, this urban strategy raises questions about the implications for older people. In Table 2, we have outlined some of the key criteria behind the strategies for making Milan a 15mC and, more importantly, the potential advantages and disadvantages for older citizens.

Highlighting the findings from the GIS-spatial analysis, it is evident that Milan generally provides good accessibility to essential services for senior citizens, particularly in its central areas. However, many peripheral neighbourhoods face challenges, experiencing longer walking times for services such as social services, cultural activities, and healthcare. On a positive note, public green spaces and supermarkets are among the most accessible services throughout the city, benefiting elderly residents. To better support its ageing population, the city could enhance access to healthcare and social services in the outer districts.

Overall, Milan offers a high level of accessibility to senior citizens, with most services within a 15-minute walking distance. Nevertheless, significant disparities exist between neighbourhoods, highlighting the need for targeted urban planning and policies to address the specific needs of underserved areas and create a more equitable environment for senior residents. Data shows that while some services are widely accessible, others are more unevenly distributed across the NILs. This variation indicates disparities in service access that city planners can address. In

Table 2 Criteria for 'Milan, the 15-minute city'

Key criteria	
Accessibility	Proximity to services such as shops, hospitals, parks, and public transport.
Sustainability	Promotion of eco-friendly transportation and development of pedestrian areas.
Density	Concentration of services and housing to encourage community life and social interaction.
Connectivity	Integration of transport networks and digital infrastructure to facilitate mobility and access to information.
Benefits for older people	
Improved quality of life	Easy access to services and activities can enhance the quality of life for older people, allowing them to maintain an active and independent lifestyle.
Reduction of social isolation	The proximity of public areas and meeting spots can help reduce social isolation and promote community participation.
Safety	A well-lit and accessible environment can increase older people's sense of security, reducing the risk of accidents.
Disadvantages and challenges for older people	
Physical accessibility	Elderly individuals with limited mobility may encounter difficulties in reaching certain parts of the city if not adequately equipped.
Costs	Renovating and upgrading existing infrastructure can entail high costs, which may not be sustainable in the long term.
Gentrification	The attractiveness of well-served areas could lead to rising property prices, excluding low-income elderly residents.
Training and awareness	It is necessary to provide training and support to help older people familiarise themselves with new technologies and available services.

Source: Authors' compilation

particular, the neighbourhoods of Affori, Bande Nere, and Vigentina stand out for their excellent access to multiple services, including healthcare, sports facilities, and supermarkets. In contrast, Adriano and Barona consistently exhibit below-average access to key services, particularly in social services, sports facilities, and cultural activities.

Future empirical research on the topic of X-minute city (whether X is 5, 10, or 20 minutes) should also consider that the age group of +65 years is not homogeneous, and subdivision of age groups is necessary while exploring mobility needs and travel patterns. For instance, one can distinguish between two groups, that is, young elderly (aged 65–75 years) and old elderly (over 75 years) (Alsnih & Hensher, 2003). Other factors, such as gender and socio-economic patterns of individuals, as well as a deeper analysis of the neighbourhood layout, are crucial in conducting future studies.

To make Milan a truly inclusive city for older residents, continued investment in infrastructure, social policies, and awareness programmes is essential, ensuring seniors live independently and actively. Key areas for improvement include accessibility to services, mobility, public spaces, and social engagement.

Accessibility to Services: Milan has a strong healthcare network and social services, but some peripheral areas lack easy access to essential services.

Additionally, service hours are not always flexible for seniors. To address this, a 'mixed zoning' approach that integrates residential, commercial, and service areas is needed, along with mobile services like libraries and assistance centres to improve access.

Mobility and Transport: Milan's public transport is extensive, but congestion and inadequate pedestrian infrastructure create challenges for older residents. Strategies to improve this include creating more comfortable, accessible public transport, promoting alternatives to cars (e.g., pedal-assisted bicycles), and enhancing pedestrian infrastructure with wide, obstacle-free pavements and rest areas.

Public Spaces and Socialisation: While Milan has many parks and events for seniors, not all public spaces are easily accessible, and some older adults face social isolation. More inclusive public spaces should be developed, with a focus on accessibility and green spaces that promote social interaction and environmental sustainability.

Technology and Innovation: Milan is integrating digital services and telemedicine, but many older people lack the skills to use these tools. Expanding digital literacy programmes and teleassistance services would help seniors access healthcare and public services more easily.

Public Participation: Although older adults contribute to community life through volunteering, they could be more involved in decision-making processes regarding urban planning and services. Greater participation in these areas would ensure services better meet their needs and promote social integration.

In conclusion, the 15mC has become one of the most prominent and appealing concepts in modern urban planning. However, while it offers significant potential, it should not be viewed as a stand-alone solution. Instead, it must be integrated with other planning strategies to fully realise its benefits and address the unique challenges of each urban environment. Combining the 15mC model with smart densification can enhance public transport services and create accessible, welcoming public spaces. Furthermore, incorporating social inclusion policies ensures that everyone—regardless of age, income, or background—can access the opportunities and services within their communities. Striking a balance between innovation and inclusion is key to creating a truly accessible and welcoming city for all.

Appendix

Results of the spatial analysis: percentage of NIL area within 15-minute walking distance of each service. In this table, each row represents a distinct NIL, and the columns indicate the percentage of the area within a 15-minute walking distance of different types of services.

Neighbourhood (NIL)	ID_ NL	Cultural Activities	Healthcare Services	Places of Worship	Social Services	Public Green Spaces	Sports Facilities	Super markets	Average
ADRIANO	17	51.36%	79.39%	99.99%	0%	100%	8.82%	66.00%	57.94%
AFFORI	80	81.68%	91.62%	100%	53.03%	100%	68.37%	100%	84.96%
BAGGIO	55	61.12%	77.40%	96.93%	78.03%	99.52%	20.38%	80.85%	73.46%
BANDE NERE	52	69.11%	100%	100%	62.32%	100%	53.72%	100%	83.59%
BARONA	46	97.48%	79.41%	82.89%	4.63%	100%	14.51%	91.92%	67.26%
BICOCCA	15	99.63%	100%	95.36%	0%	100%	83.10%	100%	82.58%
BOVISA	77	85.45%	63.90%	99.98%	83.95%	100%	48.67%	88.99%	81.56%
BOVISASCA	81	34.84%	24.07%	100%	38.87%	100%	27.03%	98.95%	60.54%
BRERA	2	100%	100%	100%	100%	100%	32.42%	100%	90.35%
BRUZZANO	83	60.85%	87.54%	100%	9.10%	100%	45.10%	100%	71.80%
BUENOS AIRES—VENEZIA	21	100%	100%	100%	90.26%	100%	77.07%	100%	95.33%
CANTALUPA	47	8.59%	1.42%	18.73%	0.20%	37.95%	0%	12.02%	11.27%
CASCINA TRIULZA—EXPO	73	51.23%	24.01%	44.56%	0.34%	46.40%	3.20%	14.42%	26.31%
CENTRALE	10	100%	100%	100%	90.34%	100%	78.23%	100%	95.51%
CHIARAVALLE	34	0%	0%	99.99%	0%	6.85%	0%	0%	15.26%
CITTA' STUDI	22	100%	100%	100%	91.25%	100%	96.87%	99.47%	98.23%
COMASINA	82	15.94%	86.14%	100%	53.97%	100%	29.45%	100%	69.36%
CORSICA	25	77.60%	98.90%	100%	49.53%	100%	29.59%	100%	79.37%
DE ANGELI—MONTE ROSA	58	100%	97.41%	100%	74.57%	100%	96.36%	100%	95.48%
DERGANO	79	94.01%	100%	100%	100%	100%	98.00%	100%	98.86%
DUOMO	1	100%	100%	100%	94.95%	100%	56.78%	100%	93.10%
EX OM—MORIVIONE	37	58.41%	100%	100%	66.78%	100%	53.77%	100%	82.71%
FARINI	78	100%	100%	100%	90.05%	100%	73.26%	100%	94.76%
FIGINO	63	0%	0%	82.25%	0%	72.72%	0%	0%	22.14%
FORZE ARMATE	56	26.68%	56.98%	96.15%	30.95%	100%	42.19%	77.89%	61.55%
GALLARATESE	65	49.55%	80.36%	97.68%	29.01%	100%	3.73%	73.49%	61.97%
GARIBALDI REPUBBLICA	9	100%	100%	100%	100%	100%	63.79%	100%	94.83%
GHISOLFA	70	100%	100%	100%	76.50%	100%	51.50%	100%	89.71%
GIAMBELLINO	49	78.71%	64.59%	100%	63.59%	100%	56.76%	100%	80.52%
GIARDINI PORTA VENEZIA	3	100.00%	100%	100%	100%	100%	63.82%	100%	94.83%
GRATOSOGLIO—TICINELLO	41	47.32%	2.82%	72.35%	34.57%	97.25%	0%	61.20%	45.07%
GRECO	13	93.32%	100%	93.52%	51.81%	100%	86.34%	100%	89.28%
GUASTALLA	4	100%	100%	100%	100%	100%	76.75%	100%	96.68%
ISOLA	11	100%	100%	100%	100%	100%	95.92%	100%	99.42%
LAMBRATE	23	78.37%	54.29%	81.09%	52.29%	100%	36.26%	35.04%	62.48%
LODI—CORVETTO	35	37.71%	84.39%	84.87%	48.06%	91.33%	41.33%	84.36%	67.44%
LORENTEGGIO	53	23.07%	44.81%	89.44%	24.91%	100%	97.33%	95.28%	67.83%

Neighbourhood (NIL)	ID_ NL	Cultural Activities	Healthcare Services	Places of Worship	Social Services	Public Green Spaces	Sports Facilities	Super markets	Average
LORETO	20	96.98%	100%	100%	70.94%	100%	87.37%	100%	93.61%
MACIACHINI—MAGGIOLINA	12	93.55%	100%	100%	94.67%	100%	53.39%	100%	91.66%
MAGENTA—S. VITTORE	7	100%	100%	100%	50.64%	100%	99.25%	100%	92.84%
MAGGIORE—MUSOCCO	72	26.37%	42.27%	75.91%	37.13%	98.24%	0.53%	69.21%	49.95%
MECENATE	30	33.03%	81.69%	78.66%	34.17%	100%	30.63%	43.37%	57.36%
MUGGIANO	54	0%	0%	100%	0%	100%	0%	10.07%	30.01%
NAVIGLI	44	100%	100%	100%	100%	100%	99.81%	100%	99.97%
NIGUARDA—CA' GRANDA	14	71.00%	96.08%	100%	38.13%	100%	68.37%	93.04%	80.95%
ORTOMERCATO	29	74.22%	100%	94.33%	25.13%	100%	0%	86.98%	68.67%
PADOVA	19	56.91%	100%	100%	38.35%	100%	81.97%	100%	82.46%
PAGANO	68	100%	100%	100%	58.78%	100%	75.29%	100%	90.58%
PARCO AGRICOLO SUD	87	0%	1.03%	24.71%	1.99%	29.80%	0%	13.96%	10.21%
PARCO BOSCO IN CITT…	88	8.85%	11.28%	46.13%	11.89%	77.76%	12.02%	8.46%	25.20%
PARCO DEI NAVIGLI	86	11.25%	0.37%	27.51%	0.14%	32.65%	4.85%	3.15%	11.42%
PARCO DELLE ABBAZIE	85	4.76%	1.49%	30.65%	2.44%	33.75%	0.26%	6.29%	11.38%
PARCOFORLANINI—ORTICA	24	18.04%	9.34%	38.17%	4.14%	100%	15.31%	30.79%	30.83%
PARCO LAMBRO—CIMIANO	18	19.15%	56.14%	74.97%	4.75%	99.23%	42.47%	73.24%	52.85%
PARCO MONLUE'—PONTE LAMBRO	31	1.29%	2.31%	58.43%	17.81%	77.29%	20.27%	6.00%	26.20%
PARCO NORD	84	73.13%	78.00%	100%	42.63%	100%	28.68%	86.77%	72.74%
PARCO SEMPIONE	8	100%	100%	100%	100%	100%	100%	100%	100.00%
PORTA ROMANA	27	100%	100%	100%	71.83%	100%	23.43%	100%	85.04%
PORTELLO	67	100%	100%	100%	79.85%	100%	100%	100%	97.12%
QT 8	66	20.22%	37.60%	100%	90.09%	100%	66.67%	36.78%	64.48%
QUARTO CAGNINO	61	92.51%	0.60%	91.50%	0%	100%	45.85%	0%	47.21%
QUARTO OGGIARO	76	60.74%	0%	100%	3.54%	100%	70.25%	95.52%	61.44%
QUINTO ROMANO	62	0%	0%	64.07%	0%	79.30%	1.60%	0%	20.71%
QUINTOSOLE	39	0%	0%	72.26%	0%	58.39%	0%	15.17%	20.83%
RIPAMONTI	38	72.97%	62.32%	90.72%	0.06%	99.84%	7.87%	82.82%	59.51%
ROGOREDO	33	0%	96.47%	77.51%	0%	100%	54.55%	97.06%	60.80%
RONCHETTO DELLE RANE	40	0%	0%	100%	20.45%	26.57%	0%	0%	21.00%
RONCHETTO SUL NAVIGLIO	48	22.64%	58.52%	80.55%	9.34%	84.12%	52.28%	73.75%	54.46%
S. CRISTOFORO	45	72.44%	100%	100%	88.97%	100%	68.64%	100%	90.01%

Neighbourhood (NIL)	ID_ NL	Cultural Activities	Healthcare Services	Places of Worship	Social Services	Public Green Spaces	Sports Facilities	Super markets	Average
S. SIRO	60	12.62%	32.55%	72.86%	13.52%	100%	34.03%	34.79%	42.91%
SACCO	74	16.75%	32.81%	98.91%	8.36%	98.62%	0.21%	33.30%	41.28%
SARPI	69	100%	100%	100%	100%	100%	93.19%	100%	99.03%
SCALO ROMANA	36	56.04%	96.84%	97.31%	39.04%	100%	65.35%	97.72%	78.90%
SELINUNTE	57	98.50%	100%	100%	40.71%	100%	100%	100%	91.32%
STADERA	42	75.51%	79.15%	85.74%	28.72%	99.23%	26.11%	80.05%	67.79%
STEPHENSON	75	77.54%	0%	67.47%	0%	100%	0.39%	71.44%	45.26%
TIBALDI	43	100%	100%	100%	58.48%	100%	52.79%	100%	87.32%
TICINESE	6	100%	100%	100%	100%	100%	96.92%	100%	99.56%
TORTONA	50	100%	100%	100%	54.55%	100%	81.42%	100%	90.85%
TRE TORRI	59	100%	100%	100%	100%	100%	72.05%	100%	96.01%
TRENNO	64	1.31%	83.05%	99.87%	0%	100%	0%	83.11%	52.48%
TRIULZO SUPERIORE	32	0%	70.90%	73.84%	0%	86.05%	32.49%	69.67%	47.56%
UMBRIA—MOLISE	28	100%	100%	100%	52.81%	100%	4.20%	100.00%	79.57%
VIALE MONZA	16	81.05%	99.04%	98.15%	23.88%	100%	95.98%	98.80%	85.27%
VIGENTINA	5	100%	100%	100%	100%	100%	95.27%	100%	99.32%
VILLAPIZZONE	71	82.06%	85.11%	100%	48.46%	100%	12.02%	99.85%	75.36%
WASHINGTON	51	100%	100%	100%	15.52%	100%	65.97%	100%	83.07%
XXII MARZO	26	100%	100%	100%	91.91%	100%	32.72%	100%	89.23%
Average		62.65%	69.48%	89.27%	45.66%	92.42%	46.40%	75.92%	68.83%

References

Achuthan, K., Titheridge, H., & Mackett, R. (2010). Mapping accessibility differences for the whole journey and for socially excluded groups of people. *Journal of Maps, 6*(1), 220–229. https://doi.org/10.4113/jom.2010.1077

Alberti, F., & Radicchi, A. (2022). The proximity city: A comparative analysis between Paris, Barcelona and Milan. *TECHNE—Journal of Technology for Architecture and Environment, 23*, 69–77. https://doi.org/10.36253/techne-12151

Allam, Z., Bibri, S. E., Chabaud, D., & Moreno, C. (2022). The theoretical, practical, and technological foundations of the 15-minute city model: Proximity and its environmental, social and economic benefits for sustainability. *Energies, 15*(16), 6042. https://doi.org/10.3390/en15166042

Alsnih, R., & Hensher, D. A. (2003). The mobility and accessibility expectations of seniors in an aging population. *Transportation Research Part A: Policy and Practice, 37*(10), 903–916.

Arango, J., & Montufar, J. (2008). Walking speed of older pedestrians who use canes or walkers for mobility. *Transportation Research Record, 2073*(1), 79–85.

Banister, D., & Bowling, A. (2004). Quality of life for the elderly: The transport dimension. *Transport Policy, 11*(2), 105–115.

Boeing, G. (2018). A multi-scale analysis of 27,000 urban street networks: Every US city, town, urbanized area, and Zillow neighborhood. *Environment and Planning B: Urban Analytics and City Science, 47*(4), 590–608. https://doi.org/10.1177/2399808318784595

Bonotti, R., Rossetti, S., Tiboni, M., & Tira, M. (2015). Analysing space-time accessibility towards the implementation of the light rail system: The case study of Brescia. *Planning Practice and Research, 30*(4), 424–442. https://doi.org/10.1080/02697459.2015.1028254

Buffel, T., Phillipson, C., & Scharf, T. (2012). Ageing in urban environments: Developing 'age-friendly' cities. *Critical Social Policy, 32*(4), 597–617. https://doi.org/10.1177/0261018311430457

Buliung, R. (2024). 15-minute cities, 'walkability' and last millimeter problems. *Disability & Society, 40*, 1–5. https://doi.org/10.1080/09687599.2024.2385919

C40. (2022). *The 15-minute city: International experiences*. Research Reports. November 2022. https://www.c40knowledgehub.org/s/article/The-15-minute-city-International-experiences

Casarin, G., MacLeavy, J., & Manley, D. (2023). Rethinking urban utopianism: The fallacy of social mix in the 15-minute city. *Urban Studies, 60*(16), 3167–3186. https://doi.org/10.1177/00420980231169174

Comune di Milano. (2020). *Strade Aperte. Strategie, azioni e strumenti per la ciclabilità e la pedonalità, a garanzia delle misure di distanziamento negli spostamenti urbani e per una mobilità sostenibile*, Milano. https://bicycleinfrastructuremanuals.com/manuals5/Comune-Milano-Strade-Aperte-2020.pdf

Comune di Milano. (2022). *Mi15 Spazi e Servizi per Milano a 15 minuti*. https://economiaelavoro.comune.milano.it/progetti/mi15#:~:text=5-,Mi15%20è%20il%20nuovo%20bando%20del%20Comune%20di%20Milano%20che,%22Città%20a%2015%20minuti%22

Comune di Milano. (2023). *Piazze aperte. Un programma per lo spazio pubblico di Milano*. https://www.comune.milano.it/aree-tematiche/quartieri/piano-quartieri/piazze-aperte

Comune di Milano. (2024). *Mi15 Spazi e Servizi per Milano a 15 minuti*. https://servizi.comune.milano.it/en/dettaglio-contenuto/-/asset_publisher/pqxq/content/mi15-spazi-e-servizi-per-milano-a-15-minuti

Delafontaine, M., Neutens, T., & Weghe, N. (2012). A GIS toolkit for measuring and mapping space–time accessibility from a place-based perspective. *International Journal of Geographical Information Science, 26*(6), 1131–1154. https://doi.org/10.1080/13658816.2011.635593

Dovey, K., Woodcock, I., & Pike, L. (2017). Isochrone mapping of urban transport: Car-dependency, mode-choice and design research. *Planning Practice and Research, 32*(4), 402–416. https://doi.org/10.1080/02697459.2017.1329487

Duim, E., Lebrão, M. L., & Antunes, J. L. F. (2017). Walking speed of older people and pedestrian crossing time. *Journal of Transport & Health, 5*, 70.

Elldér, E. (2024). The 15-minute city dilemma? Balancing local accessibility and gentrification in Gothenburg, Sweden. *Transportation Research Part D: Transport and Environment, 135*, 104360. https://doi.org/10.1016/j.trd.2024.104360

Ferrer-Ortiz, C., Marquet, O., Mojica, L., & Vich, G. (2022). Barcelona under the 15-minute city lens: Mapping the accessibility and proximity potential based on pedestrian travel times. *Smart Cities, 5*(1), 146–161. https://doi.org/10.3390/smartcities5010010

Florida, R., Rodríguez-Pose, A., & Storper, M. (2023). Critical commentary: Cities in a post-COVID world. *Urban Studies, 60*(8), 1509–1531. https://doi.org/10.1177/00420980211018072

Foster, S., Lynch, B., & Lowe, M. (2023). What are the lessons from covid-19 for creating healthy, sustainable, resilient future cities? *NPJ Urban Sustainability, 3*(1), 29. https://doi.org/10.1038/s42949-023-00107-y

Gorrini, A., Presicce, D., Federico, M., & Choubassi, R. (2023). Walkability for children in Bologna: Beyond the 15-minute city framework. *Journal of Urban Mobility, 3*, 100052. https://doi.org/10.1016/j.urbmob.2023.100052

ISTATA—The Italian National Institute of Statistics (NA). (n.d.). https://www.istat.it/en/

Khavarian-Garmsir, A. R., Sharifi, A., Abadi, M. H. H., & Moradi, Z. (2023). From garden city to 15-minute city: A historical perspective and critical assessment. *Land, 12*(2), 512. https://doi.org/10.3390/land12020512

Levasseur, M., Généreux, M., Bruneau, J., Vanasse, A., Chabot, É., Beaulac, C., et al. (2015). Importance of proximity to resources, social support, transportation and neighborhood security

for mobility and social participation in older adults: Results from a scoping study. *BMC Public Health, 15*(1), 503. https://doi.org/10.1186/s12889-015-1824-0
Logan, T. M., Hobbs, M. H., Conrow, L. C., et al. (2022). The x-minute city: Measuring the 10, 15, 20-minute city and an evaluation of its use for sustainable urban design. *Cities, 131*, 103924.
Louail, T., Lenormand, M., Picornell, M., Cantú, O., Herranz, R., Frías-Martínez, E., et al. (2015). Uncovering the spatial structure of mobility networks. *Nature Communications, 6*(1), 6007. https://doi.org/10.1038/ncomms7007
Mobasheri, A., Sun, Y., Loos, L., & Ali, A. (2017). Are crowdsourced datasets suitable for specialized routing services? Case study of openstreetmap for routing of people with limited mobility. *Sustainability, 9*(6), 997. https://doi.org/10.3390/su9060997
Moreno, C. (2024). *The 15-minute city: A solution for saving our time & our planet.* John Wiley & Sons.
Moreno, C., Allam, Z., Chabaud, D., Gall, C., & Pratlong, F. (2021). Introducing the "15-minute city": Sustainability, resilience and place identity in future post-pandemic cities. *Smart Cities, 4*(1), 93.
Mouratidis, K. (2024). Time to challenge the 15-minute city: Seven pitfalls for sustainability, equity, livability, and spatial analysis. *Cities, 153*, 105274. https://doi.org/10.1016/j.cities.2024.105274
Nieboer, A. P., & Cramm, J. M. (2017). Age-friendly communities matter for older people's well-being. *Journal of Happiness Studies, 19*(8), 2405–2420. https://doi.org/10.1007/s10902-017-9923-5
Pantelaki, E., Maggi, E., & Crotti, D. (2021). Mobility impact and well-being in later life: A multidisciplinary systematic review. *Research in Transportation Economics, 86*, 100975. https://doi.org/10.1016/j.retrec.2020.100975
Perry, C. A. (1929). The neighborhood unit. In *Regional plan of New York and its environs* (Vol. VII).
Pinto, F., & Akhavan, M. (2018). Promoting urban mobility for an ageing society. Exploring Policies and Initiatives in the European context. *Urbanistica Informazioni, vol.* Special issue, marzo—aprile 2018, pp. 6–12, ISSN: 0392-5005.
Pinto, F., & Akhavan, M. (2022a). Scenarios for a post-pandemic city: Urban planning strategies and challenges of making "Milan 15-minutes city". *Transportation Research Procedia, 60*, 370–377. https://doi.org/10.1016/j.trpro.2021.12.048
Pinto, F., & Akhavan, M. (2022b). Spazio e tempo nella città post covid. La "x minute city" per una rinnovata idea di prossimità. In R. Gerundo (Ed.), *Città e piani del rischio pandemico*. ESI.
Pinto, F., & Sufineyestani, M. (2018). Key characteristics of an age friendly neighbourhood. *Tema. Journal of Land Use, Mobility and Environment, 11*, 331.
Rosso, A. L., Auchincloss, A. H., & Michael, Y. L. (2011). The urban built environment and mobility in older adults: A comprehensive review. *Journal of Aging Research, 2011*, 1–10. https://doi.org/10.4061/2011/816106
Song, G., He, X., Kong, Y., Ke, L., Song, H., Zhai, S., et al. (2022). Improving the spatial accessibility of community-level healthcare service toward the '15-minute city' goal in China. *ISPRS International Journal of Geo-Information, 11*(8), 436. https://doi.org/10.3390/ijgi11080436
Ulloa-León, F., Correa, J. A. A., Vergara-Perucich, F., Cancino-Contreras, F., & Núñez, C. A. A. (2023). "15-minute city" and elderly people: Thinking about healthy cities. *Smart Cities, 6*(2), 1043–1058. https://doi.org/10.3390/smartcities6020050
Willberg, E., Fink, C., & Toivonen, T. (2023). The 15-minute city for all?—Measuring individual and temporal variations in walking accessibility. *Journal of Transport Geography, 106*, 103521. https://doi.org/10.1016/j.jtrangeo.2022.103521
Wong, R., Szeto, W., Yang, L., Li, Y., & Wong, S. C. P. (2017). Elderly users' level of satisfaction with public transport services in a high-density and transit-oriented city. *Journal of Transport & Health, 7*, 209–217. https://doi.org/10.1016/j.jth.2017.10.004

Yamamoto, S., Matsunaga, A., Wang, G., Hoshi, K., Kamiya, K., Noda, C., et al. (2014). Effect of balance training on walking speed and cardiac events in elderly patients with ischemic heart disease. *International Heart Journal, 55*(5), 397–403. https://doi.org/10.1536/ihj.14-017

Zhang, S., Wu, W., Xiao, Z., Wu, S., Zhao, Q., Ding, D., et al. (2023). Creating livable cities for healthy ageing: Cognitive health in older adults and their 15-minute walkable neighbourhoods. *Cities, 137*, 104312. https://doi.org/10.1016/j.cities.2023.104312

Zysk, E., & Zalewska, K. (2024). The methodology for assessing the 15 minute age-friendly walkability (AFW) of urban public spaces. *Sustainability, 16*(15), 6406. https://doi.org/10.3390/su16156406

Cycling in Low-Density Areas and Outskirts: Enablers and Barriers to Achieve an Equitable 15-Minute City

Isabel Cunha

1 Introduction

The decades of car-oriented planning have affected the evolution of urban areas, land use and travel behaviour, creating a solid symbolic attachment to this mode of transport (Lyons & Davidson, 2016), which has been profoundly path-dependent and increasingly disconnected from broader social and environmental policy goals (Vigar, 2017). Accordingly, studies suggested that in motorised societies, people with no access to the dominant mode of transport are likely to struggle to access opportunities to fulfil basic needs (i.e. employment, education, health assistance), ultimately leading to the phenomenon of transport poverty (Lucas, 2012) and transport-related social exclusion, among socioeconomically disadvantaged people (Martens, 2017; Martens & Golub, 2012; Pereira et al., 2017).

To overcome this conundrum, research and policy have sought to explore potential avenues for advancing sustainable transport modes, entailing a balance between economic development, environmental preservation and social equity goals (Lucas et al., 2013; Manaugh et al., 2015), which requires a broad understanding of a myriad of social, technical, cultural and economic factors (Goldman & Gorham, 2006).

Cities worldwide are employing measures and policies to promote sustainable transport systems to mitigate climate change burdens and spawn carbon-neutral, accessible, liveable, and environmentally friendly urban landscapes. To address this agenda, the 15-minute city concept has been advocated as a suitable framework that combines components of density, diversity, proximity and digitalisation to give

I. Cunha (✉)
Urban Planning, Economics and Transport Laboratory (LAET), Graduate School of Civil, Environmental and Urban Engineering (ENTPE), University of Lyon, CNRS, Vaulx-en-Velin, France
e-mail: isabel.cunha@entpe.fr

L. Bouzouina et al. (eds.), *15-Minute Cities*, Footprints of Regional Science,
https://doi.org/10.1007/978-3-032-20840-8_9

people access to essential services and daily needs opportunities through active travel (Moreno et al., 2021).

Accordingly, promoting cycling for daily travel has a potential impact on fostering this agenda, since the bicycle carries significant economic, environmental and social benefits (Banister, 2011; Hickman et al., 2011; Temenos et al., 2017; Van Nunen et al., 2011). This transport mode is considered affordable, low-carbon and flexible enough to increase access to opportunities for the whole community, democratising mobility and ensuring social diversity and inclusion (Golub & Martens, 2014; Karner, 2016; Pereira et al., 2017). Likewise, in urban areas, the bicycle can cover a significant part of the utilitarian journeys of up to 30 min or 5 km currently undertaken in motorised vehicles (Lanzendorf & Busch-Geertsema, 2014; Nielsen & Skov-Petersen, 2018).

Nevertheless, there is little evidence on the role of the bicycle in the 15-minute city model regarding spatial segregation and accessibility inequalities in car--dominated cities in metropolitan areas. Scholars argue that an equitable shift towards sustainable mobility and active travel should consider the mobility challenges beyond the neighbourhood limits and examine the metropolitan-level access disparities. This framework is helpful for exploring discrepancies in terms of unequal access to jobs, education facilities and healthcare for disadvantaged and minority groups. Previous studies indicate that inequalities tend to accumulate, with high levels of spatial segregation correlating with high levels of mobility inequality (Tammaru et al., 2023).

Likewise, scholars argue that the cycling benefits (i.e. accessibility, health gains, infrastructure coverage) are unevenly distributed across distinct societal groups in cities (Cunha & Silva, 2022) despite the efforts to normalise the bicycle as a fully recognised daily transport mode (Aldred et al., 2016), suggesting that the equity implications are often neglected during the planning and decision-making processes (Manaugh & El-Geneidy, 2011). Accordingly, empirical evidence indicates that central and socio-economically privileged areas have better cycling accessibility than peripheral areas and disadvantaged neighbourhoods (Cunha et al., 2024).

In order to examine this phenomenon in a metropolitan region and explore bicycle accessibility disparities across socio-economic groups located in high-density and low-density zones, this study applied the tool TIRE (Cunha & Silva, 2023a) and spatial analysis to assess the equity impact of the new bicycle master plan (BMP) in implementation in the metropolitan area of Lyon, France, which comprises 58 municipalities. Acknowledging that the 15-minute concept was hardly addressed outside high-density areas, this study looks beyond the inner cities to discuss the mobility challenges that car-dependent inhabitants face in the outskirts and the peri--urban regions.

Hence, the main goals of this study are: (1) to quantitatively measure and visualise the cycling accessibility conditions in central and peripheral areas and identify distributive discrepancies in a metropolitan region; (2) to assess the equity implications of a metropolitan bicycle master plan; (3) to examine if the selected plan aligns with the 15-minute city framework and (4) to propose a set of policy and planning recommendations to ensure fairness in planning and decision-making processes.

After this introduction, the second section presents a literature review encompassing the 15-minute city conceptual framework focused on active mobility. Afterwards, the methodological section illustrates the case study and how TIRE tool was calibrated and operationalised to assess the metropolitan area of Lyon in France. The fourth section discusses the main results of the spatial and quantitative analysis and the implications for planning practice and policymaking. The last chapter summarises the main research outcomes, limitations and future research pathways.

2 Active Travel and the 15-Minute City

Contemporary uncertainties encompassing rapid urban expansion, climate change and the COVID-19 pandemic have pushed for new ways of thinking and creating cities. In response, the 15-minute city concept is a model that has been gaining momentum in recent years as an approach to reducing car dependence and promoting active transport (Logan et al., 2022). The concept, first introduced by Moreno et al. (2021), was promoted by the C40 Cities agenda for post-pandemic Green and Just Recovery (C40 Cities, 2020; Moreno et al., 2021) and endorsed by many policymakers worldwide. In general terms, this framework expects that residents should be able to meet most of their needs within 15 minutes from their homes by walking or cycling. Some studies have also adapted this travel-time threshold (i.e. x-minute city) when assessing metropolitan regions (Knap et al., 2023; Logan et al., 2022; Lu & Diab, 2023). Previous research has examined the relationship between the 15-minute city's underlying principles (Khavarian-Garmsir et al., 2023), encompassing the dimensions of human scale, urban design, density, diversity, flexibility, proximity, digitalisation and connectivity. Scholars claim this model can potentially drift planning approaches from monocentric city strategies to polycentric structures, increasing the availability and diversity of local urban amenities (Graells-Garrido et al., 2021).

With regard to the transport systems, alternative and sustainable mobility solutions such as cycling infrastructure provision and bike-sharing services (BSS) are promoted strategies within the 15-minute city vision. However, scholars claim that the 15-minute city model overlooks particular challenges facing non-urban car-dependent neighbourhoods (Poorthuis & Zook, 2023). Sustainable mobility solutions are not equally accessible in all urban communities and for all socioeconomic groups, particularly those in less-dense suburban fringes. For instance, concerning BSS distribution, a study disclosed that such services and other mobility innovations tend to reinforce rather than mitigate existing mobility inequalities unless equity concerns related to infrastructure distribution for cycling are explicitly addressed in the planning process (Caggiani et al., 2020).

The present review (Table 1) identified that the most recent research on this topic focuses on walkability within 15 minutes of travel time, and less attention is paid to cycling and the role of public transport (Celbiş & Bouzouina, 2025). Moreover, most empirical evidence encompasses the European context, with most of the

Table 1 Summary of the literature review, methodology, transport modes (pedestrians, bike users, and public transport users) and geographical zones

Country	Methodological approach	Ped.	Bike	PT	References
Australia	Content analysis of documents	x*	x*		(Lu & Diab, 2023)
Canada	Cumulative opportunity accessibility measure Mobility survey, multilevel binary logistic model	x*	x*		(Birkenfeld et al., 2023; Hosford et al., 2022)
China	Floating catchment accessibility Walk score metric Spatial regression	x*	x*		(Liu et al., 2022; Weng et al., 2019; Zhang et al., 2023)
Colombie	Proximity analysis Contour measures	x*			(Guzman et al., 2021)
Denmark	Mobility survey Ordinary least-square regression models		x		(Schneider et al., 2023)
Finland	Contour measure, place-based accessibility	x*			(Willberg et al., 2023)
Germany	Mobility survey Ordinary least-square regression models		x		(Schneider et al., 2023)
Italy	Isochrones accessibility, proximity index, space syntax, regression analysis	x*			(Murgante et al., 2024; Olivari et al., 2023; Staricco, 2022)
Netherlands	Standardised gravity-based accessibility Mobility survey Ordinary least-square regression models Origin-destination matrix		x	x*	(Knap et al., 2023; Poorthuis & Zook, 2023; Schneider et al., 2023)
New Zealand	Proximity analysis	x*			(Logan et al., 2022)
Pakistan	Awareness survey Ordinal regression model	x	x		(Abdullah et al., 2022)
Spain	Origin-destination flows and local accessibility, distance-based accessibility, negative binomial regression model	x			(Ferrer-ortiz et al., 2022; Graells-Garrido et al., 2021)
USA	Proximity analysis, content analysis of documents	x*			(Logan et al., 2022; Lu & Diab, 2023)

* equity considerations

findings concentrated in urban areas. Still, there is an increasing interest in the equity implications of this vision, considering specific target areas and population groups. To illustrate, previous research suggested that the 15-minute city framework is not a one-size-fits-all model, which provides goals that are hardly reachable in the context of the global south and sprawling cities. In these geographic zones, few households can conduct all their daily travel near their home by active travel (Birkenfeld et al., 2023).

Previous research conducted in Europe suggested that in dense cities, the 15-minute threshold should not be considered the main target since many services are already accessible by walking or cycling within this time by the majority of the population (Staricco, 2022). Accordingly, scholars suggest that the 15-minute city framework could be modified to fit the culture of the location under study and its inhabitants. Nevertheless, although there is a greater focus on European cases, there is a lack of research on the identification of a set of destination types for different contexts and user preferences (Knap et al., 2023).

Focusing on walking, a study proposed a 15-minute walkable city index to assess the walkability levels of 121 European metropolitan areas, considering seven amenity types. The data were equally combined using a multicriteria approach to assign a final score and present a ranking of 15-minute cities. This study demonstrates European-level disparities and indicates that most of the leading cities are in Germany, while most of the worst performers are in the UK. This study also revealed a statistically significant difference in the 15-minute city performance due to their GDP per capita disparities (Bartzokas-Tsiompras & Bakogiannis, 2023).

The assessed literature employed quantitative frameworks to explore the extent to which cities are approaching the 15-minute vision. Accordingly, scholars have employed accessibility metrics to assess how essential activities and opportunities are reachable in 15-minute walking and cycling travel time. Accessibility analysis often results in 2D spatial maps, which illustrate accessibility disparities across zones and identify zones for further interventions (Cunha & Silva, 2023).

Considering accessibility metrics, previous research conducted a catchment area analysis to evaluate accessibility to parks and spatial equity for both walking and cycling scenarios. Results suggested an unequal accessibility distribution with low-access areas mainly distributed on the urban fringes (Zhang et al., 2023). Another study evaluated accessibility to grocery stores by walking and cycling in Vancouver using a cumulative opportunity measure to count the number of grocery stores accessible within a 15-minute walk and cycling from people's homes. Results indicate that 99% of the city has good accessibility conditions for cycling (Hosford et al., 2022).

In Barcelona, Spain, it was identified that most residents of this dense and compact city live in areas with proximity to services within a 15-minute travel time. Yet, accessibility shortfalls in peripheral areas were identified (Ferrer-ortiz et al., 2022). In Finland, another study indicated that 34% of the older population walk for 15 minutes to access the closest grocery store compared to a baseline of 93%. This study highlights the importance of considering the variation in people's walking abilities, road conditions and service networks when modelling walking accessibility (Willberg et al., 2023).

Through a longitudinal approach, a study explored 16 years of travel surveys in the Netherlands to discuss differences in car dependency in urban and non-urban areas as well as differences in extra travel time across the urban–rural continuum.

Results indicate that the 15-minute city model may overlook the particular challenges facing non-urban neighbourhoods (Poorthuis & Zook, 2023).

In Bogotá, Colimbia, a study conducted a geographical proximity analysis of non-work services and facilities within the 15-minute city framework, also addressing the challenges associated with the COVID-19 pandemic. This study found inequalities among income groups regarding access to essential services (Guzman et al., 2021). In China, highly walkable communities are primarily concentrated in the central areas, and poorly walkable communities are dispersed in rural areas (Weng et al., 2019).

All in all, there is a lack of empirical evidence on cycling, especially considering equity analysis within both high- and low-density zones.

3 Methodology

3.1 Case Study

This study explores the equity implications of the cycling plan under implementation in the Lyon Metropolitan Region (LMR), France, encompassing 58 municipalities spread over 538 km2 (Annex 1). Over the half-century, the urban area of Lyon experienced a population increase of almost 50%. Yet, as in most European and American cities, this evolution was not uniform due to the increased suburbanisation process (Bouzouina et al., 2021).

Since the 1990s, transport planning authorities have made efforts to encourage a shift from private cars to alternative modes (Systral, 2017). The modal split in the LMR has been gradually changing over the years, with a slight decrease in motorisation and an increase in active travel and public transport usage. Nevertheless, the automobile remained the preferred mode of transport in 2014 (43.9%), compared to a residual share of 2.2% of bicycle usage (Table 2). Despite planning efforts, LMR has not yet reached the European average cycling level of 7%, whereas the bicycle share in France remains low (5%) compared to other European countries (Havet & Bouzouina, 2024).

Local authorities have been implementing several measures to increase diversity in mobility. Among new sharing practices, strategies have been focusing on redistributing road space and providing new dedicated infrastructure to encourage walking and cycling to improve accessibility and create liveable urban spaces. These policies have also helped to improve road safety in the conurbation. On a broader level, intermodality has been placed at the heart of the initiative to connect the peripheral zones and the city centre (SYSTRAL, 2017).

Concerning the population density (Fig. 1), LMR is mostly populated in its core cities (i.e. Lyon, Villeurbanne), where there is a more diverse land use scheme. According to the OECD, most job opportunities are concentrated in the centre of LMR, well connected by public transport (ITF, 2020). Yet, according to this report, over 150,000 people enter Lyon city centre every morning, which is expected to

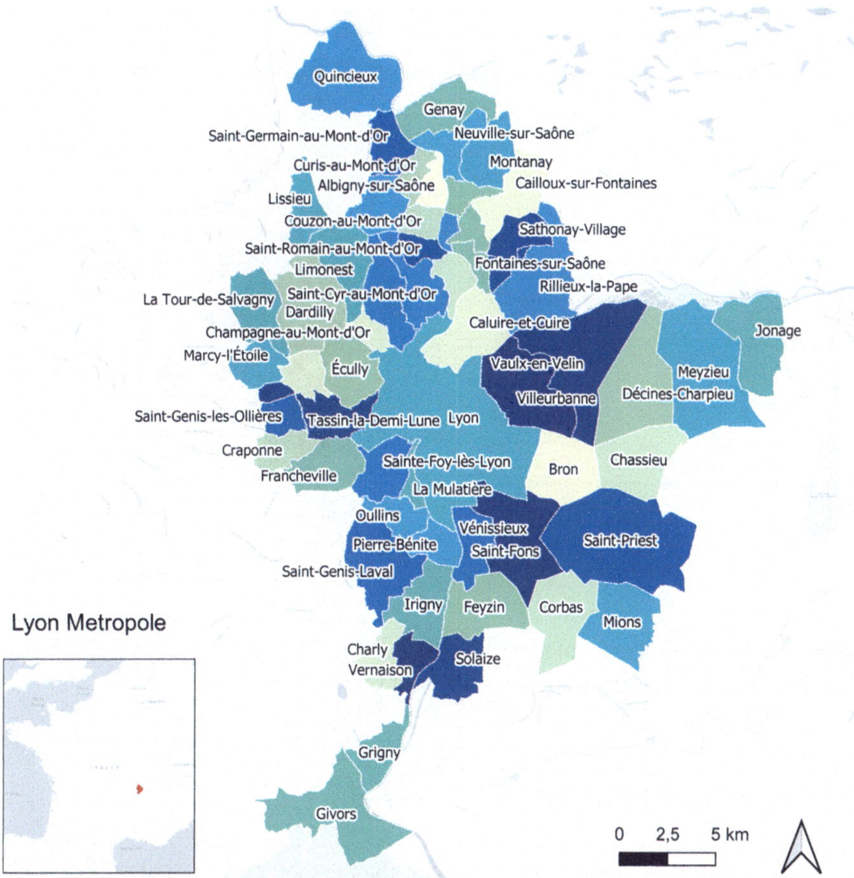

Annex 1 Lyon metropole region

Table 2 Overview of the modal split changes in the Lyon metropolitan region

Period	Walking	Cycling	Public transport	Driving
1995	31.5%	1.3%	14.1%	53.1%
2006	32.5%	2.2%	16.1%	49.2%
2015	34.1%	2.2%	19.8%	43.9%

grow in the future. These commuting trips are mainly performed in private cars (34%) or public transport (39%).

To better explore the accessibility distribution of the LMR population, this study redistributed the population count in a hexagonal grid (i.e. the H3 geospatial index system) at resolution 9 to facilitate the spatial analysis of such a large case study, reducing sampling bias due to edge effects (Cunha & Silva, 2023; Pereira et al.,

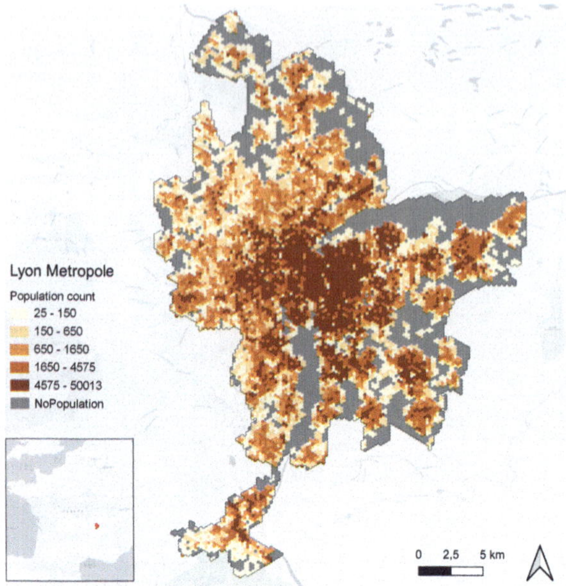

Fig. 1 Population count in the Lyon metropolitan region

2021). Accordingly, the circular structure of a hexagonal grid enables more natural data distribution and representation compared to square grids.

3.2 *Accessibility Measurement*

To measure the accessibility implications of the cycling strategy in the LMR, this study adapted a bicycle accessibility indicator proposed by a previous study (Cunha & Silva, 2023). As used in the TIRE tool, this study used a contour measure to assess the number of accessible activities by bike within a threshold of 15 minutes for each origin under analysis.

By following the proposed method, points of interest (POIs) were collected using OpenStreetMap. This model considered only six basic services (i.e. education, health, social services, public transport, culture and park). In this study, grocery shopping was excluded from the analysis due to its over-representation compared to other activities. Afterwards, the cycling network was calibrated, considering the effects of slope and road hierarchy on cycling speed, following a previous work (Silva et al., 2021).

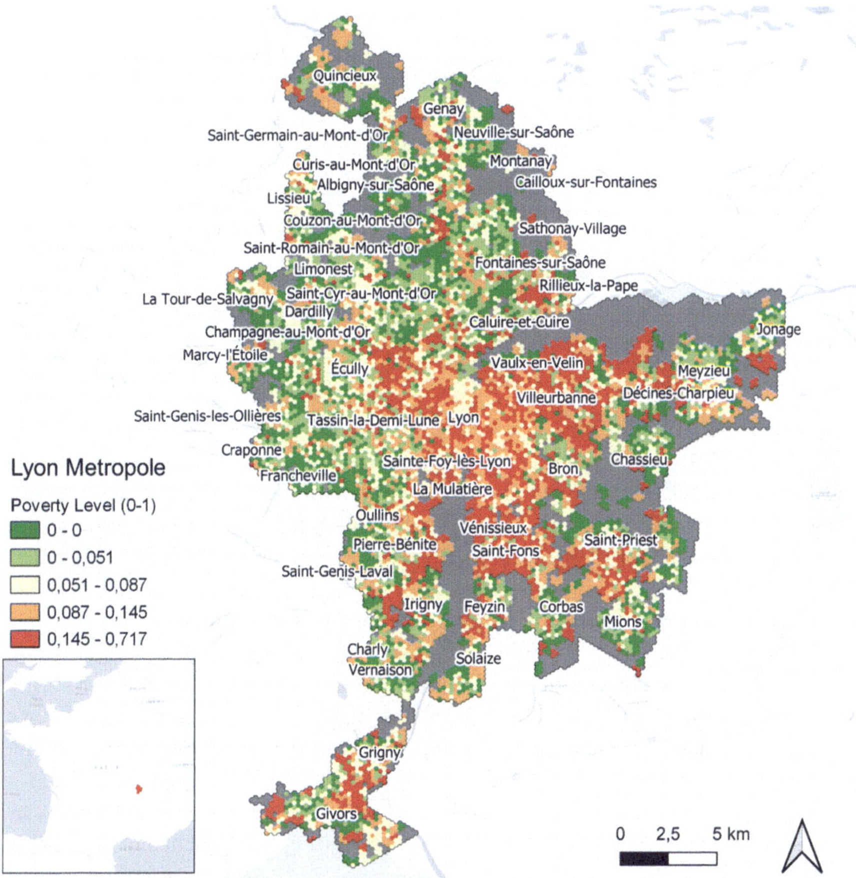

Annex 2 Poverty level in the metropolitan area of Lyon (based on INSEE data for 2019)

3.3 Socioeconomic Index

The Lyon agglomeration is divided into a poorer east and a richer west (Annex 2). Lyon's socio-spatial structure is rooted in its history as an industrial city, where a dominant economic elite resided in the city centre. Following rapid population growth after the war, large social housing estates emerged on the outskirts, especially in the eastern suburbs. These areas have since become home to predominantly working-class, low-income and ethnic minority communities, while middle- and upper-class residents are more concentrated in the city centre and the western districts (Carpenter & Verhage, 2014). As represented in the Annex 2, the poverty index, based on the French Statistics Institute (INSEE, 2021), ranges from 0 to 1, with the last one indicating the highest poverty level rate. According to the national survey, the poverty rate rose in France between 2020 and 2021, from 13.6% to 14.5%, whereas poverty intensity increased from 18.7% to 20.2%.

Table 3 LMR SE index

Case study	Socio-economic variables (%)	Weights*
Lyon metropole	Poor housing units	*−0.87*
	Individuals owning a house	*0.78*
	Single-parent household	*−0.20*
	Social housing units	*−0.75*
	Average income	*0.82*
	Households with 5 family members	*Excluded*
	One-person household	*Excluded*

Kaiser–Meyer–Olkin adequacy measure: 0.819. Bartlett's sphericity test: 0.000. 1st λ 3.122 (62.45% of total variance) and 2nd λ 0.727 (14.55% of total variance)
* Rotation method: Varimax with Kaiser normalisation

To create the SE index, this study relied on the available socioeconomic data (i.e. local census 2019) to explore the phenomenon of socioeconomic advantage and disadvantage, including the number of poor households, individuals owning a house, single-parent households, households with 5 family members, one-person households, the number of social housings and the income rate tailored to the metropolitan average.

The variables were converted to proportions and then standardised using Z-scores to prevent variables with larger ranges from disproportionately influencing the final index. Then, the principal component analysis (PCA) was conducted using SPPS to obtain the loadings for each component. As proposed by previous studies, we removed correlated variables with loadings less than 0.3 since such values are not considered strong indicators of relative advantage and disadvantage (Pink, 2013).

Table 3 depicts the final list of variables aggregated into the final SE index, as well as the respective weights derived from each loading component and eigenvalue resulting from the PCA analysis.

3.4 *Bivariate Analysis*

This study developed bi-dimensional hexagonal-grid maps using Geographic Information Systems (GIS) to spatially represent the BA and SE indicators. In the spatial bivariate analysis, both indicators were intersected to illustrate the distribution of different accessibility levels (i.e. high and low) across population groups based on the socio-economic status (i.e. advantage or disadvantage).

Finally, TIRE illustrates the equity impact, depicting four distributive clusters described in a previous research (Cunha & Silva, 2023), with cluster I identifying an equitable outcome and cluster II zones to be prioritised in future equity-oriented projects. The scheme below shows how the tool is operationalised in terms of data input and the expected outputs (Fig. 2).

In order to further explore the mismatches between SE status, population density and the corresponding bicycle accessibility, a bivariate local Moran's I analysis was

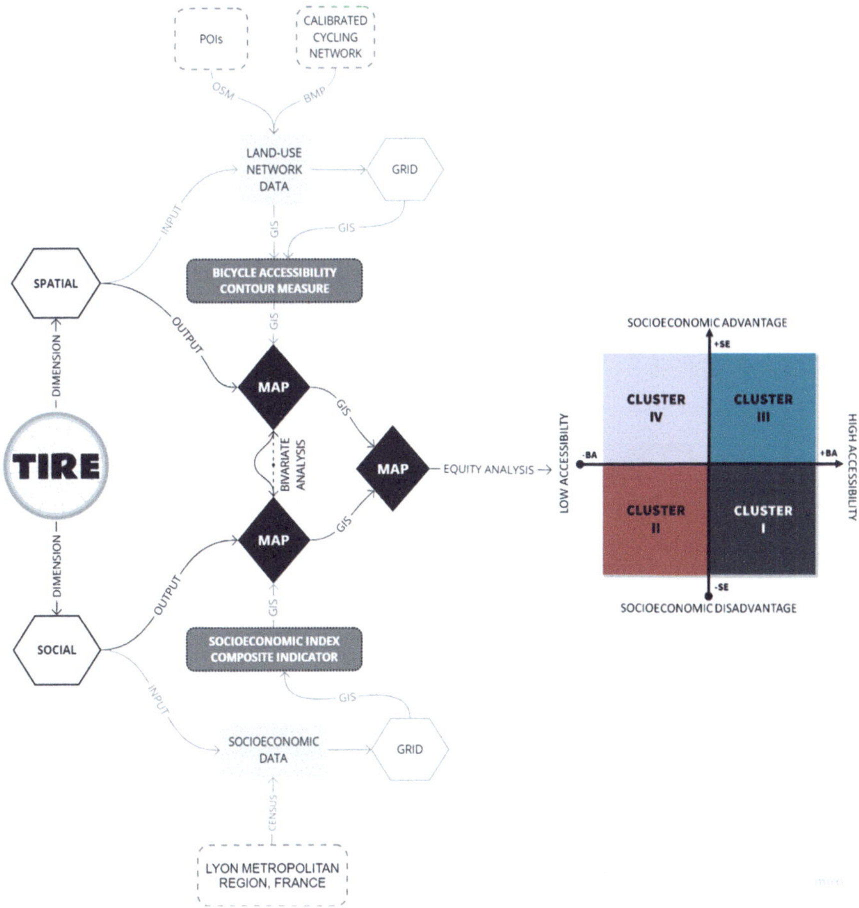

Fig. 2 TIRE operationalisation for the LMR—bivariate analysis. (Source: Cunha, I. B. da. (2023))

conducted using GeoDA software to foster discussion around the 15-minute city and the role of cycling in deprived areas and urban fringes. Moran's I is a measure of spatial autocorrelation, which evaluates the value of x at observation i with its spatial neighbours value of y. This study assessed the correlation between BA and income levels, as well as between BA and population count. By assessing Moran's I value, the pattern points in the scatterplot and the significance, it is possible to infer whether the spatial relationship between the two variables is clustered, dispersed or random (Field, 2013; Washington et al., 2011).

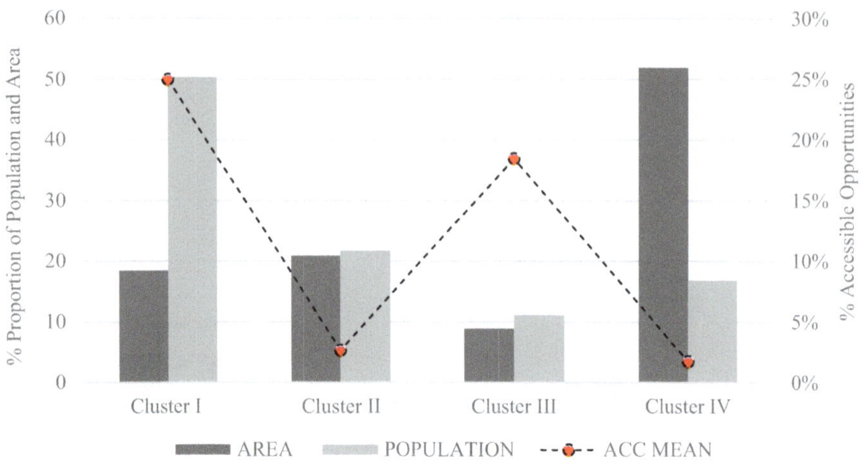

Fig. 3 Area and population proportion per cluster and % of accessible opportunities by bike (mean)

4 Results and Discussion

4.1 Equity Impact of the Metropolitan Cycling Plan

The application of the TIRE tool to the LMR allowed the assessment and geographic distribution of bicycle accessibility levels across distinct socioeconomic groups, considering a travel time threshold of 15 minutes (Fig. 5). The accessibility analysis disclosed the percentage of accessible activities per cluster, highlighting important equity gaps across the territory. The core portion of LMR has a higher concentration of cycling infrastructure and diverse activities than the municipalities on the outskirts and peri-urban areas.

To illustrate, Fig. 3 depicts the sizes of all four clusters in terms of population, area and average accessible opportunities. The first cluster aggregates zones with both higher accessibility levels and higher presence of socioeconomically disadvantaged groups. This distribution represents an equitable outcome of the cycling infrastructure allocation. As represented in Fig. 5, this cluster is mostly located in the core of the metropolitan region, which presents not only a higher diversity of activities but is also well served by the network conceived to facilitate the flows between each origin and destination. This cluster has better access to all evaluated destinations, and it is characterised as a very dense urban form, where 50% of the total metropole citizens live distributed in over 19% of the total area.

The second cluster, on the other hand, aggregates zones with a higher concentration of disadvantaged groups with low accessibility levels. This cluster is mostly scattered over the peri-urban and outskirts of LMR. As illustrated in Fig. 3, there is a huge gap between the number of accessible opportunities the residents in these areas can reach. For instance, in these areas, residents have access to only 1.78% of educational facilities and 2.35% of health care within a 15-minute bike ride.

Considering access to public transport options, cultural activities and green areas, this cluster also has limited bicycle accessibility compared to the first and third clusters. This cluster represents 22% of the total population living through 19% of the total territory.

With regard to the advantaged groups, this study revealed a higher share of residents living in peri-urban areas with lower bicycle accessibility. Indeed, the third cluster aggregating socioeconomically advantaged groups with high bicycle accessibility levels is mainly located at the core of LMR, but in a relatively small dimension. This cluster represents only 11% of the population distributed in 9% of the total territory. On the other hand, cluster IV, aggregating advantaged groups in the outskirts, encompasses 17% of the population dispersed in 52% of the total territory. This figure implies that residents in these areas are likely to rely on motorised modes of transport, living in low-density zones and single-family homes. Within 15-minute travel time, the residents of this cluster can hardly access education (0.084%), healthcare (1.08%) and social services (1.78%).

By comparing the average bicycle accessibility conditions across the different clusters (Fig. 4), this study suggests that the LMR cycling strategy has a general equitable outcome, as socioeconomically disadvantaged profiles located at the core of the metropolitan region can access a higher number of opportunities by bicycle. In other words, this group has access to 25% of all opportunities distributed in the entire region, compared to 3% found in Cluster II, 18% in Cluster III and 2% in Cluster IV. This study suggests that 21% of the territory, inhabited by 22% of the population, should be prioritised in future interventions to reduce the current equity gap.

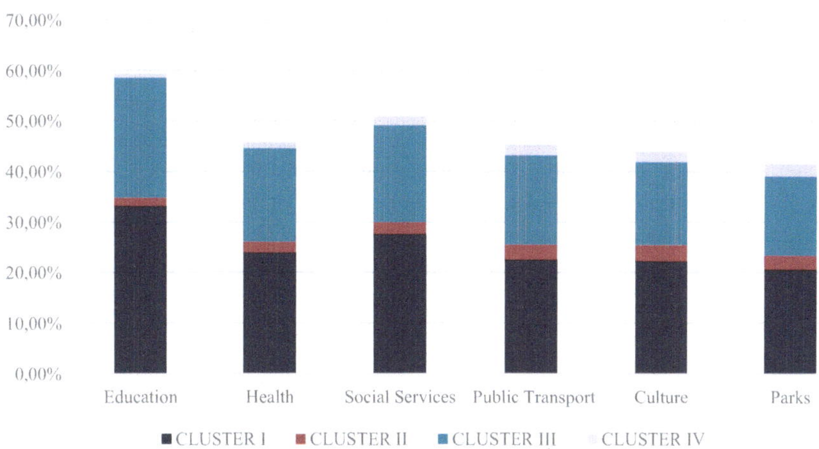

Fig. 4 Percentage of accessible opportunities per cluster

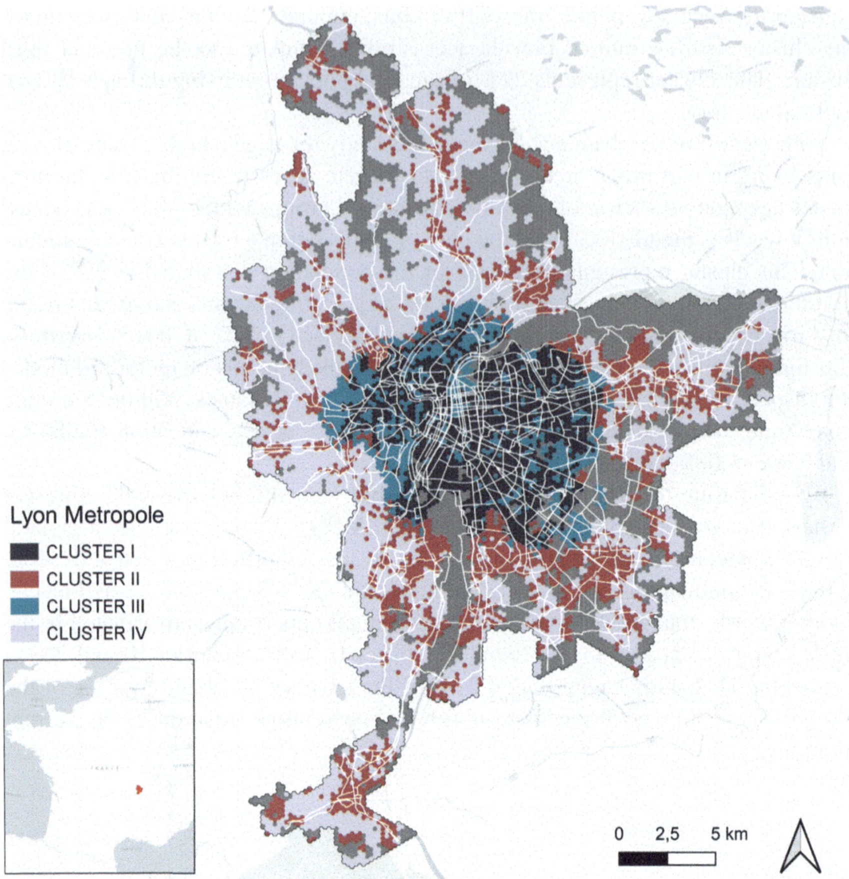

Fig. 5 TIRE application to Lyon metropolitan region

4.2 Bridging the Gap with the 15-Minute City Framework

This study revealed that bicycle accessibility in low-density, suburban areas and peripheral zones is disproportionately low compared to the high-density core of LMR (Fig. 6). Particularly, residents outside Lyon have residual access to essential opportunities provided in the territory by active travel. Whereas the centre has a higher density of both cycling infrastructure and opportunities, in the outskirts, there is a lack of land use diversity and an adequate cycling network. People living in this zone are most likely to rely on motorised modes of transport to reach opportunities and basic services. As illustrated in Fig. 6, people living in the core of the municipality have more than one thousand opportunities reachable by bicycle within 15 minutes of travel time, whereas people living in the fringes of LMR have near zero access.

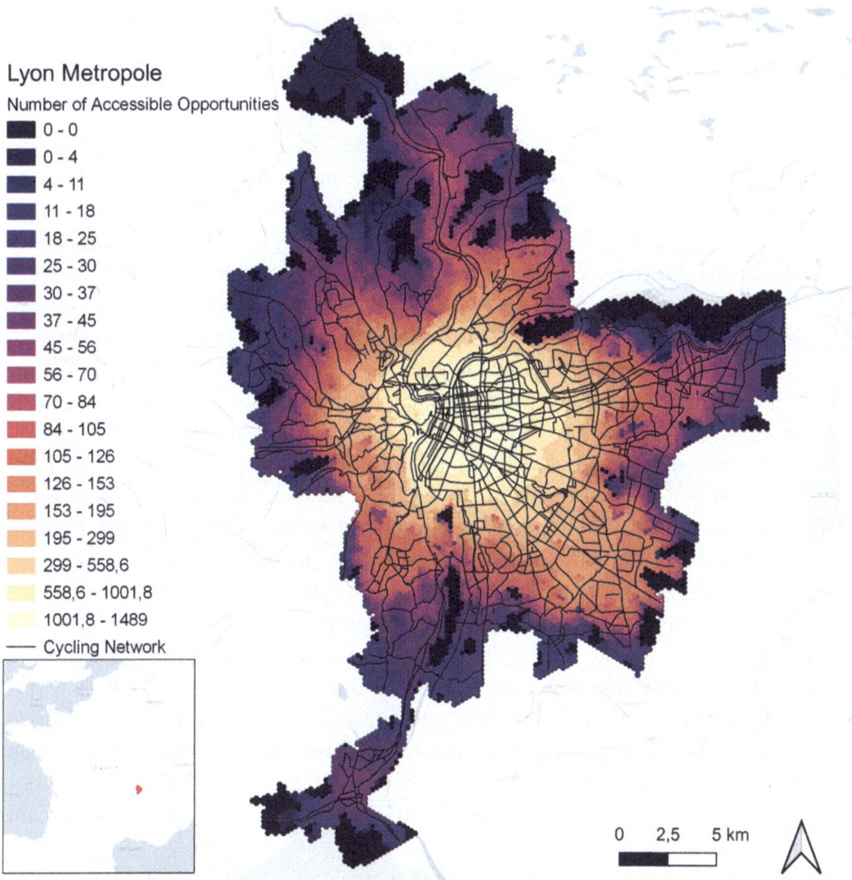

Fig. 6 Number of accessible opportunities in LMR

The findings of this study suggest that the principles of inclusiveness and ubiquity encompassing the 15-minute city framework are challenging to address on a metropolitan scale. The former expects that all residents are able to move safely in public spaces and access services with sustainable modes, irrespective of their individual capabilities. The latter expects that a given territory should equally provide proximity to essential services, public transport, mixed land use, walkable and cyclable streets, and liveable public spaces. Yet, in terms of policy-making and decision-making processes, stakeholders often make trade-offs between measures and prioritise a current political agenda. Often, measures target high-density zones to have more visibility and acceptability (Cunha et al., 2023).

To improve bicycle accessibility in low-density areas and cope with the principles underlying the 15-minute city model, this research highlights the need to integrate land use and transport planning through a set of hard-soft and push-pull measures. In terms of hard measures, implementing bicycle superhighways could be desirable to better connect origins and destinations across the metropolitan area, fostering shorter, more comfortable and safer travel experiences. Soft measures could include awareness campaigns, incentives and education programmes to raise awareness about the environmental and economic benefits of cycling.

To better distribute the cycling benefits to those living on the outskirts, it is crucial to consider strategies to foster multimodality and better connect bicycle users with local public transport. Strategies like bike-ride and bike-sharing systems services have been proven to improve the attractiveness of bicycles for utilitarian and commuting trips (Martens, 2007). Likewise, land-use patterns in the context of low--density and peri-urban areas differ in terms of purpose and scale. Henceforth, to improve bikeability in these areas and foster active travel, it is paramount to advance a polycentric development and mixed land use so that residents can access a variety of activities and fulfil their daily needs close to their homes.

In order to further explore these issues and measure the spatial correlation between accessibility levels and the population's characteristics, this study conducted local Moran's I analyses (Fig. 7). Figure 7 represent the spatial relationship between two variables: the first one focused on the accessibility levels and population density, and the second one on the accessibility and income levels.

Results from Local Moran's I correlation analysis revealed a positive spatial correlation between high population density and high bicycle accessibility levels. The p-value of 0.614 indicates spatial clustering of similar values, meaning that high values of one variable (i.e. population count) tend to be associated with high levels of the other variable (i.e. accessibility), and similarly for low values. The positive slope represented in the graph represents a strong positive autocorrelation. The map illustrates a significant correlation between population and accessibility levels, considering the cluster low–low (2159 cases) and high–high (898 cases). No significant correlation was found in 5843 cases.

On the other hand, the analysis returned a negative spatial correlation between income and accessibility, with a p-value of −0.167, suggesting spatial clustering of dissimilar values (i.e. high-income levels are associated with lower levels of accessibility). Yet, as in this case, Moran's I is close to zero, and there is no clear and significant spatial relationship between the two variables. Indeed, the shallow slope communicates weak or no spatial autocorrelation.

5 Conclusions

This study aimed to quantitatively measure and visualise the cycling accessibility conditions in central and peripheral areas and identify distributive discrepancies in a European metropolitan region. The Lyon Metropolitan Region (LMR) served as the testbed in this empirical research. Based on GIS and spatial correlation analysis, the equity impact of the LMR's bicycle master plan (BMP) was examined, considering the principles of the 15-minute city framework.

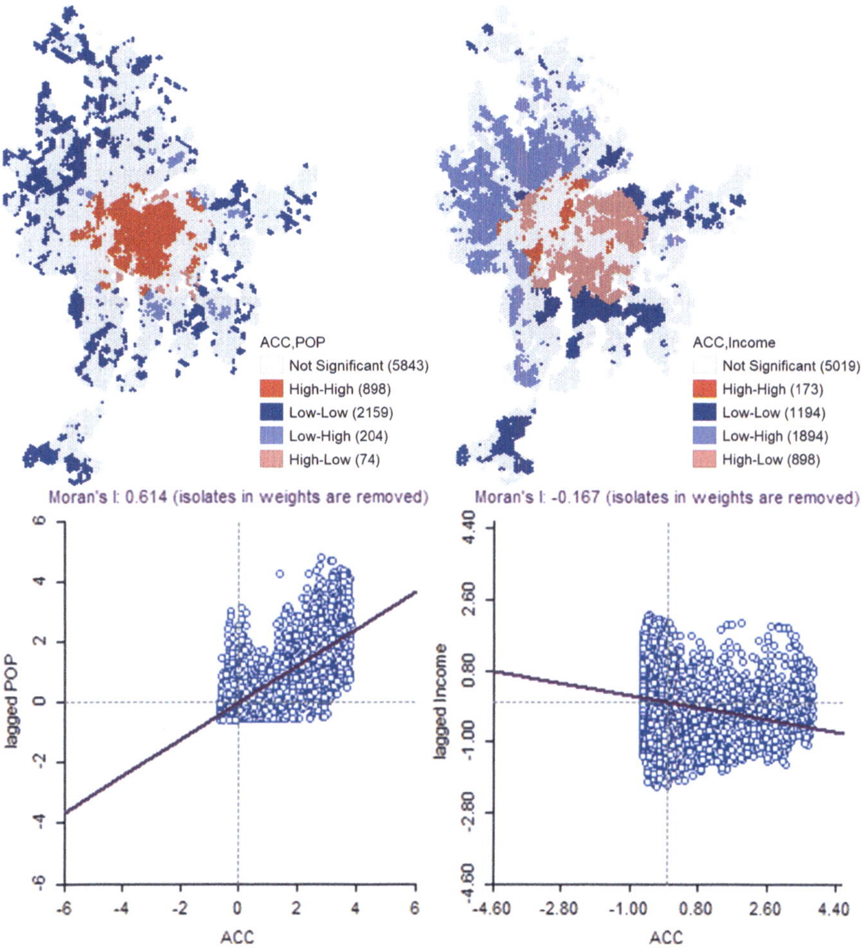

Fig. 7 Local Moran's I analysis. Left side: spatial correlation analysis between the population density and bicycle accessibility levels and the respective scatterplot. Right side: spatial correlation analysis between income level and bicycle accessibility levels and the respective scatterplot

Through the application of the tool TIRE, this study revealed that the assessed BMP is, to a certain extent, equitably distributed in the metropolitan core. Most of the cycling infrastructure and activities are concentrated in this zone compared to the peri-urban municipalities. The first cluster aggregates zones with a higher concentration of disadvantaged groups and accessibility levels, which communicates an equitable outcome. This cluster has better access to all evaluated destinations and is characterised as a very dense urban form, where 50% of the total metropole citizens live in over 19% of the total area. Moreover, this study identified areas to be prioritised in future equity-oriented interventions based on the second cluster. This cluster aggregates zones with a higher concentration of disadvantaged groups with low accessibility levels and is mainly scattered over the peri-urban and outskirts of LMR, encompassing 22% of the total population living through 19% of the total territory.

This study revealed that bicycle accessibility in low-density, suburban areas and peripheral zones is disproportionately low compared to the high-density core of LMR, and, therefore, the principles of inclusiveness and ubiquity encompassing the 15-minute city framework are challenging to be achieved. Also, results from Moran's I correlation analysis revealed a positive spatial correlation between high population density and high bicycle accessibility levels. This outcome suggests that the planning strategies focus on central zones to ensure public acceptability and increase cycling attractiveness. Conversely, this research found no significant spatial correlation between bicycle accessibility and income level. Indeed, the TIRE tool disclosed high incomers living in the west portion of LMR in low-density municipalities. People living in this zone are mostly likely to rely on motorised modes of transport to reach opportunities and basic services.

In order to advance the discussion around the adaptability of the 15-minute city framework on a metropolitan scale, this research discussed potential measures to improve bicycle accessibility in low-density, peri-urban and outskirts areas. Nevertheless, this study has some caveats that need to be addressed in future research. Although the research focused on a quantitative approach based on an open-source dataset, the model lacks information on education level and car ownership that could shed new evidence on mobility patterns in low-density areas. Future research could examine the perceived accessibility in these areas through a qualitative research approach involving local residents, planning practitioners and policymakers in order to foster a discussion about the equity implications of cycling in deprived zones, considering the 15-minute city vision.

References

Abdullah, M., Ali, N., Javid, M. A., & Hussain, Q. (2022). Awareness and knowledge levels of engineering and planning students and practitioners about the 15-minute city concept in a developing country. *Journal of Urban Mobility, 2*, 100037.

Aldred, R., Woodcock, J., & Goodman, A. (2016). Does more cycling mean more diversity in cycling? *Transport Reviews, 36*, 28–44.

Banister, D. (2011). Cities, mobility and climate change. *Journal of Transport Geography, 19*, 1538–1546.

Bartzokas-Tsiompras, A., & Bakogiannis, E. (2023). Quantifying and visualizing the 15-minute walkable city concept across Europe: A multicriteria approach. *Journal of Maps, 19*, 1–9.

Birkenfeld, C., Victoriano-Habit, R., Alousi-Jones, M., Soliz, A., & El-Geneidy, A. (2023). Who is living a local lifestyle? Towards a better understanding of the 15-minute-city and 30-minute-city concepts from a behavioural perspective in Montréal, Canada. *Journal of Urban Mobility, 3*, 100048.

Bouzouina, L., Baraklianos, I., Bonnel, P., & Aissaoui, H. (2021). Renters vs owners: The impact of accessibility on residential location choice. Evidence from Lyon urban area, France (1999–2013). *Transport Policy, 109*, 72–84.

Caggiani, L., Colovic, A., & Ottomanelli, M. (2020). An equality-based model for bike-sharing stations location in bicycle-public transport multimodal mobility. *Transportation Research Part A Policy and Practice, 140*, 251–265.

Carpenter, J., & Verhage, R. (2014). Lyon city profile. *Cities, 38*, 57–68.

Celbiş, M. G., & Bouzouina, L. (2025). To what extent walking and biking are substitutes or complements to public transport? Interpretable machine learning findings from the University of Lyon, France. *Journal of Transport Geography, 123*, 104071.

Cunha, I., & Silva, C. (2022). Equity impacts of cycling: Examining the spatial-social distribution of bicycle-related benefits. *International Journal of Sustainable Transportation, 0*, 1–19.

Cunha, I., & Silva, C. (2023). Assessing the equity impact of cycling infrastructure allocation: Implications for planning practice. *Transport Policy, 133*, 15–26.

Cunha, I., Silva, C., & Büttner, B. (2023). Practitioners' perspectives on cycling equity: Bridging the gap between planning priorities. *Transportation Research Part D: Transport and Environment, 123*, 103902.

Cunha, I., Silva, C., Büttner, B., & Toivonen, T. (2024). Pursuing cycling equity? A mixed-methods analysis of cycling plans in European cities. *Transport Policy, 145*, 237–246.

da Cunha, I. B. 2023. *Bicycle accessibility planning: Towards an equitable approach* (Doctoral thesis, Faculdade de Engenharia, Universidade do Porto). Repositório Aberto da Universidade do Porto. https://hdl.handle.net/10216/153196

Ferrer-ortiz, C., Marquet, O., Mojica, L., & Vich, G. (2022). Barcelona under the 15-Minute City lens: Mapping the accessibility and proximity potential based on pedestrian travel times. *Smart Cities, 5*, 146–161.

Field, A. (2013). *Discovering statistics using IBM SPSS statistics* (4th ed.). SAGE Publications.

Goldman, T., & Gorham, R. (2006). Sustainable urban transport: Four innovative directions. *Technology in Society, 28*, 261–273.

Golub, A., & Martens, K. (2014). Using principles of justice to assess the modal equity of regional transportation plans. *Journal of Transport Geography, 41*, 10–20.

Graells-Garrido, E., Serra-Burriel, F., Rowe, F., Cucchietti, F. M., & Reyes, P. (2021). A city of cities: Measuring how 15-minutes urban accessibility shapes human mobility in Barcelona. *PLoS One, 16*, 1–21.

Guzman, L. A., Arellana, J., Oviedo, D., & Moncada Aristizábal, C. A. (2021). COVID-19, activity and mobility patterns in Bogotá. Are we ready for a '15-minute city'? *Travel Behaviour and Society, 24*, 245–256.

Havet, N., & Bouzouina, L. (2024). Bicycle use in the university community: Empirical analysis using MobiCampus-UdL data (Lyon, France). *Journal of Transport and Land Use, 17*, 299–320.

Hickman, R., Ashiru, O., & Banister, D. (2011). Transitions to low carbon transport futures: Strategic conversations from London and Delhi. *Journal of Transport Geography, 19*, 1553–1562.

Hosford, K., Beairsto, J., & Winters, M. (2022). Is the 15-minute city within reach? Evaluating walking and cycling accessibility to grocery stores in Vancouver. *Transportation Research Interdisciplinary Perspectives, 14*, 100602.

INSEE. (2021). *La France et ses territoires*.

ITF. (2020). *Shared mobility simulations for Lyon case-specific policy analysis*.

Karner, A. (2016). Planning for transportation equity in small regions: Towards meaningful performance assessment. *Transport Policy, 52*, 46–54.

Khavarian-Garmsir, A. R., Sharifi, A., & Sadeghi, A. (2023). The 15-minute city: Urban planning and design efforts toward creating sustainable neighborhoods. *Cities, 132*, 104101.

Knap, E., Ulak, M. B., Geurs, K. T., Mulders, A., & van der Drift, S. (2023). A composite X-minute city cycling accessibility metric and its role in assessing spatial and socioeconomic inequalities—A case study in Utrecht, The Netherlands. *Journal of Urban Mobility, 3*, 100043.

Lanzendorf, M., & Busch-Geertsema, A. (2014). The cycling boom in large German cities--empirical evidence for successful cycling campaigns. *Transport Policy, 36*, 26–33.

Liu, D., Kwan, M. P., Kan, Z., & Wang, J. (2022). Toward a healthy urban living environment: Assessing 15-minute green-blue space accessibility. *Sustainability (Switzerland), 14*, 1–12.
Logan, T. M., Hobbs, M. H., Conrow, L. C., Reid, N. L., Young, R. A., & Anderson, M. J. (2022). The x-minute city: Measuring the 10, 15, 20-minute city and an evaluation of its use for sustainable urban design. *Cities, 131*, 103924.
Lu, M., & Diab, E. (2023). Understanding the determinants of x-minute city policies: A review of the north American and Australian cities' planning documents. *Journal of Urban Mobility, 3*, 100040.
Lucas, K. (2012). Transport and social exclusion: Where are we now? *Transport Policy, 20*, 105–113.
Lucas, K., Marsden, G., Brooks, M., & Kimble, M. (2013). An assessment and critique of capabilities for examining the long-term social sustainability of transport and land-use strategies. *Transportation Research Record: Journal of the Transportation Research Board, 2013*, 30–37.
Lyons, G., & Davidson, C. (2016). Guidance for transport planning and policymaking in the face of an uncertain future. *Transportation Research Part A Policy and Practice, 88*, 104–116.
Manaugh, K., Badami, M. G., & El-Geneidy, A. M. (2015). Integrating social equity into urban transportation planning: A critical evaluation of equity objectives and measures in transportation plans in north america. *Transport Policy, 37*, 167–176.
Manaugh, K., & El-Geneidy, A. M. (2011). Who benefits from new transportation infrastructure? Using accessibility measures to evaluate social equity in transit provision. In K. Geurs, K. Krizek, & A. Reggiani (Eds.), *For accessibility and transport planning: Challenges for Europe and North America* (pp. 1–24). Edward Elgar.
Martens, K. (2007). Promoting bike-and-ride: The Dutch experience. *Transportation Research Part A Policy and Practice, 41*, 326–338.
Martens, K. (2017). Seeking transportation justice. In *Transport justice designing fair transportation systems* (pp. 151–163). Routledge.
Martens, K., & Golub, A. (2012). *A justice-theoretic exploration of accessibility measures.* Accessibility Analysis and Transport Planning: Challenges for Europe and North America, 195–210.
Moreno, C., Allam, Z., Chabaud, D., Gall, C., & Pratlong, F. (2021). Introducing the "15-Minute City": Sustainability, resilience and place identity in future post-pandemic cities. *Smart Cities, 4*, 93–111.
Murgante, B., Valluzzi, R., & Annunziata, A. (2024). Developing a 15-minute city: Evaluating urban quality using configurational analysis. The case study of Terni and Matera, Italy. *Applied Geography, 162*, 103171.
Nielsen, T. A. S., & Skov-Petersen, H. (2018). Bikeability—Urban structures supporting cycling. Effects of local, urban and regional scale urban form factors on cycling from home and workplace locations in Denmark. *Journal of Transport Geography, 69*, 36–44.
Olivari, B., Cipriano, P., Napolitano, M., & Giovannini, L. (2023). Are Italian cities already 15-minute? Presenting the next proximity index: A novel and scalable way to measure it, based on open data. *Journal of Urban Mobility, 4*, 100057.
Pereira, R. H. M., Braga, C. K. V., Servo, L. M., Serra, B., Amaral, P., Gouveia, N., & Paez, A. (2021). Geographic access to COVID-19 healthcare in Brazil using a balanced float catchment area approach. *Social Science & Medicine, 273*, 113773.
Pereira, R. H. M., Schwanen, T., & Banister, D. (2017). Distributive justice and equity in transportation. *Transport Reviews, 37*, 170–191.
Pink, B. (2013). *Socio-economic indexes for areas (SEIFA).* Australian Bureau of Statistics.
Poorthuis, A., & Zook, M. (2023). Moving the 15-minute city beyond the urban core: The role of accessibility and public transport in The Netherlands. *Journal of Transport Geography, 110*, 103629.
Schneider, F., Jensen, A. F., Daamen, W., & Hoogendoorn, S. (2023). Empirical analysis of cycling distances in three of Europe's most bicycle-friendly regions within an accessibility framework. *International Journal of Sustainable Transportation, 17*, 775–789.

Silva, C., Marques, J. S., Lopes, M., & Dias, A. M. (2021). The gross potential for cycling: Planning for human scale urban mobility. *Transport in Human Scale Cities, 10*, 157–168.

Staricco, L. (2022). 15-, 10- or 5-minute city? A focus on accessibility to services in Turin, Italy. *Journal of Urban Mobility, 2*, 100030.

SYSTRAL. (2017). *Plan de Déplacements Urbains Agglomération Lyonnaise 2017–2030.*

Tammaru, T., Sevtsuk, A., & Witlox, F. (2023). Towards an equity-centred model of sustainable mobility: Integrating inequality and segregation challenges in the green mobility transition. *Journal of Transport Geography, 112*, 103686.

Temenos, C., Nikolaeva, A., Schwanen, T., Cresswell, T., Sengers, F., Watson, M., & Sheller, M. (2017). Theorizing mobility transitions: An interdisciplinary conversation. *Transfers, 7*, 113–129.

Nunen, J. Van, Huijbregts, P., & Rietveld, P. (2011). *Introduction to transitions towards sustainable mobility*. In Transitions Towards Sustainable Mobility.

Vigar, G. (2017). The four knowledges of transport planning: Enacting a more communicative, trans-disciplinary policy and decision-making. *Transport Policy, 58*, 39–45.

Washington, S., Karlaftis, M., & Mannering, F. (2011). *Statistical and econometric methods for transportation data analysis*. Taylor & Francis.

Weng, M., Ding, N., Li, J., Jin, X., Xiao, H., He, Z., & Su, S. (2019). The 15-minute walkable neighborhoods: Measurement, social inequalities and implications for building healthy communities in urban China. *Journal of Transport and Health, 13*, 259–273.

Willberg, E., Fink, C., & Toivonen, T. (2023). The 15-minute city for all?—Measuring individual and temporal variations in walking accessibility. *Journal of Transport Geography, 106*, 103521.

Zhang, D., Ma, S., Fan, J., Xie, D., Jiang, H., & Wang, G. (2023). Assessing spatial equity in urban park accessibility: An improve two-step catchment area method from the perspective of 15-mintue city concept. *Sustainable Cities and Society, 98*, 104824.

Everyday Mobility and the 15-Minute City in Sub-Saharan Africa: Insights from Dakar, Senegal

Pascal Pochet and Gaële Lesteven

1 Introduction

Introduced in 2016 by Moreno, the concept of the '15-Minute City' gained prominence during the COVID-19 pandemic. Its success lies in its simple formula of the proximity-based planning principle, which has been widely discussed in public debates (Mouratidis, 2024) and in scientific literature. The concept refers to a situation where residents are able to access basic needs—'living, working, commerce, healthcare, education, and entertainment'—within 15 minutes, either on foot or bicycle, from their home (Moreno et al., 2021, p. 100).

Three additional urban planning principles come with the proximity principle, forming the four pillars of the 15-Minute City: the principle of density (of population), the principle of diversity (of urban functions), and the principle of digitalisation (Allam et al., 2022). Proponents of the concept often foreground its potential to reduce car dependency, decrease fossil fuel consumption, and improve air quality (Allam et al., 2022; Moreno et al., 2021). They add public health implications, encouraging active modes of transport to mitigate obesity risks (Weng et al., 2019). However, depending on the urban context, the 15-minute limit should be interpreted with some flexibility. Some authors prefer to consider the '20-minute city' (Calafiore et al., 2022; Capasso Da Silva et al., 2020) or even the '30-minute city' (Levinson, 2020). Another variant incorporates public transport (PT) alongside active modes (Calafiore et al., 2022; Liu et al., 2024).

The implementation of the 15-Minute City raises questions regarding social and spatial inequalities and equity in access to urban resources. Is the 15-Minute City inclusive or does it primarily benefit affluent populations residing in urban centres? Moreover, it raises questions about the transferability of the concept, which has

P. Pochet (✉) · G. Lesteven
LAET, ENTPE, Vaulx-en-Velin, FRANCE
e-mail: pascal.pochet@entpe.fr

L. Bouzouina et al. (eds.), *15-Minute Cities*, Footprints of Regional Science,
https://doi.org/10.1007/978-3-032-20840-8_10

been developed for European cities, with their densely populated centres and car-dependent suburbs. How relevant is the concept in the context of cities in the Global South?

Through a review of 98 studies on 15-Minute City practices across the world, Teixeira et al. (2024) demonstrate that the concept is highly context-dependent. Among the case studies, 58 were based in Europe, 22 in North America, 8 in Asia and Oceania, 2 in Latin America, and none in Africa. Recently, Dumedah et al. (2024) investigated the case of a low-income suburb community in Ghana and assessed its alignment with the access condition of the 15-Minute City. The geographic distribution of the case studies illustrates the Western-centric nature of the concept, which seeks proximity as a countermeasure to the dominant automobile norm. It reflects the worldwide diversity of modal shares at the urban scale (Lesteven, 2023). If the private car is predominant in European and North American cities, walking remains the prevailing norm in many cities in the Global South, particularly in sub-Saharan Africa (Diaz Olvera et al., 2013; Porter et al., 2020).

Sub-Saharan African cities are growing fast, and planning policies remain too limited to ensure balanced urban development. Urban peripheries are diverse in terms of social composition, housing, and accessibility. In areas where auto-construction is predominant, access to basic services is often challenging (Meth et al., 2024). Informal economy and local communities have enabled the development of markets, shops, and services of all kinds. Despite these developments, essential public services, such as education and healthcare, are insufficient to meet the needs of residents of the peripheries in both quantity and quality, especially for low-income populations, as Stokenberga et al. (2025) showed through an accessibility analysis of ten African cities. Job opportunities are still largely concentrated in central areas, which limits their accessibility. Residents in peripheral areas frequently endure long, arduous commutes on congested urban roads (Bertrand, 2010; Carmichael et al., 2024; Lesteven et al., 2023; Melbye et al., 2015; Salon & Gulyani, 2019). Some residents commute by private car or motorcycle. But for the vast majority, limited incomes compel them to rely on PT or walk.

Walking is the primary mode of transport on the African continent (Benton et al., 2023). It is often the most relevant mode of travel, and can be as efficient as, or even more efficient than, PT (Stokenberga et al., 2025), particularly in traffic congestion. However, frequently, walking is a default choice. Many residents have to walk long distances, sometimes while carrying goods or children, under challenging conditions due to a lack of infrastructure and difficult climatic conditions. Walking is a physical effort that soils clothing due to sweat, dust, and rain (Cooke et al., 2022). In most large African cities, active modes are limited to walking, as bicycle use remains marginal. Cycling is considered dangerous and carries a negative image, as the bicycle is often perceived as a transport mode for the poor (Mendiate et al., 2022). When they can afford it, people opt for PT. The cost of PT is thus a significant issue for low-income people, whether they are workers, students, or unemployed individuals (Diaz Olvera et al., 2013; Venter, 2011). This unequal access to modes of transport reveals major economic and social inequalities in African cities (Boyer et al., 2016; Diaz Olvera et al., 2013).

The chapter evaluates the relevance of the 15-Minute City concept to urban mobility and everyday life practices in the West African city of Dakar, Senegal. The originality of this research lies both in its case study of an African city and in its methodology. As previously noted, the number of studies focusing on the 15-Minute City in the Global South is limited. Moreover, most of these studies rely on GIS tools to estimate accessibility to essential services (Dumedah et al., 2024; Guzman et al., 2021; Guzman et al., 2024; Liu et al., 2024; Nieto-Combariza et al., 2024). This study adopts a different approach by specifically focusing on actual daily mobility and access, using data from a household travel survey to analyse the proportion of trips for out-of-home activities that are compatible with the 15-Minute City. We propose four variants of the 15-Minute City in order to take into consideration the major role of walking and to include PT. The first variant considers all trips made on foot within 15 minutes, the second all trips made on foot and by PT within 15 minutes, the third all trips made on foot within 30 minutes, and the last one all trips made on foot and by PT within 30 minutes.

The rest of the chapter is structured as follows. We first present the context of the Dakar region and our methodology based on four variants of the 15-Minute City. We then highlight what daily activities align with the 15-Minute City concept and analyse the related mobility patterns of different population categories. Finally, we conclude with a discussion about the main results of the study, showing that the 15-Minute City is a de facto reality for many residents in the Dakar region. In particular, the mobility of women, unemployed individuals, and the low-income population is circumscribed to their neighbourhoods, even when it is not their preference. This Dakar case underscores the significant challenges in fostering local activities and improving living conditions in the underserved outskirts of African cities.

2 Context and Methodology

This study is based on the case of Dakar. The Senegalese capital is located in the western part of the country on the Cape Verde Peninsula. Home to 4 million inhabitants (ANSD, 2025), the Dakar region covers an area of 547 km^2 from west to east. The tip of the peninsula constitutes the central department of Dakar and houses nearly one-third of the population. The remaining population resides in the nearby suburbs in the Guédiawaye department, and in the more distant suburbs of Pikine, Keur Massar, and Rufisque departments (ANSD, 2025). Nearly six out of ten jobs are located in the department of Dakar (Diaz Olvera et al., 2024). This separation between places of residence and places of work results in long commuting distances (Sakho, 2014). As household ownership of private cars and motorcycles remains limited (Diaz Olvera et al., 2020), people rely on motorised PT services. PT accounts for 24% of daily trips, compared to 5% by private modes (Sitrass-Curem, 2016).

The Dakar region presents a diverse PT system, encompassing both formal and informal modes, as it is common in African cities. Formal modes include the Dakar Dem Dikk public bus company, along with AFTU privately owned buses and a fleet of yellow-black taxis. Informal modes consist of Cars Rapides minibuses, Ndiaga Ndiaye 40-seater buses, and Clandos shared taxis. The transport system is undergoing rapid transformation, marked by the opening of a regional express train line in 2021 and a BRT line in 2024, alongside the expansion of informal motorcycle taxis since the 2010s.

Walking is the main mode of daily travel, with a modal share of 70%. It often takes place under poor conditions due to inadequate infrastructure and poor management of roads and public spaces. Issues include street hawkers, lack of pavements, insufficient lighting, rubbish, and flooding during the rainy season (Sitrass-Curem, 2016). In the most peripheral department of Rufisque, walking share is the highest. It accounts for 79% of trips, compared to 63% in the Dakar department (Lesteven et al., 2022). Conversely, car use is highest in the central Dakar department, where incomes are also the highest (Diaz Olvera et al., 2020). Additionally, gender inequality is significant, as private cars and motorised two wheelers are used by men predominantly (Pochet & Lesteven, 2025).

The study draws on data collected through a Household Travel Survey (HTS) carried out in the Dakar region in 2015, in which one of the authors participated (Sitrass-Curem, 2016). The sample includes 13,415 individuals aged 11 years and older from 3176 households surveyed in 419 census zones. After weighting, the sample is representative of the population.[1] The questionnaire comprises several sections aimed at gathering information on household and individual characteristics, as well as for each individual, characteristics of all the trips made on the day before—including short trips entirely on foot—, such as origin and destination, activity at destination, transport mode, travel distance, and duration. It is important to note that not all daily activities involving travel are performed on a daily basis. Some regular activities may occur once or several times a week, while others are more occasional but not less essential to livelihoods. Surveys have been completed on weekdays and Saturdays. A set of 38 semi-structured interviews conducted with Dakar residents in 2022 and 2023 (Lesteven et al., 2023) completes the household travel survey. The interviews are focused on daily activities residents consider important, mobility challenges they encounter, strategies they develop to overcome these difficulties, and the modes of transport they use.

To evaluate the relevance of 15-Minute City principles to the Dakar context, we aggregate data from the household travel survey according to: (1) time distances and transport modes and (2) activities.

(1) We consider the original definition of the 15-Minute City, which includes all trips made on foot or by bicycle within a 15-minute walking or cycling distance from home. Since bicycle use is rare and walking represented 99% of trips made by active modes in the Dakar region, we refer to it as the *15-minute walking city*.

To provide a more comprehensive perspective, we consider three variants of the original 15-Minute City. The first variant expands the original definition to include

[1] All the figures presented in the tables and text are from the weighted sample.

trips made by PT within a 15-minute distance from home, referred to as the *15-minute walking and PT city*.

Given the high prevalence of walking, we extend the time limit to 30 minutes, proposing two additional variants: one considering trips lasting up to 30 minutes on foot, called the *30-minute walking city*, and the other including PT trips, which we name the *30-minute walking and PT city*. The 30-minute time limit for both walking and PT trips excludes the longest and most arduous trips on foot or by PT, ensuring a scale for daily travels that people are likely to consider acceptable in terms of quality of life.

To differentiate the concept of the 15-Minute City from the four variants we propose, we use capital letters for the generic form.

(2) We group daily activities into three main categories. The first category, which we call *Ordinary life,* refers to all activities related to 'living', 'commerce', and 'healthcare', which are three of the six basic needs, as classified by Moreno et al. (2021). It refers to all trips made for food and non-food purchases, water and firewood supplies, administrative tasks, healthcare and service purposes, religion, and accompanying other people (children, etc.).

The second category, named *Social life,* adds activities related to 'entertainment', another of the six basic needs, such as leisure and sporting activities. It also includes activities related to 'living' that consist of visiting relatives, family and neighbours, community meetings, and participation in social events, especially traditional ceremonies. In West African cities, social life is mainly structured around these activities. They are important markers of sociability as well as solidarity within the extended family and the community, as institutional social protection is almost non-existent.

Finally, the third category, which we synthesise as *Working life,* consists of all activities related to 'working' and 'education', the last two basic needs defined by Moreno et al. (2021). It groups all home-to-work and home-to-school trips and other trips generated by professional or scholarly activities.

3 An Urban Life in the Vicinity of Home… and Beyond

Almost half of the daily basis trips in the whole Dakar region are dedicated to working life. Trips related to ordinary life represent a little less than a third, and the remaining 20% corresponds to trips for social life (Table 1).

This hierarchy is reversed when trips within a 15-minute walking distance are considered. Trips related to ordinary life, such as commerce, healthcare, prayer,[2] and administrative tasks, are more frequent, accounting for 45% of trips, compared to 30% across the entire region. The proportion of social life trips within a

[2]The Muslim religion is dominant in the Dakar region (94% of the population compared to 4% being Christian). Although religious practice is slightly less widespread than in the rest of Senegal (ANSD, 2013), prayers are a significant reason for travel, particularly for men.

Table 1 Distribution of daily trips per category of activities, for each definition of the 15-Minute City (%)

	(a) Ordinary life	(b) Social life	(c) Working life	Other	All trips
(i) 15-minute walking trips	45	25	29	1	100
(ii) 15-minute walking & PT trips	44	24	31	1	100
(iii) 30-minute walking trips	41	24	34	1	100
(iv) 30-minute walking & PT trips	38	23	37	1	100
All trips	30	21	48	1	100

Source: Dakar 2015 HTS; calculations: authors
(a) Ordinary life: trips for food and non-food purchases, water and firewood supplies, administrative tasks, healthcare and service purposes, religion, accompanying other people (children, etc.); (b) Social life: trips for visits to relatives, friends, and neighbours, for taking part in ceremonies, association meetings, leisure, and sporting activities. (c) Working life: home to work and home to school trips and other trips generated by professional or scholarly activities

15-minute walking distance is also more important, accounting for 25% of trips, compared to 21% at the metropolitan scale. This finding indicates that Dakar residents fulfil most of their basic daily needs in the vicinity of their homes. The daily routine of a 50-year-old millet porridge seller illustrates this pattern: 'My travels are limited to my workplace, the market, and the mill. […] They are not far from home, just about a two-minute walk'.

However, the situation may differ for trips related to work and education. These account for less than 30% within the 15-minute walking city and gradually increase when considering longer time distances. Many residents must travel beyond their home vicinity to access employment or educational opportunities, often commuting to the city centre, where job opportunities are concentrated. A 60-year-old self-employed trader living in the suburbs testified: 'If I don't go to the city centre to work, I won't be able to meet the daily expenses. Sometimes, I feel like staying home to rest, but if I realize that the daily expenses are not covered, I am forced to do what it takes to go to work'.

We now consider how compatible the three categories of activities are with the different definitions of the 15-Minute City (Table 2). All activities considered, half of the total volume of daily trips in the Dakar region falls within the 15-minute walking city. These proportions logically increase when the modal and proximity constraints are relaxed, reaching 70% of all trips being compatible with the 30-minute walking and PT limit.

Almost three-quarters of the total trips related to ordinary life and 60% of trips related to social life occur within a 15-minute distance on foot. The situation is even more explicit with the 30-minute walking and PT limit, as it covers almost nine out of ten trips for ordinary life activities, and eight out of ten for activities devoted to social life. This situation is enabled through a well-developed PT system, particularly informal transport, which serves many areas across the region. A 68-year-old retired teacher explained that, thanks to the proximity of shared taxi ranks, he could

Table 2 Share of trips compatible with different definitions of the 15-Minute City, for different categories of activities (%)

	(i) 15-min walking trips	(ii) 15-min walking and PT trips	(iii) 30-min walking trips	(iv) 30-min walking and PT trips
(a) Ordinary life	73	78	79	88
(b) Social life	60	64	69	80
(c) Working life	30	35	41	55
All trips	50	54	58	70

Source: Dakar 2015 HTS; calculations: authors
Reading: 73% of all trips related to the ordinary life of Dakar residents are within a 15-minute walk from home. This proportion rises to 88% within a 30-minute distance from home on foot or by PT

Table 3 Share of trips compatible with different definitions of the 15-Minute City, for specific trip purposes (%)

	(i) 15-min walking trips	(ii) 15-min walking and PT trips	(iii) 30-min walking trips	(iv) 30-min walking and PT trips
Administrative tasks (*ordinary life*)	17	26	21	42
Healthcare (*ordinary life*)	25	35	29	61
Visiting family (*social life*)	36	42	44	60
Attending ceremonies (*social life*)	24	26	28	35
Seeking a job (*working life*)	11	17	26	45

Source: Dakar 2015 HTS; calculations: authors

access all local amenities he needed: 'We have no difficulty accessing these various services because the Clando [informal shared taxi] ranks are very close, and we can rely on them at any time'.

Based on these findings, one can consider that the 15-Minute City de facto exists in the Dakar region regarding a large part of activities related to ordinary and social life. However, less than a third of work commutes and education-related activities are within the 15-minute walking city, and slightly more than half fall within the broader definition of the 30-minute walking and PT city. Many workers have to make long trips to reach their workplace, as is the case for many high school or university students.

Table 3 presents activities that play an important role in terms of economic and social integration, even if they are less frequent and occur on a weekly basis or occasionally.

Although often compulsory, administrative tasks depend on the location of public administrations and services, which are most of the time centrally situated and far from residential areas. They require long travel. Similarly, trips for healthcare are, in general, not compatible with the 15-minute or 30-minute walking city. Nevertheless, their frequency doubles when including PT within a 30-minute distance. A 35-year-old tailor shared his experience of relying on PT to take his sister to the hospital: 'She couldn't move, so I went to find a car [informal shared taxi], as the walking distance was estimated to be more than 20 minutes'.

In contexts where institutional social protection systems provide little or no support to most of the population, maintaining connections with one's social network is crucial (Lindell & Utas, 2012). Part of social life unfolds within the neighbourhood, fostering local social ties essential for everyday life. However, this sociability in the neighbourhood cannot address all difficulties that households face in their daily routine. Visiting relatives and family members, who may live throughout the city (Gough et al., 2015), is critical for preserving relationships, obtaining support, or resolving problems. Attending ceremonies, such as weddings, birth celebrations, or funerals, is a significant aspect of social life, especially for women. An 80-year-old fruit seller living in northern Dakar explained: 'Some of us [from the family] stayed in the city centre. I often go there when there is a death—may Allah protect us—but also for weddings and other events'. These events reinforce kinship relations and involve the exchange of gifts and large sums of money (Moya, 2015).

Interviews with Dakar residents highlight how much visiting relatives and attending ceremonies are considered vital and worthy of preservation, despite the time, physical, and financial costs they entail, even when walking, due to sweat, dust, or rain, soiling the clothing, and can hinder participation in a ceremony (Cooke et al., 2022). For this man, a 60-year-old self-employed trader, these trips are definitely part of his mandatory trips: 'The first [the most important] is going to work, followed by attending funeral ceremonies'. Visits to the extended family often exceed the limits of the 15-Minute City, except in its broader version of the 30-minute walking and PT city. Participation in ceremonies, though less frequent, usually requires longer trips, with only one-third of such trips falling within the 30-minute walking and PT city. As a 25-year-old skilled worker emphasised: 'Trips for courtesy visits and ceremonies are made on a regional level, they are not limited to the neighbourhood'. Similarly, job seeking requires long travels, making these trips even less compatible with the 15- or 30-minute city than commutes for work or school.

Even if they are not frequent, these activities are among the most important out-of-home activities and require trips that are less compatible with the limits of the 15-Minute City compared to other routine activities.

4 The 15-Minute City for Whom?

Work and education-related activities often impose spatial constraints that necessitate long commutes, making those not in employment or education the least mobile group. They make the fewest trips daily, and these trips are the closest to home (Table 4). This is the category for which the trips are most compatible with the principles of the 15-Minute City.

The population in education has a high commute rate and aligns with the 15-Minute City concept (Table 4). However, significant differences based on age and level of education exist within this category, as training institutions become increasingly scarce and geographically dispersed as students progress to a higher level of education. Primary and middle school children usually travel within the 15-minute walking city. A 70-year-old retired accountant explained the situation for children in school in her neighbourhood located in the outer department of Rufisque: 'If you live in the neighbourhood, it's not necessary [to take PT] unless you're coming from other neighbourhoods'.

In contrast, high school students travel further, with nearly three out of four of their trips completed within the 30-minute limit. The situation is even more pronounced for university students, who need to travel even further and sometimes use multiple means of transport to reach their destination: 45% of the trips made by university students exceed the 30-minute limit. A 25-year-old student from the outer suburbs of Keur Massar describes her complex commute to school in Dakar: 'Yesterday, I went to school. As usual, I took a bus to the station, then a *Ndiaga Ndiaye* [bus] to Patte d'Oie [a main junction], where I waited for [AFTU] bus 44'.

A similar situation applies to working people (Table 4). They represent half of Dakar's metropolitan population aged 11 and over in 2015. However, not all working people face long commutes: 16% work from home and 39% walk to their workplace. Nevertheless, employment imposes a significant spatial constraint on

Table 4 Frequency of trips (in number) and share of trips compatible with different definitions of the 15-Minute City, for specific trip purposes (%)

	Mean number of trips	(i) 15-min walking trips	(ii) 15-min walking and PT trips	(iii) 30-min walking trips	(iv) 30-min walking and PT trips
Population in education	4.1	56	59	68	77
Primary school	*4.4*	*71*	*73*	*84*	*87*
Middle school	*4.2*	*62*	*65*	*77*	*84*
High school	*3.9*	*48*	*52*	*61*	*74*
University	*3.5*	*33*	*39*	*38*	*55*
Population in work	3.5	45	49	54	64
Unemployed population	2.4	64	69	72	84
Total population	3.3	50	54	58	70

day-to-day mobility. As shown in Table 4, working people make 45% of their trips within the 15-minute walking city. This is 11% lower than that of students, regardless of their age, and 19% lower than that for unemployed individuals. The difference in mobility persists among these three groups when considering the 30-minute walking and PT city. This indicates that, in general, working people make more trips outside their neighbourhood compared to other groups.

Table 5 shows that gender-related differences also exist. About 55% of trips made by women align with the 15-minute walking city, compared to 45% of trips made by men. In African societies, which are largely traditional, women are typically responsible for household activities, such as going to the market to buy groceries for the day. A 60-old-year man reported: 'Women take a long walk to go to the market [...], or if you have the money, you can take a Clando every day to go to the market'. This also explains why women, if they work, tend to work closer to home on average than men. Overall, the mobility of women is more compatible with the 15-Minute City, especially in its broadest definition, compared to men.

Table 5 also presents differences in terms of position in the household. African households are often large, with an average of 6.6 persons. In the Dakar region, household heads are typically men (69%), most of them have a job (69%), and when a household owns a private vehicle, they are usually its primary users (Diaz Olvera et al., 2020). Household responsibilities often require mobility patterns that are less centred around the home and involve greater movement across the city. As shown in Table 5, only 37% of household heads have a daily mobility pattern compatible with the 15-minute walking city, and 56% align with the broadest definition of the 30-minute walking and PT city. The remaining 44% of their trips use private motorised modes, or on foot or by PT for trips exceeding 30 minutes. The situation differs for other household members, who, regardless of their position, make more than half of their daily trips within the 15-minute walking city.

When looking at incomes (Table 6), it is evident that, similar to the unemployed, the poorest individuals tend to stay close to their homes, and their daily mobility patterns align with the 15-Minute City definition. In contrast, the wealthiest households are the furthest from the 15-minute target. The income effect is progressive

Table 5 Share of trips compatible with the different definitions of the 15-Minute City, depending on gender and position in the household (%)

	(i) 15-min walking trips	(ii) 15-min walking and PT trips	(iii) 30-min walking trips	(iv) 30-min walking and PT trips
Gender				
Male	45	49	54	64
Female	55	60	63	77
Household position				
Household head	37	42	42	56
Partner/spouse	56	63	63	79
Child	51	55	61	72
Other member	55	59	63	75
Total population	50	54	58	70

Table 6 Share of trips compatible with different definitions of the 15-Minute City, depending on the per capita household income (%)

	(i) 15-min walking trips	(ii) 15-min walking and PT trips	(iii) 30-min walking trips	(iv) 30-min walking and PT trips
1st quintile*	58	61	69	78
2nd quintile	54	58	63	73
3rd quintile	49	54	58	71
4th quintile	45	51	52	67
5th quintile	39	44	44	58
Total population	50	54	58	70

* first quintile: the 20% of households with the lowest per capita income, and so on

Table 7 Share of trips compatible with different definitions of the 15-Minute City, depending on the department of residence (%)

	(i) 15-min walking trips	(ii) 15-min walking and PT trips	(iii) 30-min walking trips	(iv) 30-min walking and PT trips
Dakar	44	50	50	66
Guédiawaye	47	53	54	68
Pikine	52	56	61	71
Keur Massar	54	57	65	74
Rufisque	58	62	69	77
Total population	50	54	58	70

and consistent across each variant of the 15-Minute City. The differences, between the first and fifth quintiles, are more pronounced for the 15-minute and 30-minute walking cities compared to those that also include PT. Many poor people do not always have enough money to take PT. A 50-year-old seller responded that he mostly walks, as he lived far from PT and had a minimal income: 'I am far from public transport, which makes my travel complicated, and finding money every day is not easy'. As household income increases, trips outside the neighbourhood become more frequent, along with better access to public and private motorised transport modes.

As Table 7 underlines, people living in the outlying departments of Pikine, Keur Massar, and Rufisque have mobility patterns that are more compatible with the 15-Minute city than those of residents in the Dakar department and its nearby suburbs of Guédiawaye. Differences between the city centre and its outer suburbs are particularly marked for the 15-minute walking city, and these differences diminish when PT trips are included. In Dakar, and to a lesser extent in Guédiawaye, using PT is easier due to a combination of factors: the transport supply is better, it is more affordable because of the higher incomes of their residents, and urban resources are more accessible. However, differences between the centre and the outskirts are also linked to the lower proportion of working people among the residents of Pikine, Keur Massar, and Rufisque than those in Dakar or Guédiawaye.

Finally, as it is shown in the previous tables, women, the unemployed, the poor, and those living in the outer peripheries present mobility patterns that are the most compatible with the 15-minute walking city. In the Dakar region, one can consider that the 15-Minute City is a de facto reality for many residents, whether it is their choice or not.

5 Discussion and Conclusion

Based on analyses of the Dakar household travel survey, completed through interviews conducted with Dakar residents, this study shows that the 15-Minute City is a de facto reality for many residents. A significant proportion of daily trips made in the Dakar region align with the principles of the 15-Minute City, especially trips related to ordinary and social life, and, to a lesser extent, to working life. Informal businesses and local development ensure the presence of local amenities, such as shops, markets, and mosques. Guzman et al. (2024) presented similar findings based on an accessibility index for Bogota in Colombia.

The 15-Minute City by default, even in its broadest definition of the 30-minute city, including PT, does not fulfil all the aspirations of its inhabitants. For better job opportunities, universities, or specific services such as healthcare or administrative tasks, people need to travel longer distances. Similarly, in order to maintain social networks and ensure social protection, people visit extended family and attend ceremonies, sometimes far from home.

Inequalities regarding access to essential services are pronounced in the Dakar region as in other developing contexts, whether in terms of income (Guzman et al., 2021) or gender (Nieto-Combariza et al., 2024). Most of the daily trips that are made by women, unemployed, and poor people occur in the vicinity of their home, whereas men, employees, university students, and members of affluent households largely travel beyond the neighbourhood to reach their destinations.

Dakar's metropolitan development is characterised by a concentration of formal jobs and higher-tier services in the city centre, alongside urban sprawl in the peripheries, marked by informal settlements and infrastructure deficits. This organisation is characteristic of cities in West Africa and more generally of the Global South. It can lead to urban fragmentation with, on the one hand, people who stay in the neighbourhood, and on the other hand, highly mobile people who make long, tiring, and costly journeys from the periphery (Andreasen et al., 2024).

Addressing the issue of social and territorial equity in access to essential services and employment emerges as central to discussions about the 15-Minute City. For more than a decade, the Senegalese government has made significant efforts to develop a metropolitan mass transit network and create new employment hubs on the peripheries. The challenge lies in integrating metropolitan development with the promotion of a desirable neighbourhood-scale lifestyle as outlined in the principles of the 15-Minute City (Moreno et al., 2021) and not only as a default choice.

To achieve this, several recommendations in terms of urban development and transport planning can be considered. A primary focus is on endowing neighbourhoods with collective services. Neighbourhoods, especially in older peripheral areas of Dakar, appear relatively well provided with elementary schools, mosques, and small clinics. However, intermediate urban functions, such as middle and high schools or hospitals, should be further developed to reduce the need for long-distance travel.

A second key issue relates to mobility policies, with a particular emphasis on recognising walking as a transport mode per se in policies and planning documents. This is especially important given that walking remains the primary mode of transport in the Dakar region. Therefore, there is an urgent need to improve walking infrastructure (Guzman et al., 2024; Oviedo et al., 2021) and to redefine how public authorities value this fundamental mode (Benton et al., 2023). Another recommendation could be to promote bicycle use, which requires public efforts to change perceptions among both the population and policymakers, invest in visible and dedicated infrastructure, and implement a comprehensive bicycle policy. These are the first steps towards rehabilitating this mode of transport (Guzman et al., 2024) and increasing the proportion of trips within a 15-minute distance. As for walking, this is a complex and long-term effort that requires strong public commitment.

Besides active modes, PT plays a crucial role within the 30-minute limit. Over the past two decades, the Senegalese government has intensified its efforts to improve Dakar's PT system, particularly by developing mass transit. However, while mass transit serves the metropolitan region, it does not effectively cover the needs of a 15-Minute City, contrary to informal transport that provides access to essential services (Nieto-Combariza et al., 2024). Public policies should recognise the importance of these modes, especially shared taxis, serving medium-distance routes in the peripheries (Lesteven et al., 2022), and better integrate them into the metropolitan transport system.

Lastly, and most importantly, designing a metropolis accessible to the entire population is crucial, as it enables everyone to make occasional metropolitan-scale trips to meet their social and economic needs. This requires managing car use, linking mass transit with local shared transport options, and regulating PT fares to ensure they remain affordable for the population.

To conclude, this research provides an opportunity to explore the relevance of the 15-Minute City concept in West Africa. By focusing on Dakar, it fills a critical research gap, as a review found no previous studies based in Africa. Based on data about access to daily activities on foot and by PT from a household travel survey, it provides empirical evidence of how residents actually move and access services, rather than just what could be accessed. It thus demonstrates that the 15-Minute City is a de facto reality in the African context, and often a default choice, notably due to limited household ownership of private cars and motorcycles and the important role of informal PT modes. To reverse this paradigm, several recommendations in urban planning and mobility are suggested, focusing on developing local amenities, improving walking infrastructure, considering cycling as a relevant mode, and integrating informal PT modes, while making the city accessible to the entire population.

The research presents some limitations. The available quantitative data do not include the latest urban and mobility transformations. The methodology focuses on questions of daily mobility and access without incorporating an accessibility index. Furthermore, digitalisation—the fourth pillar of the 15-Minute City, alongside proximity, density, and diversity—is not yet included in the analysis.

Future research perspectives could include making a new household travel survey to examine urban changes over the past decade to understand how the proportion of 15- and 30-minute trips on foot or by PT has evolved. The new survey would also assess the impact of digital technology on daily mobility practices, whether through the use of smartphones or the development of new mobility services. It would also be valuable to cross-reference the findings in terms of access to essential services with accessibility analyses conducted in other contexts of the Global South, such as Ghana (Acheampong & Asabere, 2022) or Colombia (Guzman et al., 2024; Nieto-Combariza et al., 2024).

Finally, applying the methodology developed in this chapter, based on household travel survey data and interviews, to other metropolitan contexts, whether in the Global South or North, would be valuable. This approach would allow for an exploration of urban equity issues in Western cities, comparing the 15-Minute City model, designed for affluent urban residents in city centres, to the daily reality of poorer residents, who often have no choice but to remain in their peripheral neighbourhoods.

References

Acheampong, R. A., & Asabere, S. B. (2022). Urban expansion and differential accessibility by car and public transport in the greater Kumasi city-region, Ghana—A geospatial modelling approach. *Journal of Transport Geography, 98*, 103257.

Allam, Z., Bibri, S. E., Chabaud, D., et al. (2022). The theoretical, practical, and technological foundations of the 15-minute city model: Proximity and its environmental, social and economic benefits for sustainability. *Energies, 15*(16), 6042.

Andreasen, M. H., Agergaard, J., Oteng-Ababio, M., et al. (2024). High mobility lifestyles: Unpacking travel behavior in Accra's rapidly expanding periphery. *Cities, 155*, 105471.

ANSD. (2013). *Deuxième enquête de suivi de la pauvreté au Sénégal (EPSII-2011)*. Rapport définitif. https://www.ansd.sn/sites/default/files/2022-11/Rapport_ESPS-2011.pdf

ANSD. (2025). *Situation économique et sociale de la région de Dakar 2022–2023*. https://www.ansd.sn/sites/default/files/2025-05/SES-Dakar_2022-2023.pdf

Benton, J., Jennings, G., Walker, J., et al. (2023). "Walking is our asset": How to retain walking as a valued mode of transport in African cities. *Cities, 137*, 104297.

Bertrand, M. (2010). De l'accès au logement à la relation domicile-travail: enjeux sociaux et spatiaux des mobilités dans la région du grand Accra (Ghana). *Revue Tiers Monde, 1*, 87–106.

Boyer, F., Gouëset, V., & Delaunay, D. (2016). Les mobilités quotidiennes, un révélateur des inégalités sociospatiales à Ouagadougou. *Autrepart, 80*, 47–68.

Calafiore, A., Dunning, R., Nurse, A., et al. (2022). The 20-minute city: An equity analysis of Liverpool City region. *Transportation Research Part D Transport and Environment, 102*, 103111.

Capasso Da Silva, D., King, D. A., & Lemar, S. (2020). Accessibility in practice: 20-minute city as a sustainability planning goal. *Sustainability, 12*(1), 129.

Carmichael, F., Darko, C. K., Daley, P., et al. (2024). Time poverty and gender in urban sub-Saharan Africa: Long working days and long commutes in Ghana's Greater Accra metropolitan area. *Journal of International Development, 36*(1), 343–364.

Cooke, S., Ryseck, B., Siame, G., et al. (2022). Proximity is not access: A capabilities approach to understanding non-motorized transport vulnerability in African cities. *Frontiers in Sustainable Cities, 4*, 811049.

Diaz Olvera, L., Plat, D., & Pochet, P. (2013). The puzzle of mobility and access to the city in sub-Saharan Africa. *Journal of Transport Geography, 32*, 56–64.

Diaz Olvera, L., Plat, D., & Pochet, P. (2020). Access to the car in the cities of sub-Saharan Africa: Practices and users in Dakar. *Flux, 1*(119–120), 73–89.

Diaz Olvera, L., Plat, D., & Pochet, P. (2024). Changes in daily mobility and new public transport services supply in Dakar (2000–2015). *Case Studies in Transport Policy, 16*, 101214.

Dumedah, G., Iddrisu, S., Kpevu, G. A., et al. (2024). Do informal urban communities in Ghana qualify as a 15-Minute City based on access to socioeconomic activities? *African Geographical Review, 44*(1), 70–86.

Gough, K., Esson, J., Andreasen, M. H., et al. (2015). City dynamics: Mobility and livelihoods of urban residents, FP7 project RurbanAfrica. *Deliverable, 3*, 2.

Guzman, L. A., Arellana, J., Oviedo, D., et al. (2021). COVID-19, activity and mobility patterns in Bogotá. Are we ready for a '15-minute city'? *Travel Behaviour and Society, 24*, 245–256.

Guzman, L. A., Oviedo, D., & Cantillo-Garcia, V. A. (2024). Is proximity enough? A critical analysis of a 15-minute city considering individual perceptions. *Cities, 148*, 104882.

Lesteven, G. (2023). Getting around the city: Overview of urban mobility around the world. In G. Lesteven (Ed.), *Urban mobility Systems in the World* (pp. 3–22). ISTE-Wiley.

Lesteven, G., Cissokho, D., Pochet, P., et al. (2022). Daily mobility in urban peripheries. The role of clandestine taxis in Dakar, Senegal. *Sustainability, 14*(11), 6769.

Lesteven, G., Diongue, M., Pochet, P., et al. (2023). S'adapter face aux difficultés de mobilité ? Eclairages d'une métropole africaine, Dakar. *Espace populations sociétés, 2*, eps.13899.

Levinson, D. (2020). *The 30-Minute City: Designing for access*. Network Design Lab.

Lindell, I., & Utas, M. (2012). Networked city life in Africa: Introduction. *Urban Forum, 23*, 409–414.

Liu, D., Kwan, M. P., & Wang, J. (2024). Developing the 15-Minute City: A comprehensive assessment of the status in Hong Kong. *Travel Behaviour and Society, 34*, 100666.

Melbye, D. C., Møller-Jensen, L., Andreasen, M. H., et al. (2015). Accessibility, congestion and travel delays in Dar Es Salaam—A time-distance perspective. *Habitat International, 46*, 178–186.

Mendiate, C. J., Nkurunziza, A., Soria-Lara, J. A., et al. (2022). Cycling in sub-Saharan African cities: Differences and similarities with developed world cities. *IATSS Research, 46*(3), 398–410.

Meth, P., Charlton, S., Goodfellow, T., et al. (2024). Introduction. In P. Meth, S. Charlton, T. Goodfellow, & A. Todes (Eds.), *Living the urban periphery: Infrastructure, everyday life and economic change in African City-regions* (pp. 1–45). Manchester University Press.

Moreno, M., Allam, Z., Chabaud, D., et al. (2021). Introducing the "15-Minute City": Sustainability, resilience and place identity in future post-pandemic cities. *Smart Cities, 4*(1), 93–111.

Mouratidis, K. (2024). Time to challenge the 15-minute city: Seven pitfalls for sustainability, equity, livability, and spatial analysis. *Cities, 153*, 105274.

Moya, I. (2015). Unavowed value: Economy, comparison, and hierarchy in Dakar. *Hau Journal of Ethnographic Theory, 5*(1), 151–172.

Nieto-Combariza, M., Arellana, J., López-Fabra, M. F. et al. (2024). *The 15-minute informal settlement: Problematising the idea of hyper proximity in marginalised contexts*. Available at SSRN 4892457.

Oviedo, D., Okyere, S. A., Nieto, M., et al. (2021). Walking off the beaten path: Everyday walking environment and practices in informal settlements in Freetown. *Research in Transportation Business and Management, 40*, 100630.

Pochet, P., & Lesteven, G. (2025). The spread of motorcycles in sub-Saharan Africa: Dynamics and public issues. *Transportation Research Procedia, 82*, 3237–3250.

Porter, G., Abane, A., & Lucas, K. (2020). *User diversity and mobility practices in sub-Saharan African cities: Understanding the needs of vulnerable populations*. The Volvo Research and Educational. The state of knowledge and research.

Sakho, P. (2014). *La production de la ville au Sénégal: entre mobilités urbaines, migrations internes et internationales*, Vol. 1. Rapport de synthèse. Thèse de doctorat d'Etat sur travaux. Université Cheikh Anta Diop de Dakar.

Salon, D., & Gulyani, S. (2019). Commuting in urban Kenya: Unpacking travel demand in large and small Kenyan cities. *Sustainability, 11*(14), 3823.

Sitrass-Curem. (2016). *Enquête sur la mobilité, le transport et l'accès aux services urbains dans l'agglomération de Dakar*. Rapport d'Analyse final, pour le compte du CETUD. CETUD.

Stokenberga, A., Saïsset, E., Kerzhner, T., et al. (2025). Connecting through public transport: Accessibility to health and education in major African cities. *Area Development Policy, 10*, 210–227.

Teixeira, J. F., Silva, C., Seisenberger, S., et al. (2024). Classifying 15-minute cities: A review of worldwide practices. *Transportation Research Part A: Policy and Practice, 189*, 104234.

Venter, C. (2011). Transport expenditure and affordability: The cost of being mobile. *Development Southern Africa, 28*(1), 121–140.

Weng, M., Ding, N., Li, J., et al. (2019). The 15-minute walkable neighborhoods: Measurement, social inequalities and implications for building healthy communities in urban China. *Journal of Transport and Health, 13*, 259–273.

Land Use and Urban Form

The City of Proximity as an Inspiration for New Concepts of Development of Kraków and Its Agglomeration

Aleksander Noworól, Piotr Kopyciński, Jeremiasz Salamon, and Paweł Hałat

1 Introduction

The concept of a 15-minute city, or a city of short distances, has become one of the main orientations in the design and modernisation of cities and metropolitan areas. The situation is no different in Polish cities, where the concept is increasingly incorporated into urban development strategies. This process has been reinforced since 2018 through the ongoing reform of Poland's territorial development management system, which encompasses all planning levels, including local, metropolitan, regional, and national. In this respect, legislative proposals presuppose the integration of socio-economic planning with spatial planning, creating an opportunity to include cities with short distances in development concepts.

Kraków is an example of an attempt to adapt and implement this concept. The conclusions of the report entitled 'Can Kraków be a 15-minute city' (Noworól et al., 2021), which was also published in an international publication (Noworól et al., 2022), were the subject of a vibrant discussion on the future model of the city's development and were reflected in the 2022–2023 draft's 'Update of the Kraków Development Strategy. This is where I want to live. Kraków 2030/2050' (from now on, the draft Update of Strategy) and the proposals for the directions of spatial development of the Kraków agglomeration (Noworól et al., 2024). The process of drafting the most important document from the city's development perspective, which the strategy undoubtedly is, was formatted in a way that ensured the actual involvement of the residents and entities representing their interests, which was in line with the postulate of bottom-up formulation of public policies inherent in the concept of a city of short distances (Pozoukidou & Chatziyiannaki, 2021).

A. Noworól (✉) · P. Kopyciński · J. Salamon · P. Hałat
Krakow University of Economics, Krakow, Poland
e-mail: noworola@uek.krakow.pl; salamonj@uek.krakow.pl

L. Bouzouina et al. (eds.), *15-Minute Cities*, Footprints of Regional Science,
https://doi.org/10.1007/978-3-032-20840-8_11

The documents mentioned above propose a way of understanding the concept of a city with short distances on a large city's scale, along with its surroundings. A significant development challenge in Kraków is the touristification of the historic (relatively small) city centre, which fits perfectly into the model of a short-distance city but is losing its residents because of gentrification and the difficulties associated with tourism development (Szromek et al., 2019; Kowalczyk-Anioł, 2019, pp. 14–20). This development is accompanied by the rapid growth of new settlements on the city's periphery and, in the last two decades, also by vigorous suburbanisation in the suburban zone (Harańczyk, 2015, pp. 85–101; Sykała et al., 2023, pp. 14–27). The recent development lacks essential services, and only a few peripheral neighbourhoods have public spaces with a centre-forming character. Therefore, a network of neighbourhood, local, sub-metropolitan, and suburban centres has been proposed in the urban and metropolitan system to transform Kraków's highly centralised urban centre into a more polycentric system with better accessibility for the residents of the entire agglomeration.

The purpose of this chapter is, therefore, to present, using the example of Kraków and the Kraków metropolitan area, how the concept of a city of short distances may influence the strategy of the metropolitan core city as well as the spatial policy of the Kraków agglomeration. In this regard, the authors refer to analytical and consulting studies in which they were personally involved. Due to the scale of this publication, the description of the studies has been reduced to the necessary minimum, with references to publicly available and published resources. The conclusions draw out public policy challenges and suggest further research, particularly on the implications of a polycentric urban layout for developing a metropolitan community.

2 Is Kraków a 15-Minute City, and Can It Become One?

One of the main ideas behind the concept of a city of short distances is the assumption that its residents should be able to satisfy their daily life needs, as well as to fulfil some above-standard needs, within a few or several minutes' walking or cycling distance from their place of residence (Allam et al., 2024; Moreno et al., 2021; Pozoukidou & Chatziyiannaki, 2021; O'Gorman & Dillion-Robinson, 2021). An approach based on ensuring accessibility to urban services through proximity should guarantee a high quality of life on the one hand and minimise unnecessary and long journeys, especially those taken by car, on the other. An important consideration is the universality and provision of equal access to services and infrastructure—the ability to meet needs locally should, therefore, apply to the whole or at least most of the inhabited area of a city, not just selected neighbourhoods (Duany & Steuteville, 2021; Moreno et al., 2021, p. 100). These opportunities are determined by a reasonably even and equitable distribution of amenities in the urban space, especially those facilities that fulfil the daily needs of the residents.

After the Second World War, Kraków underwent intensive expansion and industrialisation under the conditions of a socialist economy. According to modernist development paradigms, this was initiated by the construction of a metallurgical plant and the 'ideal city' of Nowa Huta, located approximately 8 km from the city centre. Between 1945 and 1989, the city's area increased from 168 km^2 to 326 km^2, and the population more than tripled (to around 750,000). The post-war development of the city involved a shift from a compact building structure to large housing estates built in newly incorporated areas, often at a considerable distance from the centre. Housing estates were usually built according to designs that included public services and public transport systems, but service programmes were not consistently implemented comprehensively, and infrastructure development faced delays (Gorczyca, 2009, pp. 89–124; Włodarczyk, 2013, pp. 98–99; Ziobrowski et al., 2023, pp. 51–74; Gorzelak, 2024, pp. 42–45). Significant areas incorporated into the city had (in some instances, still have) a quasi-rural character and remained poorly communicated with the city centre.

The next period of intensive urban development, which began after Poland acceded to the EU (2004), took place under conditions of a spatial planning crisis in Poland, often leading to new settlements with a chaotic urban structure, lacking the provision of essential services and usually public transport (Izdebski, 2013, pp. 124–135; Pluta, 2016, pp. 159–168; Drozda, 2016, pp. 183–186). Recently, housing development in suburban areas has also been intensifying, with a notable feature of the settlement around Kraków being the high dispersion of development in rural areas (Harańczyk, 2015, p. 93; Różycka-Czas et al., 2021, p. 11; Węgrzynowicz, 2022, p. 22). As a consequence, Kraków, like other Polish cities, has experienced a sharp increase in the intensity of individual car use (Raport..., 2014, p. 5; Wójcik, 2022, p. 50).

The objective of the 2021 study of the accessibility (understood as geographical proximity) of services in Kraków was to assess the degree to which the city space meets the above assumptions. In particular, it was essential to identify areas that provide, within a 15-minute walk, access to a set of services that meet the needs of residents, as well as areas where local services are deficient. Based on the literature (Gil Solá & Vilhelmson, 2019; Weng et al., 2019; O'Gorman & Dillion-Robinson, 2021; Happach, 2021), six categories of needs were selected and implemented within a city of short distances: education and childcare, healthcare, culture, green spaces, recreation and sport, shopping and services, and public transport. The categories have been detailed about the specific types of services and facilities where they are provided. The facilities were divided into two groups: those implementing the minimum version of the '15-minute city' concept, which included essential local services, and the optimal version, which included above-standard services. For each category of facilities, the serviced areas were selected, delimited by a 15-min (10-min for bus and tram stops) equidistant range, corresponding to 1,200 m covered by a pedestrian moving at 1.34 m/s (80 m/min). The area providing 15-min proximity to the set of features in the minimum and optimal versions of the '15-minute city' was created as the geometric standard part of the buffer layers of the individual facilities (on the methodology of the study, see in more detail Noworól

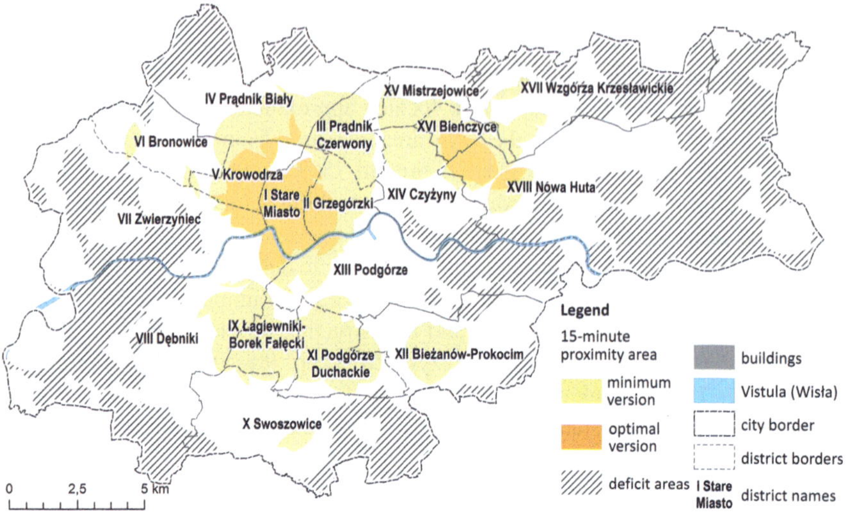

Fig. 1 Areas of Kraków located within a 15-min proximity and deficit areas. (Source: Own study)

et al., 2021; Noworól et al., 2022). Deficient areas were also delimited, which are out of reach of more than two categories of objects in the minimum version of the '15-minute city' (Fig. 1).

The area of the city within 15 minutes' proximity to essential local services (the 'minimum' version) is inhabited by approximately 66% of Krakow's residents (data included those registered for permanent or temporary residence, excluding people living without registration). The area encompasses the city centre and most of the block housing estates built in the second half of the twentieth century. Apart from this, however, a significant proportion of new housing estates are closer to the city borders. The deficit areas are primarily located in the low-populated outskirts of the city.

Only about 19% of Kraków's population inhabit the area providing complete accessibility to services. The area is mainly covered by quarters of traditional, compact buildings, a multifunctional character in the city centre, and the old part of Nowa Huta, built according to the concept of a neighbourhood unit and constituting a second (much weaker) sub-centre of Kraków. This stems from the concentration, characteristic of the Kraków metropolitan area, of most public and commercial functions in the downtown area, which constitute activity generators (Noworól et al., 2024). Areas experiencing depopulation tend to be the best supplied with services. This is particularly true of the inner city area, which is subject to intensive depopulation (the Old Town district has lost about 43% of its permanent residents in 20 years). One reason for this phenomenon is the touristification of the city centre and the replacement of the residential function by hotels or short-term leases. However, many twentieth-century settlements (e.g. Nowa Huta) are also being depopulated. Simultaneously, there is an intensive development of settlements in the outer part of the city, as well as in its immediate surroundings, often without sufficient service facilities (Fig. 2).

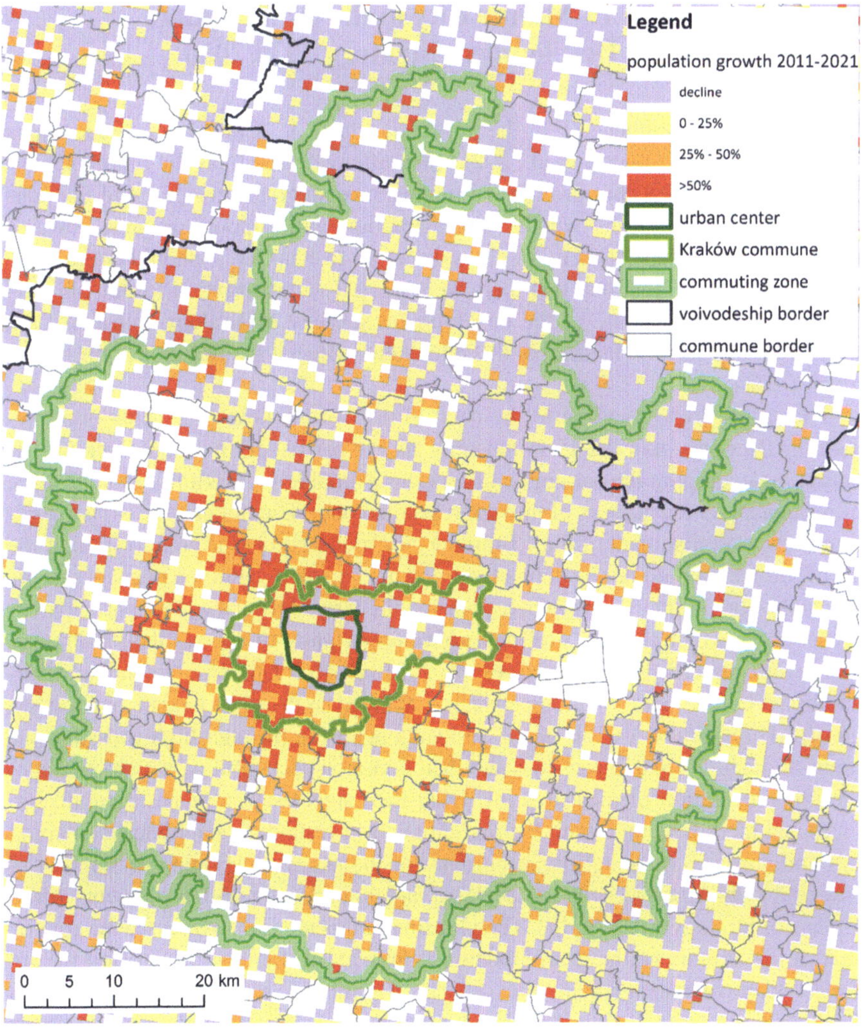

Fig. 2 Changes in population in Kraków and the metropolitan area over the last 10 years. (Source: Own study based on spatial data from the 2011 and 2023 National Population and Housing Censuses)

Considering the depopulation process and the assumption that as large a proportion of the population as possible should have access to a variety of services at their place of residence, one of the significant strategic challenges for the city is to respond to demographic changes in the urban space and to stop the population decline in the city's best-supplied neighbourhoods.

3 Concept of the 15-Minute City in the Draft Update of the Kraków Development Strategy 2030/2050

Published in November 2021, the report is entitled. 'Can Kraków become a 15-minute city?' (Noworól et al., 2021) (hereafter referred to as the report) identifies areas of the city that provide 15-minute proximity to services, and contains several recommendations whose implementation will support progress towards achieving the concept of a city of short distances in Kraków. It should come as no surprise, therefore, that many of them were not only an inspiration for the ongoing preparation of the update of the development strategy of Kraków, but above all, these were included in the draft of this document, which was published at the beginning of 2024.

The report drew attention to the need to involve Kraków's residents in implementing the concept of the 15-minute city and, in particular, to ensure their participation in developing the catalogue of services available in proximity to their place of residence. As the authors of the report pointed out: 'The importance of participation in the establishment of the 15-minute city suggests that the composition of the set of services should not be defined top-down, i.e., solely based on decisions of the city authorities supported by experts, but should derived from a widely participatory procedure in which the city's residents have opportunity to express their opinion in this regard effectively' (Noworól et al., 2021, p. 46). This recommendation was addressed in the operational objective 1.2 included in the draft Update of Strategy, entitled High quality of life of the residents achieved through high (possibly the highest) quality of public services and their common accessibility, which envisages the implementation of measures involving the categorisation of basic public services and identification—with the participation of residents, neighbourhood councils and non-governmental organisations—of the assumed 15-min accessibility of essential public services at the scale of neighbourhoods and settlements (draft Update of Strategy, 2024, p. 17). The phrasing of this measure expresses the conviction of the authors of the draft Update of Strategy of the need to develop the so-called adaptive management, the subject of which is to be the community of Kraków's residents acting for the common good (Draft Update of Strategy, 2024, p. 10).

Although operational goal 1.2 of the draft does not specify the services whose proximity is to be increased. Based on the subsequent provisions of the draft Update of Strategy, it is possible to identify its essential catalogue. The analysis of indicators, intended to be measures of success in achieving operational goal 1.2, points out that, thanks to its implementation, the residents of Kraków are expected to gain better access to services, in particular those related to:

- Childcare (increase in the percentage of Kraków's residents with access to a nursery within 500 m from their place of residence);
- Education (increase in the percentage of Kraków's residents with access to a kindergarten within 500 m from their place of residence and increase in the

percentage of Kraków's residents with access to a primary school within 750 m from their place of residence);
- Health care (increase in the percentage of Kraków's residents with access to a primary healthcare facility within 1,000 m from their place of residence);
- Culture (increase in the percentage of Kraków's residents with access to cultural facilities within 1,000 m from their place of residence);
- Sport and recreation (increase in the percentage of Kraków's residents with access to sports facilities within 1,000 m from their place of residence);
- Public transport (increase in the percentage of Kraków's residents with access to: 1) a tram or bus stop within 400 m; 2) a railway station serving the Agglomeration Rapid Rail within 1,000 m) (Draft Update of Strategy, 2024, pp. 32–33).

Furthermore, according to the draft Update of Strategy (2024, p. 15), the dissemination of e-government services is expected to increase accessibility to public services in Kraków.

The report indicates that 'effective implementation of the concept of the 15-minute city should rather focus on creating conditions in which satisfying most of the needs related to the daily functioning of people will be possible near their residence. In other words, the activities of city authorities should be focused on bringing services and goods to people in their place of residence, rather than creating opportunities for people to access services and goods outside their place of residence' (Noworól et al., 2021, p. 47). The draft Update of Strategy assumes the implementation of this postulate by creating a polycentric spatial structure of Kraków, thus making it a 'city of many cities'. Measures envisaged to achieve this goal include, among others:

- Restoring public spaces to residents by limiting access of cars to the city centre;
- Creating and/or enhancing public spaces in peripheral neighbourhoods in conjunction with material heritage sites and green areas. Acquisition of new areas for public spaces;
- Preserving the specificity of neighbourhoods and settlements based on tangible and intangible heritage and functional linkages. Counteracting the over-commercialisation of public spaces; stimulating the creation of attractive meeting, leisure, and recreation spaces in these areas, especially in squares, which are not only accessible to users of commercial facilities;
- Designing green spaces in the surroundings of urban developments and within housing estates and city centres ('pocket parks', courtyards, squares);
- Keeping excessive urbanisation and the effect of exurbanisation under control (draft Update of Strategy, 2024, pp. 27–28).

The foundation of a city of short distances in Kraków is a network of so-called local centres and neighbourhood centres (Fig. 4). The former is to constitute an essential feature of the management system for proximity to public services in the city and a primary factor shaping local identity regarding the tangible and intangible heritage of Kraków's neighbourhoods and settlements (Draft Update of Strategy, 2024, p. 45). The 27 neighbourhood centres, created based on public spaces dispersed within the city, will constitute the essential service centres on the scale of a neighbourhood or a group of settlements and ensure a high standard of public spaces, being at the same time a place of activity for the residents and a place to spend their leisure time. They are also to be places of particular symbolic and aesthetic importance, supporting the civic activity of the residents. Neighbourhood centres should be characterised by excellent accessibility by public transport (especially by rail), as well as by walking and cycling. The draft Update of Strategy only schematically identifies the scope and location of neighbourhood centres, indicating that the detailed decisions in this respect should be the subject of public consultations and workshops (Draft Update of Strategy, 2024, pp. 81–82).

Proximity is also influenced by the minimum number of spatial barriers, making it more challenging to walk and increasing walking time (Noworól et al., 2021, p. 48), especially for city residents with special needs (e.g. seniors, persons with disabilities). Therefore, the draft Update of Strategy (2024, pp. 16, 17, 25, 28) formulates measures to eliminate barriers and ensuring balanced access to public services for residents from groups at risk of being excluded; adaptation of infrastructure to ensure proper parameters for vulnerable traffic participants, including people with special needs; the application of the so-called ‘universal design’, i.e. barrier-free access to public services or developing and adapting public services, technical infrastructure and information systems to the needs of the senior citizen and people with disabilities.

The authors of the report considered balancing housing policy, providing housing with different living areas and affordability across all city neighbourhoods, as essential aspects of the 15-minute city concept (Noworól et al., 2021, p. 48). Therefore, it should not come as a surprise that the draft of Kraków’s strategic development plan formulates a set of measures aimed at ensuring universal access to housing in the city. These measures notably include: improving access to housing for those with incomes which makes it impossible to purchase or rent a flat on a commercial basis; striving to ensure the availability of premises and buildings in the municipal resource for people with special needs; expanding the real estate resource for municipal housing; increasing the possibility of satisfying the housing needs of people threatened with social exclusion given limited income or complicated personal situation, or implementing support programmes for housing communities and cooperatives (Draft Update of Strategy, 2024, p. 19). The draft Update of Strategy (2024, pp. 82–83) assumes the territorialisation of activities in this area. For this purpose, the Area of Strategic Intervention (ASI) for low- and moderate-rent housing, including communal housing, was established. This initiative responds to the growing demand for housing available to people and families with incomes that do not allow them to purchase or rent a flat on a commercial basis. The housing complexes created in these areas should, in particular, be characterised by a high-quality urban composition and public spaces; provide access to essential public services, green areas and public transport; and be shaped in a way that enables the integration of their residents, diversifies the social structure and prevents the creation of homogeneous, isolated communities.

Several recommendations in the report refer to urban mobility. According to the authors, municipal authorities striving to implement the concept of the 15-minute city should, among other things, engage in creating conditions encouraging residents to make the broadest possible use of various forms of micro-mobility, take measures to develop rail transport (trams and railways) and, when infeasible, ensure the smooth flow of bus traffic and strive for closer integration of the components of the urban transport system (Noworól et al., 2021, pp. 51–53). This category of postulates has also been addressed within the framework of the draft Update of Strategy (2024, pp. 24–25), which envisages the development of sustainable, integrated, reliable and environmentally friendly mobility, to be achieved notably through:

- Further territorial and qualitative development of public transport systems;
- Integration of transport systems with priority given to rail transport;
- Further favouring public transport in urban traffic and ensuring better accessibility and adequate frequency of public transport services;
- Completing the implementation of adopted solutions for road transport with emphasis on the integration of systems through multimodal nodes;
- Developing environmentally neutral micro-mobility, cycle paths, rented bicycles and scooters;
- Supporting electromobility at the scale of the city and Kraków Functional Area;
- Intensifying the use of the Agglomeration Rapid Rail.

Under the assumptions adopted in the draft Update of Strategy, the implementation of the measures mentioned above is expected to contribute to the following division of transport tasks: pedestrian traffic—30%; cycling traffic—10%; public transport—40% and individual transport—20% (Draft Update of Strategy, 2024, p. 58).

4 Assumptions of the 'City of Short Distances' in the Spatial Policy Model Frameworks of the Kraków Metropolitan Area

The authors believe that the 15-minute city—a city of short distances—can be interpreted at various spatial scales. It applies particularly to the metropolitan scale, to which the study published in 2024 under the title 'Spatial policy in the Kraków agglomeration in the light of metropolisation and suburbanisation processes' (Noworól et al., 2024) was dedicated.

Building on extensive studies on delimitation and management of metropolitan areas (Allmendinger & Haughton, 2009; d'Albergo & Lefèvre, 2018; Kudłacz & Markowski, 2017; Mikuła, 2019), the authors aimed to determine the premises and directions of spatial policy development in the widely understood Kraków agglomeration (Kraków metropolitan area). The study included analyses of planning documents created in local and sub-regional systems, as well as monitoring of actual processes taking place in the agglomeration. Research materials, including descriptions of demographic processes, existing and planned transportation systems, and insights into land-use issues, were collected. Both actual investment flows and spatial development plans, implemented at regional, sub-regional, and local levels, were included. Two categories of conclusions were identified—the first concerns the confirmation of the variable territorial scale of the Kraków metropolitan area. The second calls for the formulation of spatial policy directions along with conclusions for managers of territorial development processes. In each case, a study was conducted on the concentration of services, residential development and the availability of local service centres, taking into account the dominant role of Kraków.

The study confirmed the dynamic nature of the delimitation of the Kraków metropolitan area. The spatial record of processes in the proper sphere concerning demographics and population migration, transport corridors and flows in the broader Kraków area, and actual investment processes, especially those related to housing development, revealed various activity locations generating traffic and construction activity. Each of the studies mentioned above indicates a slightly different scope of what is defined here as the Kraków metropolitan area. This area is a volatile spatial construct in two respects. Firstly, the metropolis is growing with the so-called 'urban sprawl' phenomenon, which refers to different types of suburbanisation processes. Secondly, for various functions and phenomena, it is possible to delimit the borders of the metropolis slightly differently. Considering the two extreme delimitation criteria, the Kraków agglomeration (as of 2023) ranges from 15 municipalities and approximately 1.14 million residents to 58 municipalities and approximately 1.68 million residents (considering the commuting zone). This is illustrated in Fig. 3.

Several conclusions were drawn regarding the rationale for spatial policy in the metropolitan area. It was noted that investment activity is concentrated along the main road transport corridors, particularly for road connections and individual transport. This means, notably, that population growth is scarcely linked to the rail network and tends to relate to roads out of Kraków (even where a rail network exists).

The study identified the following prerequisites for urbanisation processes:

- Location of the areas in relation to transport corridors within a commuting ~45-min isochrone;
- Morphology and scenic attractiveness of the areas;
- Availability of investment areas covered by local spatial development plans;
- The spatial policy of municipalities bordering Kraków, which includes encouraging settlement.

Urbanisation barriers have also been identified: in the east, the industrial belt surrounding the metallurgical plant, and in the west, the airport approach zones and restricted-use zones in the Balice airport area.

Spatial analyses lead to the following conclusions:

- At the borders of Kraków and the neighbouring municipalities, areas with residential uses are merging in a process of concentrated urbanisation;

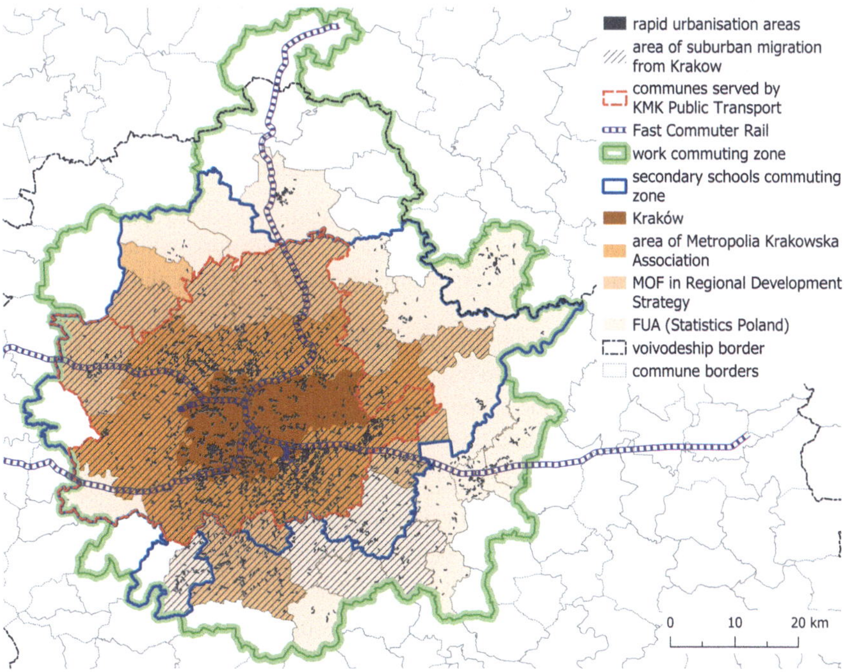

Fig. 3 Dynamic delimitation of the Kraków agglomeration. (Source: Own study)

- As a consequence, the city of Kraków is 'discreetly' enlarging in the northern and southern directions through mono-functional-residential 'cross-border' structures; these areas are functionally connected with the metropolis and operated by the transport system of Kraków, which is also a destination in terms of access to the vast majority of public, social and employment functions;
- Mono-functional 'city-bedrooms' necessitate a new approach to the formation of their spatial policy through the location and development of local metropolitan centres, which would enhance the level of service to residents in the spirit of a city of short distances, while also alleviating the burden on the metropolitan core city centre.

Considering the Polish legal framework in the sphere of integration of socio-economic planning with spatial planning, recommendations for spatial policy were formulated. It was suggested that this policy should be based on the following instruments.

Firstly, it was assumed that the lack of service centres is a weakness of the spaces surrounding Kraków, which could provide an attractive activity place for residents and visitors from the entire agglomeration area. It is the origin of the proposal to locate sub-centres of metropolitan services and suburban centres and, consequently, to deconcentrate these services in the area outside the so-called third ring road of Kraków (ring of expressways within the city). Locations of existing centres might be an example. These centres in Kraków are a contemporary 'gateway to the city', offering mainly commercial services. In developing the concept of a network of district and local centres proposed in the draft Update of Strategy (cf. text above), it was acknowledged that there is a need to extend this idea beyond the city borders, which could be based on the development of suburban centres in highly urbanised and urbanising areas.

By creating metropolitan service sub-centres, we understand the following:

- The identification of existing or potential multifunctional public spaces that can provide an attractive meeting place for residents and visitors, while also being within good (or possible) access, especially by public transport and secondarily by individual transport;
- A genuine deglomeration of public and social services of a higher standard in such a way that there is no need to travel to the centre of Kraków to access them. These are services beyond the scope of municipal tasks, such as administrative decisions, education, culture and tourism.

The deconcentration of services aims to provide them as close as possible to the place of residence, thereby reducing unnecessary trips to the city centre. Thus, public services, mainly administrative services, have a solid centralising effect. The concept is, therefore, based on a three-tier structure:

- Neighbourhood centres (described in relation to the draft Update of Strategy) within the administrative borders of the core city and suburban centres outside it, providing for the basic needs of the residents at the local level;
- Metropolitan sub-centres, located higher up in the hierarchy of services and providing services, mainly administrative services, specific to various levels and branches of administration;
- The metropolitan centre of Kraków comprises non-deconcentrated services.

The proposed polycentric arrangement of services will not only reduce the traffic load, mainly car traffic, in the centre of Kraków. However, it will also contribute to the development of the metropolitan community in social and cultural aspects.

The second key instrument in the spatial management of the Kraków agglomeration should be the concentration of urban development around public transport corridors, especially those serving public transport and rail transport, with a simultaneous increase in the degree of integration of the metropolis's public transport subsystems. Due to the metropolitan area's physiographic shape, parking facilities in the vicinity of public transport nodes and stops would be crucial in this case. The proposed new railway lines are expected to play a vital role in the spatial transformation of the agglomeration, particularly in the southern direction, which is currently underserved by rail transport. Concerning the railway corridors, we propose that the future location of further suburban centres be closer to the edge of intensive urbanisation. The instruments suggested above constitute the proposal for linking the settlement network, the main transport corridors and networks, and selected centre-forming elements of technical and social infrastructure.

The above observations are illustrated synthetically in Fig. 4.

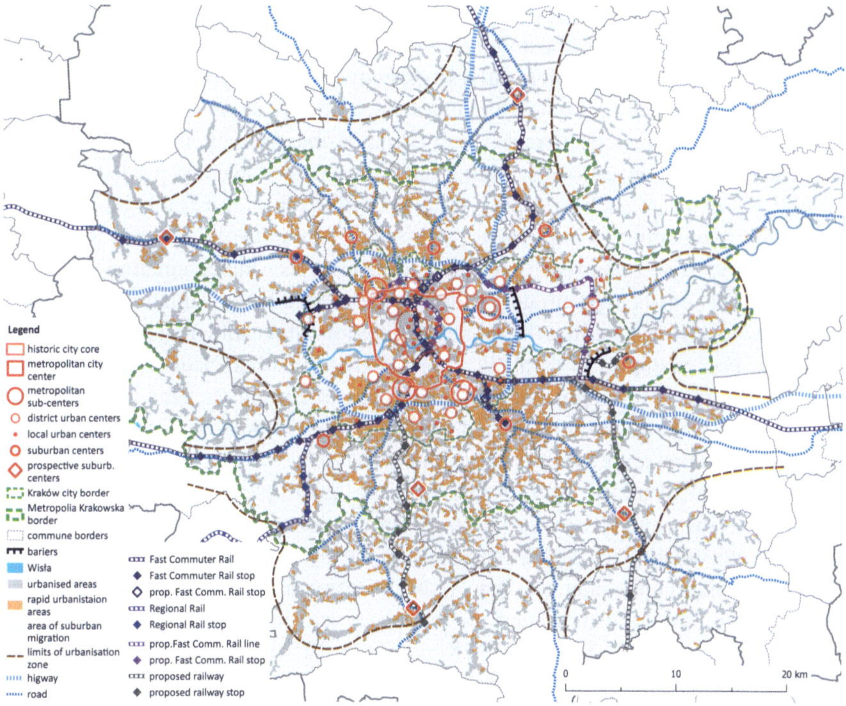

Fig. 4 Scheme of the spatial policy of the Kraków agglomeration. (Source: Own study based on Noworól et al., 2024, pp. 117–121 and draft Update of Strategy 2024, pp. 41, 65)

5 Conclusions: The Challenges for Public Policies of Regional and Local Authorities

The studies cited in this chapter, on Kraków as a 15-minute city (Noworól et al., 2021) and the Kraków agglomeration (Noworól et al., 2024), provide essential insights, both general and city-specific, which should, in turn, be reflected in the development and implementation of public policies.

First of all, the importance of implementing evidence-based policies should be emphasised. The analysis of statistical data, supplemented when necessary by mixed-methods studies, enables the identification of changes in spatial planning and the management of proximity to public services or mobility (including commuting to schools and work). Therefore, the planning of appropriate measures by local authorities (whose responsibility in Poland is the preparation of spatial development plans), as well as the coordinated action of local, regional and national authorities about appropriate transport investments, particularly in public transport (in Poland, public transport management takes place at several levels, as discussed below by way of example).

Associated with the above is a shift from short-term political thinking, limited by the term of office of the public authorities, to long-term creation of public policies. Priority plans should be implemented irrespective of political parties, and revised under new authorities (as well as the emergence of new data), but not completely abandoned. Without such an approach, it is impossible to implement, for example, long-term transport investments conditional in turn on a stable and predictable spatial policy. Thanks to such an approach, it is possible (apart from emergencies such as war or a pandemic) to implement public policy in an evolutionary (rather than revolutionary) style preferred by residents, whereby the implementation of public authority decisions is appropriately timed and preceded by consultations with residents (city-users), reconciling the interests of various stakeholders (search for compromise). For example, in Kraków, the introduction of a clean transport zone (low-emission zone) was postponed due to a lack of consensus. This initiative was initially scheduled to be implemented from mid-2024; however, according to current plans, it is now expected to occur from mid-2025.

Stakeholders of a city, who, in line with the quadruple helix concept, include public authorities, enterprises, city users, scientific units, and intermediaries, should engage in consultation to reach a consensus on the issues above (Kopyciński, 2018). The city/municipalities/metropolitan authority remains the final decision-maker. These entities should also account for the legacy and feasibility limit of proposed solutions. The concept of a city lab (Kopyciński, 2018) may be beneficial in this context. Its application may be necessary for efforts to develop alternatives to the monocentric characteristic of Kraków and its metropolitan area, which is compounded by the concentration of most public and commercial functions that generate activity in the central area of Kraków (Noworól et al., 2024). To align with the concept of the 15-minute city, Kraków and the municipalities of the Kraków agglomeration should implement the creation of a multi-centre metropolitan area, as outlined in Strategy Kraków Metropolis 2030 (2021). Thus, it is necessary to establish a network of neighbourhood centres and local centres in most of Kraków's settlements, and to develop sub-metropolitan and suburban centres in accordance with the metropolis (e.g., Niepołomice, Skawina, Wieliczka). An essential feature of the new polycentric centres, assuming their range of influence will mirror the city's metropolitan functions, will be their direct impact on residential functions and public facilities that provide these services (Draft Update of Strategy, 2024, pp. 56–57).

We emphasise the necessity to move away from fragmented measures towards integrating public policies at different levels. Referring again to the example from our study (Noworól et al., 2024), public transport in the Kraków agglomeration is currently the responsibility of the city of Kraków (buses and trams in the city and buses in the neighbouring municipalities—based on the inter-municipal agreements), the individual municipalities (which often maintain their bus lines) and the region (railways and bus feeder lines to railways). Public transport can also be organised by counties (Polish *powiat*). The public, nationwide railway company (PKP Intercity) also provides dedicated commuter services in the main rail routes of the Kraków agglomeration. Meanwhile, the lack of full integration of the public transport system is a problem in Kraków, in the agglomeration and in the region (Polish *województwo*). For example, in Wieliczka, a neighbouring city to Kraków, networks such as municipal bus lines, agglomeration bus lines (operated by agreement with the city of Kraków) and feeder bus lines (operated by agreement with the Malopolska region) are provided. Although there are positive examples of integration (such as in Wieliczka and Skawina, and the region—the Metropolitan Ticket and the Malopolska Integrated Ticket), full integration of public transport (integrated transfer systems, tariff and timetable integration), including bus and train carriers (and in Kraków also trams), is necessary for the Kraków agglomeration. The case of spatial planning is similar—although the preparation of detailed plans (and general plans by the mid of 2026 at the latest) is the prerogative of municipalities, however, given the concentration of development around public transport corridors, especially public transport and rail transport, as postulated in our study (Noworól et al., 2024), there is a need to coordinate this type of activity between these local government units (mainly at the municipal level).

Coordination of the measures mentioned above may take place through agreements between local government units (municipalities, counties, and regions) or through the statutory establishment (following the example of the Metropolis GZM (Polish *Górnośląsko-Zagłębiowska Metropolia*) already existing in Poland) of a metropolitan association grouped around Kraków. The tasks of these entities should focus primarily on the issues of the organisation (but also coordination with other local government units) of public transport. However, this entity may also consider delivering other public services over time. The flexible borders of such an association (i.e. the possibility of joining new members (municipalities) based on statistical data and other research, in accordance with evidence-based policy, and its openness to cooperation with non-member municipalities) are necessary here.

The findings presented encourage further research to be undertaken at regional and local scales. Firstly, these studies should outline the spatial arrangements of metropolitan, neighbourhood and local centres, as well as how they should be defined based on available real estate resources and existing urban and architectural contexts. Another fundamental area of research is the institutional infrastructure of the entire agglomeration and its municipalities within the most extensive metropolitan area, in the context of their ability to implement the sustainable concept of a 'short-distance metropolis'. Informal institutional linkages, as exemplified by the influence of public participation processes on spatial and transportation solutions, will play a significant role in achieving this. It is another key research area worth exploring in more depth because it is precisely the optimisation of transport systems and subsystems that promote public mobility. The conciliation and coalition capacities of local suburban and rural communities need to be analysed, as well as the capacity of transport subsystems to serve areas with extensive physiographic features.

Acknowledgements The chapter presents the results of Project no. 026/GGR/2024/POT financed from the subsidy granted to the Krakow University of Economics.

References

Allam, Z., Khavarian-Garmsir, A. R., Lassaube, U., Chabaud, D., & Moreno, C. (2024). Mapping the implementation practices of the 15-Minute City. *Smart Cities, 7*(4), 2094–2109. https://doi.org/10.3390/smartcities7040083

Allmendinger, P., & Haughton, G. (2009). Soft spaces, fuzzy boundaries, and metagovernance: The new spatial planning in the Thames gateway. *Environment and Planning A: Economy and Space, 41*, 617–633. https://doi.org/10.1068/a40208

d'Albergo, E., & Lefèvre, C. (2018). Constructing metropolitan scales: Economic, political and discursive determinants. *Territory, Politics, Governance, 6*(2), 147–158. https://doi.org/10.1080/21622671.2018.1459203

Draft Update of the Kraków Development Strategy. This is where I want to live. Kraków 2030/2050. (2024). Zarządzenie nr 519/2024 Prezydenta Miasta Krakowa z dnia 19.02.2024 r. w sprawie przyjęcia i przekazania pod obrady Rady Miasta Krakowa projektu uchwały Rady Miasta Krakowa w sprawie przyjęcia dokumentu „Strategia Rozwoju Krakowa. Tu chcę żyć. Kraków 2030/2050". Online access: https://www.bip.krakow.pl/zarzadzenie/2024/519/w_sprawie_przyjecia_i_przekazania_pod_obrady_Rady_Miasta_Krakowa_projektu_uchwalyRady_Miasta_Krakowa_w_sprawie_przyjecia_dokumentu_%E2%80%9EStrategia_Rozwoju_Krakow.html

Drozda, Ł. (2016). Transformacja w przestrzeni. Wpływ modelu deweloperskiego na urbanizację Polski po 1989 roku. *Kwartalnik Kolegium Ekonomiczno-Społecznego "Studia i Prace", 27*(3), 177–192.

Duany, A., & Steuteville, R. (2021). Defining the 15-Minute City—Urban Projectization. *Public Square. CNU Journal*. Online access: http://www.cnu.org/publicsquare/2021/02/08/defining-15-minute-city.

Gil Solá, A., & Vilhelmson, B. (2019). Negotiating proximity in sustainable urban planning: A Swedish case. *Sustainability, 11*(1), 1–18. https://doi.org/10.3390/su11010031

Gorczyca, K. (2009). Wielkie osiedla mieszkaniowe—diagnoza stanu obecnego, podejmowane działania rewitalizacyjne. In W. Jarczewski (Ed.), *Przestrzenne aspekty rewitalizacji: śródmieścia, blokowiska, tereny poprzemysłowe, pokolejowe i powojskowe* (pp. 89–123). Instytut Rozwoju Miast.

Gorzelak, G. (2024). "Diagnoza stanu gospodarki przestrzennej Polski 1981–1983"—wnioski i inspiracje. *Studia Lokalne i Regionalne, 1*(95), 40–54. https://doi.org/10.7366/1509499519503

Happach, M. (2021). Czy żyjemy lokalnie? Raporty z planowania Warszawy. *Miejska Pracownia Planowania Przestrzennego i Strategii Rozwoju Miasta*. Online access: https://architektura.um.warszawa.pl/documents/12025039/21022710/RAPORT_Z_PLANOWANIA_WARSZAWY_cz.4_Czy+żyjemy+lokalnie.pdf/60305769-0713-7bf6-d2fb-828e941eeb7e?t=1624471781098

Harańczyk, A. (2015). Procesy suburbanizacji w krakowskim obszarze funkcjonalnym. *Studia Miejskie, 18*, 85–102. https://doi.org/10.25167/sm.2439

Izdebski, H. (2013). *Ideologia i zagospodarowanie przestrzeni: doktrynalne prawno-polityczne uwarunkowania urbanistyki i architektury*. Wolters Kluwer Polska.

Kowalczyk-Anioł, J. (2019). Tourismification of the housing resources of historical inner cities. The case of Krakow. *Studia Miejskie*, 35, 9–25. https://doi.org/10.25167/sm.1014

Kopyciński P. (2018). City Lab as a Platform for Implementing Urban Innovation: the Role of Companies. In K. Wach & Maciejewski (Eds.), International Entrepreneurship as the Bridge between International Economics and International Business: Conference Proceedings of the 9th ENTRE Conference and 5th AIB-CEE Conference (pp. 257-274)(Przedsiębiorczość Międzynarodowa, vol. 4, nr 3). Uniwersytet Ekonomiczny w Krakowie.

Kudłacz, T., & Markowski, T. (2017). Miejskie obszary funkcjonalne w świetle wybranych koncepcji teoretycznych–zarys problemu. *Studia KPZK PAN, 174*, 17–30. https://doi.org/10.24425/118518

Mikuła, Ł. (2019). *Zarządzanie rozwojem przestrzennym obszarów metropolitalnych w świetle koncepcji miękkich przestrzeni planowania*. Bogucki Wydawnictwo Naukowe.

Moreno, C., Allam, Z., Chabaud, D., Gall, C., & Prationg, F. (2021). Introducing the "15-Minute City": Sustainability, resilience and place identity in future post-pandemic cities. *Smart Cities, 4*(1), 93–111. https://doi.org/10.3390/smartcities4010006

Noworól, A., Kopyciński, P., Hałat, P., Jasiński, L., Paruch, S., & Salamon, J. (2024). *Polityka przestrzenna w aglomeracji krakowskiej w świetle procesów metropolizacji i suburbanizacji*. Wydawnictwo Małopolskiej Szkoły Administracji Publicznej Uniwersytetu Ekonomicznego w Krakowie. Online access: https://politykipubliczne.pl/wp-content/uploads/2024/07/Aglomeracja_20240624_ISBN-fina.pdf

Noworól, A., Kopyciński, P., Hałat, P., Salamon, J., Baziak, J., Sypek, W., & Wirchniańska, M. (2021). *Czy Kraków może stać się "miastem 15-minutowym"?* Wydawnictwo Małopolskiej Szkoły Administracji Publicznej Uniwersytetu Ekonomicznego w Krakowie. Online access: https://politykipubliczne.pl/wp-content/uploads/2021/12/30_v2_Czy-Krakow-moze-stac-sie-miastem-15-min.pdf

Noworól, A., Kopyciński, P., Hałat, P., Salamon, J., & Hołuj, A. (2022). The 15-Minute City—The geographical proximity of Services in Krakow. *Sustainability, 14*(12), 7103. https://doi.org/10.3390/su14127103

O'Gorman, S., & Dillion-Robinson, R. (2021). *20 minute Neighbourhoods in a Scottish context, climate exchange Scotland's Centre of expertise connecting climate change and policy*. ClimateXChange. https://doi.org/10.7488/era/808

Paszkowski, Z. (2014). Tendencje w rozwoju polskiej urbanizacji. In J. Sepioł (Ed.), *Przestrzeń życia Polaków* (pp. 15–34). Stowarzyszenie Architektów Polskich. Online access: https://obserwatorium.miasta.pl/wp-content/uploads/2016/08/przestrzen_zycia_polakow_0.pdf

Pluta, M. (2016). Spatial planning in Poland in the context of 'inspire' rules and amendments to the spatial planning and development act. *Geomatics, Landmanagement and Landscape, 4*, 159–168. https://doi.org/10.15576/GLL/2016.4.159

Pozoukidou, G., & Chatziyiannaki, Z. (2021). 15-Minute City: Decomposing the new urban planning Eutopia. *Sustainability, 13*(2), 928. https://doi.org/10.3390/su13020928

Raport o ekonomicznych stratach i społecznych kosztach niekontrolowanej urbanizacji w Polsce. (2014). Fundacja Rozwoju Demokracji Lokalnej i Instytut Geografii i Przestrzennego Zagospodarowania PAN. Online access: https://frdl.org.pl/static/upload/store/import/pliki/frdl/document/zalaczniki_artykuly/Raport_Ekonomiczny_29.10.2013_calosc.pdf

Różycka-Czas, R., Czesak, B., & Staszel, A. (2021). Which polish cities sprawl the Most? *Land, 10*(12), 1291. https://doi.org/10.3390/land10121291

Strategy Kraków Metropolis 2030 (pl. Strategia Metropolia Krakowska 2030). (2021). Uchwała z dnia 14 grudnia 2021 r. Walnego Zgromadzenia Członków Stowarzyszenia Metropolia Krakowska.

Sykała Ł., Dawid M., Dawid W., Koj J., Kudłacz K., Mróz M., & Stelmaszewska N. (2023). *Procesy suburbanizacji w Polsce w świetle rozwoju budownictwa mieszkaniowego i niemieszkaniowego w strefach podmiejskich*. Warszawa–Kraków: Instytut Rozwoju Miast i Regionów. https://doi.org/10.51733/opm.2023.01

Szromek, A. R., Kruczek, Z., & Walas, B. (2019). The attitude of tourist destination residents towards the effects of Overtourism: A case study of Krakow. *Sustainability, 12*(1), 228. https://doi.org/10.3390/su12010228

Węgrzynowicz, P. (2022). Characteristics of residential areas in Kraków functional urban area. *Journal of Settlements and Spatial Planning, Special Issue, 15–24*. https://doi.org/10.24193/JSSPSI.03.CSPTER

Weng, M., Ding, N., Li, J., Jin, X., Xiao, H. Z., & Su, S. (2019). The 15-minute walkable neighbourhoods: Measurement, social inequalities and implications for building healthy communities in urban China. *Journal of Transport & Health, 13*, 259–273. https://doi.org/10.1016/j.jth.2019.05.005

Włodarczyk, M. (2013). Architektura lat 60. XX wieku. Fragment historii Krakowa i innych polskich miast. *Kwartalnik Architektury i Urbanistyki, 58*(3), 93–112.
Wójcik, S. (2022). Czynniki warunkujące wybór samochodu jako środka transportu w codziennych podróżach — przykłady miast polskich. *Studia Regionalne i Lokalne, 24*(89), 49–66. https://doi.org/10.7366/1509499538904
Ziobrowski, Z., Jaśkiewicz, M., & Kuzianik, R. (2023). *Blaski i cienie planowania rozwoju przestrzennego Krakowa. Fakty i wspomnienia*. Instytut Rozwoju Miast i Regionów.

Chronourbanism as a Driver of More Sustainable Post-Socialist Cities: Case Study—Iași City, Romania

Alexandru Bănică, Cristian-Manuel Foșalău, Alexandra Sandu, and Lucian Roșu

1 Introduction

Contemporary cities face unprecedented challenges: climate change, economic instability, global pandemics, and growing social inequalities. Agenda 2030 and the Sustainable Development Goals (SDGs) proposed by the UN in 2015 specifically recommend that policymakers bring cities closer to sustainable development (Drobniak & Stangel, 2024) by reducing pollution (including greenhouse gas emissions), using resources more wisely, and assuring social justice while maintaining economic development and enhancing the quality of life for healthier and happier people. These pressures demand new approaches to urban development, yet solutions cannot ignore the complex layering of historical urban forms and practices that shape current possibilities. In this sense, cities are true palimpsests that retain traces of their past in various forms.

This layering process is significant in Central and Eastern Europe (CEE) as cities navigate the complex transition from socialist to sustainable urban forms (Sandu, 2023). Each urban layer—from historical cores to industrial zones and brownfields, from socialist housing estates to recent commercial developments—carries its logic of spatial organisation and social practices, creating constraints and opportunities for sustainable urban development. Indeed, the socialist era left a lasting mark on the post-socialist urban form through rigid functional zoning, standardised housing estates, and

A. Bănică (✉)
Faculty of Geography and Geology, Alexandru Ioan Cuza University of Iași, Iași, Romania

Romanian Academy–Iași Branch, Geographic Research Centre, Iași, Romania

C.-M. Foșalău · L. Roșu
Faculty of Geography and Geology, Alexandru Ioan Cuza University of Iași, Iași, Romania

A. Sandu
Cañada Blanch Centre, London School of Economics and Political Science, London, UK
e-mail: a.sandu@lse.ac.uk

L. Bouzouina et al. (eds.), *15-Minute Cities*, Footprints of Regional Science,
https://doi.org/10.1007/978-3-032-20840-8_12

hierarchical service distribution. Post-1990 market-driven development introduced new patterns, among which rapid sub-urbanisation also creates both obstacles and opportunities for implementing new urban concepts and operational solutions, such as the 15-minute city, which emerges as a central concept of current policies.

Among the operational solutions that meet these goals, the 15-minute city is a concept and model proposed by Carlos Moreno in 2016. It has gained significant attention in urban policies and studies on cities since 2020, particularly amid the COVID-19 pandemic. This approach not only aims to increase accessibility to essential proximity services but also serves as a planning tool for promoting active and clean mobility (Moreno et al., 2021; Allam et al., 2022a, 2022b). Based on a long history of successive urban models and planning approaches, starting with the Garden City proposed by Ebenezer Howard (1890) and the neighbourhood units of Clarence Perry (1923) (Khavarian-Garmsir et al., 2023), the 15-minute city is one of the expressions of chronourbanism, i.e. urban planning based on time management. The 15-minute city includes proximity as a fundamental principle, allowing access in a timeframe of 15 min from residential areas to commercial, educational, health, institutions, workplaces, and recreational spaces. However, the exact reference to the 15-min time interval may seem restrictive and unrealistic given the multiple rhythms of the city and the need to access services in a differentiated manner, some being needed more frequently and in even closer proximity (e.g. 5 min). In comparison, others can be accommodated in time distances of more than 15 min. It is also worth remembering that different users will have different travel speeds, and the difference in speed between walking and cycling is significant. In this regard, the concept of the x-minute city proposed by various authors seems preferable (Logan et al., 2022). Nevertheless, to a large extent, this model relies on mixing these functions in the proximity of residential areas, thereby eliminating strictly delimited individual functional areas and creating a compact, coherent, and functional urban mosaic. This approach also reduces the need for long-distance travel and makes it possible to promote active mobility on foot or by bicycle (Gehl, 2001).

This paper explores how more efficient management of space and time can create greener and more sustainable cities better adapted to the present challenges and enhance liveability. It refers to the cities in the Eastern European Union, specifically in former communist countries, with a particular emphasis on Iaşi, the largest urban pole in Romania's Northeast Region and the country's third largest city.

2 Approaches to Urban Sustainability in Post-Socialist Cities: Advantages and Challenges of Chronourbanism

2.1 Urban Planning Models in the Post-Socialist City: Transitions and Transformations

The evolution of urban sustainability in post-socialist cities involves navigating between inherited socialist structures and market-driven urban development patterns, which have been far from straightforward due to socialist urban legacies still present in the urban morphology and infrastructure.

Socialist cities were shaped by modernist planning principles that emphasised rigid functional zoning and centralised service provision, resulting in segregated urban spaces (Coudroy de Lille, 2009). Indeed, French and Hamilton's (1979) model of a socialist city (see Fig. 1) offers insights into the urban structures that continue to influence contemporary development patterns.

The socialist city was organised into distinct functional zones: A historical core, surrounded by a central business district (CBD), with old bourgeois areas and working-class neighbourhoods combined with industrial areas. The socialist era added systematically distributed residential districts, new industrial zones on the

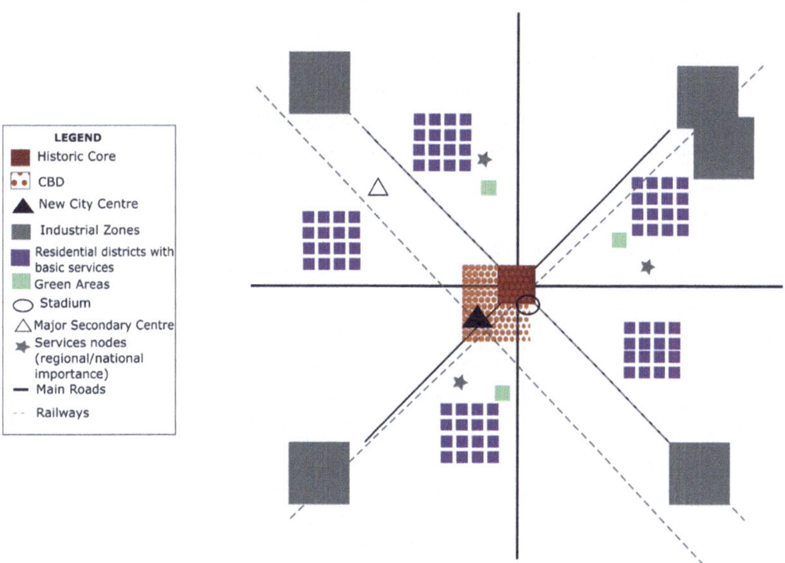

Fig. 1 The socialist city model (adapted after French & Hamilton, 1979)

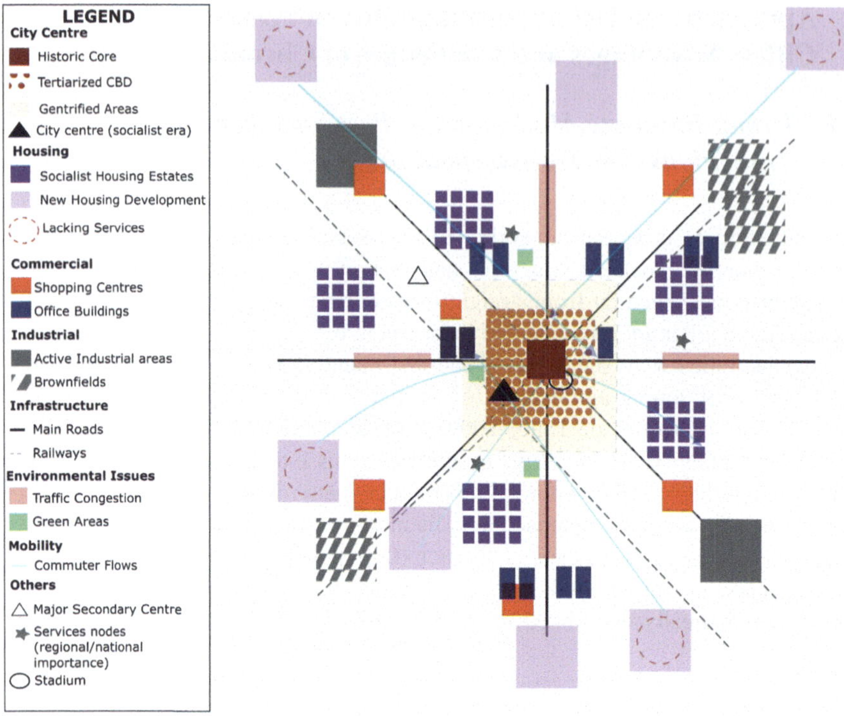

Fig. 2 The post-socialist city (adapted after Sailer-Fliege, 1999)

outskirts, and a hierarchical service centres network. This spatial organisation reflects key socialist planning priorities: a focus on industrial growth, an extensive public transport system (notably railways), and uniform service distribution. However, this rigid zoning now complicates contemporary efforts toward mixed-use development, while the centralised distribution of services—while somewhat efficient during the socialist period—struggles to adapt to today's urban population's more varied needs and demands.

While the socialist spatial organisation created specific urban challenges (Smith, 1996), the subsequent transition to market economies introduced additional complexities, clearly illustrated in Sailer-Fliege's (1999) model of the post-socialist city from CEE (see Fig. 2).

This model reveals the emergence of new spatial patterns: Fragmentation of the city centre between administrative and commercial functions, increasing socio-spatial segregation in residential areas, and a two-fold division of industrial zones (active and brownfields). The model particularly emphasises new mobility patterns and the rise of secondary commercial centres, indicating a shift toward polycentric urban development.

Nevertheless, while the transition brought numerous changes to these cities, many socialist-era structures remain, creating a pronounced path dependency that hinders systemic changes. Scholars argue that this path-dependent structure, inherited from socialist urbanism, resists transformation even decades after political shifts as socialist legacies present complex difficulties in terms of infrastructure and urban morphology (Hirt, 2013; Stanilov, 2007; Tsenkova, 2006).

Large housing blocks, for example, have created rigid urban landscapes that lack flexibility for modern mixed-use or compact city development (Stanilov, 2007). The socialist urban areas were designed with functional rigidity, prioritising industrial sectors over service-oriented or diverse-use spaces. This structural resistance—described by some scholars as "bad resilience" (Rufat, 2012) reflects a form of systemic inertia that perpetuates the spatial and functional rigidity of the socialist legacy (Bănică & Muntele, 2017; Drobniak et al., 2017; Sandu et al., 2021).

Urban planning in socialist cities operated within the rigid, top-down structures of a centralised regime that prioritised industrial productivity and ideological goals over environmental and social considerations, creating a legacy that shapes the contemporary post-socialist urban landscape (Stanilov, 2007; Tsenkova, 2006). Consequently, contemporary environmental challenges are partly rooted in these historic planning practices. Although socialist cities maintained functional segregation, the post-socialist era introduced mixed-use spaces, often unplanned and market-driven, further complicating the post-socialist urban landscape. This shift has led to a fragmentation of the previously organised service provision system, resulting in disparities in service accessibility across different parts of urban neighbourhoods (Stanilov, 2007).

A particularly problematic aspect of post-1990 development has been the chaotic expansion of residential areas. Unlike the socialist-era residential districts that were planned with accompanying services, new outer urban expansion areas frequently develop without schools, healthcare facilities, or adequate public transportation, leading to car-dependent lifestyles and increased environmental pressure (Milanovic et al., 2007; Sandu, 2023; Schmidt et al., 2015). Additionally, city centres faced their own challenges, as the post-socialist transformation has led to intense densification of central areas without corresponding increases in public spaces or green areas (Badiu et al., 2019; Sandu, 2023; Sykora, 1998). This densification of the urban core, as well as the extensive urban sprawl phenomena, have resulted in various environmental challenges, such as traffic congestion, deteriorating air quality, loss of green spaces, parking problems, and an overall reduction in quality of life (Csomós et al., 2020; Kovács et al., 2019; Miljanović et al., 2023).

In this complex post-socialist urban context, sustainable urban development must blend historical evolution with contemporary global frameworks. While the sustainable city represents the overarching framework, particularly as emphasised in Sustainable Development Goal 11 (UN, 2018), its implementation in post-socialist contexts in CEE requires a more nuanced approach than in Western contexts, one that acknowledges socialist planning legacies while incorporating the current sustainability concepts. Various urban paradigms present both opportunities and challenges: the smart city approach offers technological solutions for optimising inherited socialist infrastructure and service systems, but critics argue that its implementation must be carefully balanced against social equity concerns and the digital divide that often characterises post-socialist societies (Ibănescu et al., 2020; Kola-Bezka et al., 2016); the green city concept addresses both socialist-era environmental neglect and post-socialist environmental pressures, though its application often struggles against the path dependency of socialist-era infrastructure and the pressure of market-driven development (Grabkowska, 2022; Milovanovic et al., 2024). Similarly, the resilient city framework aims to transform "bad" resilience of socialist structures into adaptive capacity. However, its success depends heavily on overcoming institutional inertia and developing new governance models, bridging the gap between top-down socialist planning legacies and contemporary participatory approaches (Tarasevych et al., 2023). Finally, while key in humanising both socialist-era housing estates and post-socialist developments, the livable city concept faces significant challenges in balancing densification pressures with quality-of-life improvements (Bănică & Pascariu, 2023).

Therefore, the path towards urban sustainability in these contexts must navigate not only the inherited socialist structures, but also the often chaotic post-1990 market-driven development patterns. This is particularly relevant when considering the 15-minute city concept (Moreno, 2016), which paradoxically finds both potential support in the socialist-era hierarchical service system and resistance in its rigid functional segregation. Focusing on the temporal organisation of urban space and activities, this approach offers possibilities for optimising urban rhythms and mobility patterns, reducing peak-hour congestion, and promoting temporal diversity in space usage, especially in a post-COVID world (Moreno et al., 2021).

2.2 Chronourbanism in the Post-Communist City: Compatibilities and Obstacles in Urban Planning

The perception of time and space during the communist period in the Central and Eastern European states was deeply influenced by the regime's political, ideological, and socio-economic context that claimed to promote a classless society and centralised planning. Time during the communist period was par excellence collective and not personal, based on the everyday experience of the progressive march towards the ideal of the "communist future." Urban services and workplaces (especially in factories) were organised in a standardised way to level behaviours under the tutelage of state authority (and the single party) and were ways of creating an order. Therefore, time was precisely measured, especially by the interval needed to get to work, while recreational ("cultural") activities were also controlled, strongly ideologically marked and delimited; nightlife was non-existent or reduced to a minimum, as were informal activities. Through working hours, urban planning and state-organised leisure, people dictated how they spent their time and organised their space. Home-work trips were prioritised through dedicated, but often overcrowded, public transportation. Home-city centre relationships or access to public spaces were minimised precisely to reduce these trips. Thus, social life and community interactions were limited by discouraging the freedom to use "free time," restricting public spaces and meetings and inducing an atmosphere of distrust related to the lack of free access to information, freedom of expression, and the promotion of a system that encouraged duplicity and denunciation.

Housing neighbourhoods were organised using the model of Soviet districts as quasi-autonomous entities where standard services were located in proximity. Although they were promoted as an innovation of their own ("micro-rayon theory") of Soviet-inspired regimes, these districts were practically based on an adapted form of the model proposed in the 1930s by Clarence Perry ("neighbourhoods units' concept") (Becker et al., 2012; Gogishvili et al., 2017). Thus, the new housing complexes built after World War II as part of urban restructuring and development programmes included the planning of essential services (commercial, educational, proximity health) (Hirt, 2013; Kissfazekas, 2022), but to a lesser extent, they took into account access to green spaces or leisure areas. However, beyond this modular/semi-autonomous character for residential areas, communist cities became divided ensembles, in which connections with the centre were limited, the workers' neighbourhoods being organised as villages ("city villages") with standardised housing and lacking personal touch or individuality, to symbolise a society of equal people. In these collective housing, people had to spend most of their time—the only transport links, better and relatively fast, were with jobs in the new areas or industrial platforms. Many of these relationships were maintained by inertia, following the mechanisms of "path dependency" as a legacy of "bad resilience" even three decades after the fall of communism (Rufat, 2012; Ilchenko & Dushkova, 2018).

The post-socialist transition after 1990 marked a dramatic shift from the previous system's rigid organisation based on the segregation of workspace, housing, and public space to chaotic processes of *laissez-faire* based not only on the idea of freedom but, to a greater extent, on corrupt administrative systems and in which the power of influence of some individuals or groups dictated urban developments. Communist segregation was replaced by an often-dysfunctional urban mosaic in which densifications in the urban area were often made through improper constructions, which did not respect minimal urban planning norms, adding valuable services but not necessary utilities. Deindustrialisation and tertiarisation brought profound transformations to cities; the centres often became hypertrophied and crowded, and the number of cars increased exponentially, as did the trips to the city centre or the new areas where jobs were concentrated. This resulted in a noticeable increase in noise pollution and daily disturbances. Added to this was the uncontrolled expansion of the city outwards *("urban sprawl"),* which brought new challenges related to mobility and the provision of proximity services or public utilities.

The city's footprint has increasingly consumed both space and time. There has been a process of sometimes insidious urbanisation, similar to the process of percolation (Diappi, 2004), spreading into the nearby countryside. These areas are now inhabited by city workers who commute daily to the centre, primarily by personal car, causing congestion at the exits/entrances from/to the city during morning and afternoon/evening rush hours. Changing transport needs, the lack of flexibility, and reduced connectivity of public transport with infrastructure and routes inherited from the socialist period (Tuvikene et al., 2020) have emphasised the choice of individual transport as the predominant mode of travel within the city, including over short distances. However, the time spent, i.e. wasted in traffic jams or by travelling at low speeds, has impacted work productivity and general satisfaction with individuals' lives. The lack of essential public services in urban expansion areas, the poor quality of roads, and the lack of adequate services in quantity and quality have led to an even more substantial increase in this pressure. Forced collectivism and lack of essential freedoms have been replaced by an exacerbated individualism, characterised by a disregard for others' comfort, security, and freedom—creating a new form of social division. However, a certain reversal of this trend is visible in recent times, with more and more urban civic associations appearing and carrying out sustained activity to use spatial and temporal proximity in order to regenerate public space and community spirit, while also managing to put a certain pressure on decision-makers for decisions that take the public, common good into account.

This paper examines whether post-communist cities can address and potentially transform their current inequalities through several approaches: Ensuring equity in access to current proximity services, encouraging active mobility (walkability, cyclability), and implementing urban greening initiatives. These initiatives include preserving and creating new green infrastructure and "nature-based solutions" that would diminish the existing spatial inequities regarding access to green and blue spaces within the post-socialist city (Csomós et al., 2020; Bradecki et al., 2022).

Another aspect concerns the possibility of inducing a polycentric development of the post-communist city by creating local urban centres (LUCs). These LUCs are compact and multifunctional central places oriented towards providing all

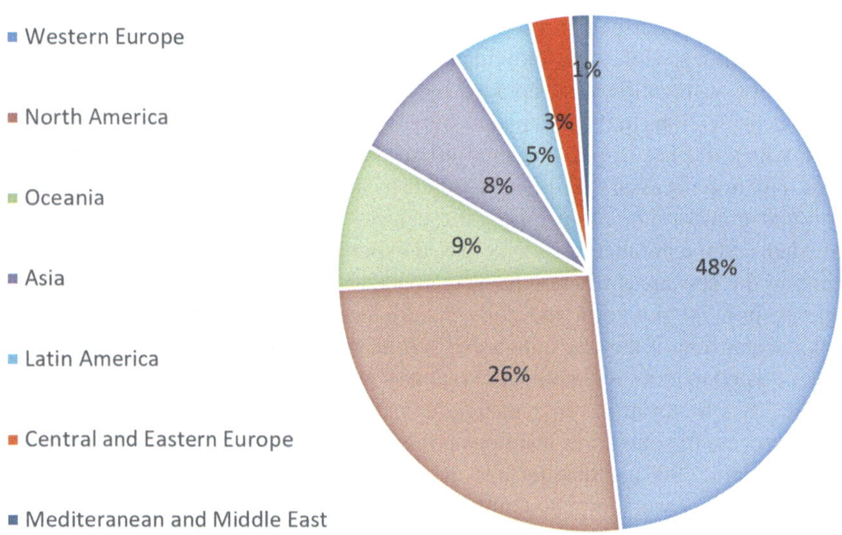

Fig. 3 The geographic distribution of 15-minute initiatives from 2008 to 2023 (Data source: Allam et al., 2022a, 2022b)

necessary services in proximity. These LUCs are hierarchically organised and interconnected by public transport, as proposed in Kosice (Drobniak & Stangel, 2024).

Although, in the last 4 years, there has been an openness and concern of many countries in CEE to discuss the 15-minute city model, there is still a relatively small number of cities in this part of Europe that have adopted the principles of the 15-minute city at the urban planning level (Fig. 3). Until 2020, there was a need to address the topic of walkability in post-communist cities, except for the Czech Republic, where this aspect has already been discussed in scientific works (Rišová, 2020). Concrete practices focused strictly on the 15-minute city have been identified in a few post-socialist cities, including Zagreb, Krakow, and Pleszew (Teixeira et al., 2024).

However, studies that discuss the appropriateness of using the 15-minute city model in the case of other cities in the Central and Eastern European Union are still emerging. These aim to present the model's advantages for the sustainable urban development in Slovakia (Mocák et al., 2022); the efficiency and liveability benefits of proximity services in Krakow, Poland (Noworól et al., 2022); how adopting the 15-minute city model can make Timişoara (Romania) more resilient to pandemics such as COVID-19 (Draghici et al., 2022); the current appreciation of urban districts built in the 1950s in Budapest according to Perry's neighbourhood unit model (Kissfazekas, 2022; Földényi, 2023); the use of LUCs in relation to district train stations for polycentric development based on the 15-minute city model in Katowice (Drobniak & Stangel, 2024); and the effects of implementing the 15-minute city approach in decarbonising urban mobility in post-socialist cities such as Tallin and Skopje (Horton, 2024).

The case study proposed in this chapter considers the city of Iaşi, the main economic driver in the northeast region of Romania. Here, the 15-minute city model could represent a transformative solution, creating sustainable and high-quality residential areas from various structures inherited from the communist period and those added or transformed in the last 30 years.

3 Case Study: Iași City

3.1 The Study Area

Iaşi is Romania's third-largest city, with 271,692 legal residents and 442,771 in its Metropolitan area (according to the National Institute of Statistics, see https://insse.ro/cms/en/). Located in the Moldavian Plateau and traversed by the Bahlui River, the city presents a tentacular development shaped by local topography and historical evolution (Sfîcă et al., 2018). Its growth was also influenced by its historical role as the capital city and central position within the historical Principality of Moldavia (1359–1859) and later by socialist industrialisation (Muntele et al., 2019). The city's structural evolution in the last 50 years can be divided into two distinct periods. Pre-1989, under the socialist regime, urban development followed a centralised approach characterised by industrialisation and high-density residential districts, with approximately 75% of the population residing in compact neighbourhoods. The post-1989 transition to a market economy triggered significant transformations in the urban fabric, leading to several challenges for the further sustainable evolution of the city.

- *Urban Sprawl*: The western and southern peripheries have experienced substantial growth, characterised by low-density residential developments lacking integrated infrastructure and services. This sprawl has created areas with reasonable environmental advantages but where basic facilities are outside walking distance, contradicting the walkable city.
- *Infrastructure Disparities*: The spontaneous peri-urban growth has resulted in environmental degradation, increased traffic congestion, and social segregation (Roşu et al., 2018), highlighting the need for better-distributed urban services and facilities as proposed in the 15-minute city model.
- *Brownfield Sites*: The deindustrialisation process has left numerous underutilised spaces within the urban core, which present opportunities for implementing 15-minute city features through strategic urban regeneration.

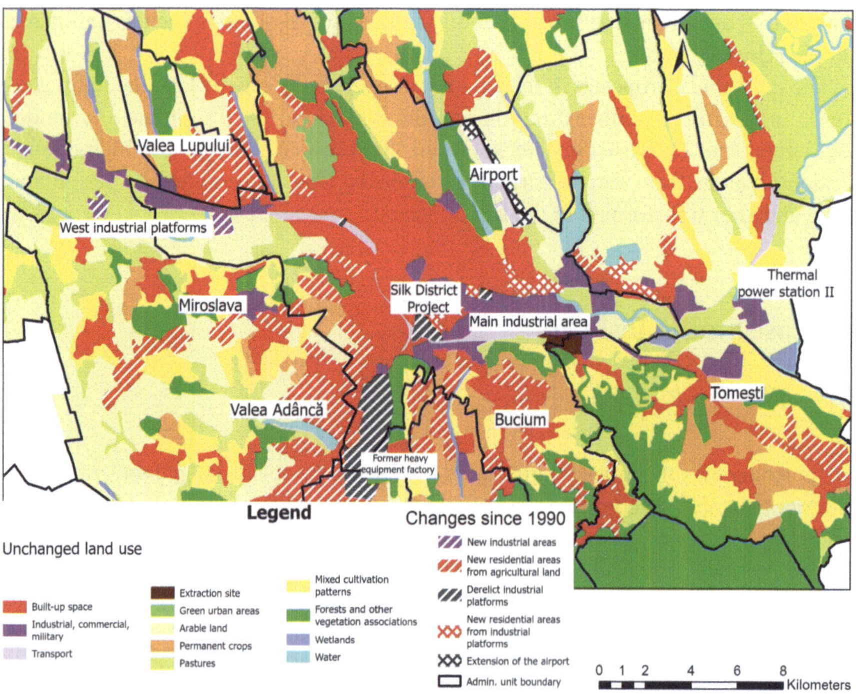

Fig. 4 Post-socialist changes of functional urban areas in Iaşi city (Own design, data sources: Urban Atlas, CLC)

The present-day pattern is often dispersed through a low-density fabric from private investors in residential complexes built by private investors within a framework largely bereft of comprehensive regulations (Fig. 4). Such unorganised expansion of the peri-urban areas facilitates environmental degradation, road congestion, and social segregation (Dinter & Roşu, 2021).

These current challenges stem from weak policy enforcement and uncoordinated urban planning post-1989. The present urban landscape reflects the complexities and contradictions inherent in modern city living. In this context, the main goal of the research is to question if the urban built area of Iaşi Municipality is ready to leap towards the concept of a 15-minute city by assessing the accessibility of essential services and amenities together with the quality of environmental conditions using specific GIS tools. This evaluation will provide insights into the feasibility of creating a more sustainable and livable urban environment that fosters urban resilience.

3.2 Access to Proximity Services: The X-minutes Iași

To achieve the aim of this research, we employ a quantitative analysis using an x-minute accessibility index that measures access to critical facilities and services. Drawing on various authors' methodologies, we considered a thorough list of services specific to the 15-minute neighbourhoods, as defined in Table 1. These align with the "six essential urban social functions" identified by Moreno et al. (2021): Working, commerce, healthcare, education, entertainment, and living.

Table 1 Services and facilities considered in the 15-minute city model for Iaşi

Service or facility	Category	No. of objects	References
Supermarkets, open-air food markets, restaurant / fast-food, pastry/confectionery, bakery	Commerce and public alimentation	910	Sun and Luo (2024), Balletto et al. (2021), Ferrer-Ortiz et al. (2022), Gaglione et al. (2022)
Pharmacies, family general practitioners	Healthcare	444	Guzman et al. (2021), Gaglione et al. (2022), Noworól et al. (2022), Song et al. (2022)
Primary schools, kindergartens	Education	168	Sun and Luo (2024), Guzman et al. (2021), Ferrer-Ortiz et al. (2022), Caselli et al. (2021)
Bus stops	Public transit	321	Sun and Luo (2024), Noworól et al. (2022), Cho et al. (2021)
Cinemas/theatres, bookstores/ libraries, churches/places of worship	Culture	241	Ferrer-Ortiz et al. (2022), Noworól et al. (2022)
Sports fields/gym/pool, parks/ green spaces, playgrounds	Sports and leisure	608	Balletto et al. (2021)

A few mentions have to be made regarding the facilities often acknowledged in other related papers and not included in this study. First, for the educational facilities, the proximity criterion is only available for kindergartens and primary schools. Romanian legislation allows secondary and high schools to be chosen by the criterion of prestige and reputation, commonly enrolling pupils through an exam organised independently by schools. Second, hospitals are institutions people do not access daily, most of which are specialised in a specific branch, so we did not include them. Third, accessibility to jobs within this framework is too difficult to assess with the present available data. Comprehensive data on job locations, commuting patterns, and resident employment status requires approaches not covered in this paper, as workplaces are rarely chosen based on proximity. Moreover, employment dynamics in a fast-developing city like Iaşi are complex; people are changing

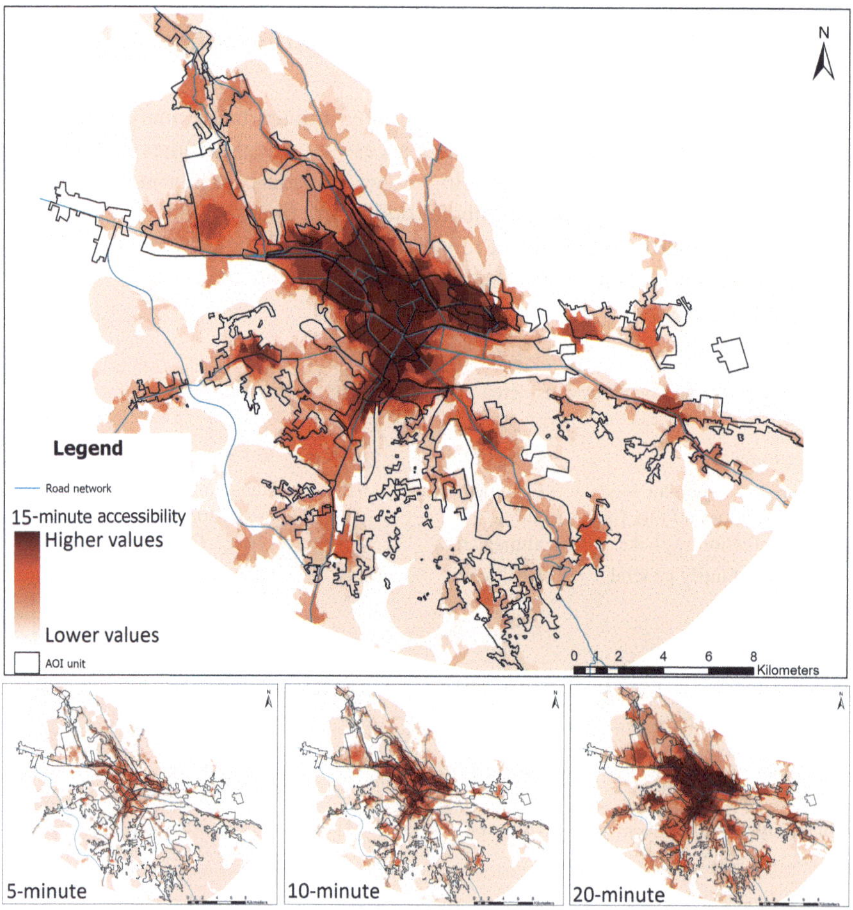

Fig. 5 The 5-, 10-, 15-, and 20-min accessibility in Iaşi (Source: own representation)

jobs throughout their lives, remote working popularity is increasing, and each job type may have different accessibility needs and attract different populations.

A spatial database was created containing point locations (based on coordinates) of urban services, and polygons for parks and green spaces. A catchment area was calculated for each facility using network analysis in ArcMap 10.4. This analysis generated isochrones for 5-, 10-, 15-, and 20-min walking distance to the chosen facilities, considering an average walking speed of 3 km/h, or 250, 500, 750, and 1000 m, respectively, for the isochrones above. Following Casseli et al.'s example (2021), we believe that 3 km/h (or 50 m/min) is a value that is inclusive for all sidewalk users, ages, and social categories, that considers the various obstacles that can slow down the walking path inside a city. The polygons delineated by these isochrones were mapped and weighted (using the AHP method) to calculate the proposed 15-min index (Foşalău, 2023). A neighbourhood where catchment areas of all the suggested services intersect has a score of 1, while areas further than x min away from any facility have a score of 0.

The 15-min accessibility index depicts the spatial location, density, and diversity of urban services in the built-up continuum of Iaşi. As depicted in Fig. 5, maximum values for the index are found in the municipality's urban core, specifically in central neighbourhoods and Tătăraşi. The latter is a typical example of a neighbourhood that was significantly developed during the systematisation process undertaken in the '60 s by the socialist regime. High scores are also found in similar situations, such as Podul de Fier, Cantemir, or Nicolina. The accessibility decreases as we move away from the city centre, with only a few peri-urban settlements getting close to the 15-minute neighbourhood model. Villages have become part of the built-up continuum and are experiencing rapid residential development, while the implementation of basic facilities tends to lag behind.

As shown in Section 2.2., one can acknowledge that in Eastern European cities, neighbourhoods developed during the socialist regime often have proximity facilities that seek to meet the needs of their residents, typically essential services such as schools, kindergartens, nurseries, basic healthcare amenities, markets, which were present in the centralised urban planning policies during that period. This type of planning was essential for complying with the high population densities resulting from vertical development. Moreover, some of these neighbourhoods still keep cultural amenities, such as community centres, parks, and recreational facilities, contributing to residents' overall quality of life. These facilities were designed to encourage public life and collectivism, equality and standardisation while considering social well-being through large open central squares, shared green spaces or collective facilities (places to eat, culture houses).

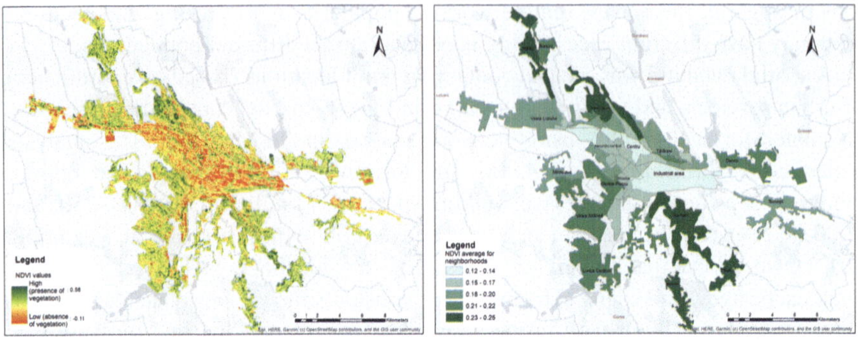

Fig. 6 The NVDI values at microscale (**a**) and urban neighbourhood (**b**) level

3.3 Iași as a Green City, Favourable to Active Mobility

Both the 15-minute city concept and its precursors, the neighbourhood unit concept, the garden city, or the new urbanism, encourage walking, biking, using public transportation, enhancing green spaces, and spending time outside public places. In this respect, a few indicators were assessed for the same study area units: Vegetation cover, air quality, and active mobility.

The key metrics of vegetation cover are vegetation indices and the normalised difference vegetation index (NDVI), which illustrates a detailed momentary situation of the presence of vegetation in the urban and peri-urban perimeters of Iași (Fig. 6). The sprawled character of the urbanisation process strongly influences the greenness of the city's outskirts through residential complexes, commercial centres, building materials, and increasingly sealed surfaces. In the context of these dynamic transformations, we used an NDVI raster computed from a Landsat 8 image from the late summer of 2020. A general look at the NDVI spatial distribution shows the areas where vegetation cover is limited, such as old industrial platforms in the eastern Iași, new counterparts from the western part, commercial centres, and thermal power plants. By contrast, peri-urban neighbourhoods naturally benefit from a wider vegetation cover due to the topography and original vegetation of the Central Moldavian Plateau in the south and the Jijia Plain in the north. A closer look offers insights regarding existent vegetation among collective buildings, neighbourhoods, urban parks, and new residential ensembles, where green facilities are commonly integrated. This phenomenon is primarily driven by investors who want to increase the appeal of newly built homes, specific to a competition-based economic system. The integration of green spaces is not aimed at solving the broader environmental issues – a problem resulting from the lack of an integrated framework from authorities.

Complementarily, air quality was assessed for each measurement station around the city. PM2.5 was monitored for about 1000 days, and an average was calculated

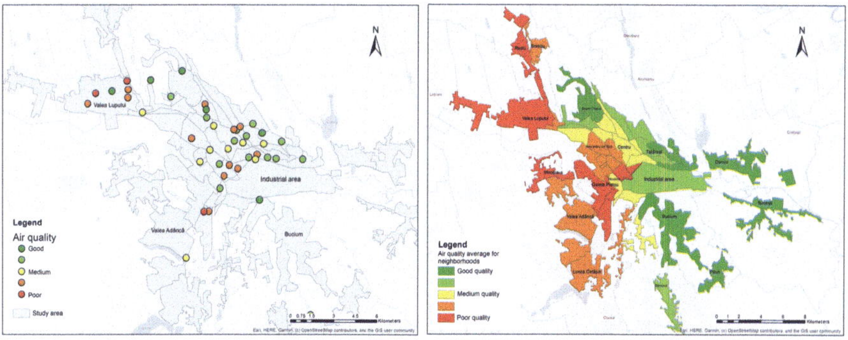

Fig. 7 Air quality indexed by monitoring points (**a**) and aggregated at urban neighbourhood level (**b**) (Data source: WAQI)

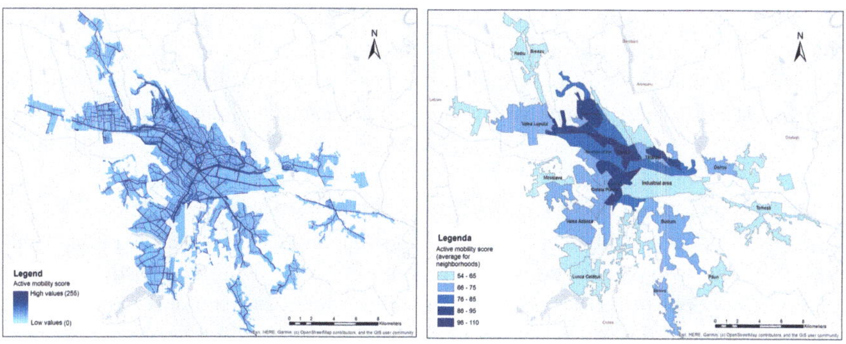

Fig. 8 The active mobility at the microscale (**a**) and urban neighbourhood (**b**) level (Data source: STRAVA app)

(Fig. 7). Their interpolation offers rough indications about the situation in the area of interest. The most polluted units appear to be the ones that are currently the most crowded: Central heavily transited neighbourhoods and high-class residential peri-urban settlements, especially in the west and south. Good air quality is constantly recorded where the topography allows, as pollutants tend to persist in lower areas, such as the valley of the Bahlui River.

Active mobility is represented by walking, jogging, and biking, and it is illustrated on the map as a network of the most frequently used routes (Fig. 8). To a large extent, it overlaps with the road network, except in peri-urban areas. The most prominent routes are along streets and boulevards that feature bike lanes, ample unobstructed sidewalks with few crosswalks, bike-friendly streets that are safe for pedestrians, and urban parks. The City Centre, Primăverii, CUG I, and the train station area stand out. At the same time, mobility is least active and most car-dependent in peri-urban villages such as Holboca, Breazu, Păun, and Miroslava.

Table 2 Correlation matrix (Pearson)

Variables	Accessibility to proximity services	Air quality	NDVI	Active mobility
Accessibility to proximity services	1			
Air quality	0,006	1		
NDVI	**−0,671**	**-0,307**	1	
Active mobility	**0,781**	0,061	**-0,658**	1

Values in bold are different from 0 with a significance level alpha = 0.05

3.4 How Green Is the X-minute Iași City?

Among the four indices calculated for 58 territorial units (neighbourhoods within the city and urbanised rural localities), several results align with expected hypotheses, such as the negative relationship between greenness (expressed through the NDVI) and air quality: the greener urban areas tend to have cleaner air (corr. Coef. −0.307). Similarly, a positive relationship between accessibility to services and mobility is observed: in areas with a high density and a variety of facilities (e.g., shops, parks, sports fields, bakeries), people are more likely to walk or bike (corr. Coef. −0.307) (Table 2).

On the other hand, two results contradict our hypotheses: the negative relationship between accessibility and NDVI suggests that areas with more and closer facilities also tend to have less consistent vegetation cover. Additionally, the negative relationship between NDVI and active mobility indicates that areas where people are more inclined to practice active mobility are not those with greener urban areas, and, on the contrary, access to green areas does not necessarily motivate the population to give up personal cars. This dynamic suggests the involvement of other factors beyond these variables that should be studied more in-depth in future research.

4 Conclusions

The 15-minute city is an urban planning model that aims to be an effective tool for transforming cities—currently significant contributors to environmental pollution and climate change and vulnerable to economic shocks and social inequalities—into greener, cleaner, and more socially cohesive systems that are economically sustainable and offer a better quality of life.

The compact neighbourhoods of the post-socialist city provide favourable conditions for partially achieving the 15-minute city goals. Following chronourbanism models (such as Perry's), urban planning strategies from the communist era ensured essential services within a neighbourhood. However, these high-density built areas, often lacking adequate vegetation coverage, are an obstacle to a sustainable city due to problems related to overcrowding, excessive standardisation, poor living conditions, and insufficient or lack of green areas.

The last 35 years of transformation induced new challenges: The new poorly planned urban mosaic and the often-chaotic expansion/sprawl have created dysfunctional, inefficient, time-consuming, polluted cities without some basic facilities and services. Among these, the uncontrolled peri-urban expansion represents a significant obstacle in implementing the urban concept of the "15-minute city," determined by the absence of a legislative framework that regulates the balanced development of facilities and mobility in these peripheral areas. The results of our case study also indicate that the lack of clear zoning policies and investment in marginal neighbourhoods perpetuates spatial inequalities, fragmenting residents' access to essential services and infrastructure.

The equitable but also flexible and adaptable ("x-minute") spatial distribution of urban services not just in the high-density urban neighbourhoods, but also in peripheral, more dispersed areas must be complemented by factors such as a robust local economy, encouragement of local producers, and social factors that foster solid communities, strengthen interactions among people, and reduce social inequalities. Regarding urban mobility, long-term solutions in Iași include investments in public transport through dedicated lanes, expanding the transport network, and fostering a mindset that understands the benefits of active mobility, ultimately being more time efficient. In the future, expansions of the 15-minute neighbourhood model could be envisioned. The growing IT industry could also be leveraged to introduce numerous smart city technologies, digitise urban networks, improve real-time data systems, monitor traffic and air pollution, and make information accessible to all citizens.

References

Allam, Z., Nieuwenhuijsen, M., Chabaud, D., & Moreno, C. (2022a). The 15-minute city offers a new framework for sustainability, liveability, and health. *Lancet Planet Health, 2022*(*6*), e181–e183.

Allam, Z., Bibri, S. E., Chabaud, D., & Moreno, C. (2022b). The '15-Minute City' concept can shape a net-zero urban future. *Humanities and Social Sciences Communications, 2022*(*9*), 126.

Badiu, D. L., Onose, D. A., Niță, M. R., & Lafortezza, R. (2019). From "red" to green? A look into the evolution of green spaces in a post-socialist city. *Landscape and Urban Planning, 187*, 156–164. https://doi.org/10.1016/j.landurbplan.2018.07.015

Balletto, G., Ladu, M., Milesi, A., & Borruso, G. (2021). A methodological approach on disused public properties in the 15-Minute City perspective. *Sustainability, 13*, 593. https://doi.org/10.3390/su13020593

Bănică, A., & Muntele, I. (2017). Urban transitions and resilience of eastern European Union cities. *The Eastern Journal of European Studies, 8*, 45–68.

Bănică, A., & Pascariu, G. C. (2023). Challenges in building more liveable cities in post-socialist European Countries: From "Cities4all" to "Cities4everyone". In L. Fusco Girard, K. Kourtit, & P. Nijkamp (Eds.), *The future of Liveable cities, footprints of regional science* (pp. 171–195). Springer International Publishing. https://doi.org/10.1007/978-3-031-37466-1_9

Becker, C., Mendelsohn, S. J., & Benderskaya, K. (2012). Russian urbanization in the Soviet and post-Soviet eras. London: Human Settlements Group, International Institute for Environment and Development.

Bradecki, T., Hallova, A., & Stangel, M. (2022). Concretization of sustainable Urban Design education in the project based learning approach—Experiences from a Fulbright specialist project. *Sustainability, 14*, 6971. https://doi.org/10.3390/su14126971

Caselli, B., Carra, M., Rossetti, S. & Zazzi, M. (2021). From urban planning techniques to 15-minute neighbourhoods. A theoretical framework and GIS-based analysis of pedestrian accessibility to public services. *European Transport \ Trasporti Europei, Issue 85*, Paper n° 10.

Cho, H., Choi, J., No, W., Oh, M., & Kim, Y. (2021). Accessibility of welfare facilities for elderly people in Daejeon, South Korea, considering public transportation accessibility. *Transportation Research Interdisciplinary Perspectives, 12*, 100514. https://doi.org/10.1016/j.trip.2021.100514

Coudroy de Lille, L. (2009). *Relire la ville socialiste*. Marne-la-Vallée (Société française d'histoire urbaine).

Csomós, G., Farkas, J. Z., & Kovács, Z. (2020). Access to urban green spaces and environmental inequality in post-socialist cities. *Hungarian Geographical Bulletin, 69*, 191–207. https://doi.org/10.15201/hungeobull.69.2.7

Diappi, L. (Ed.). (2004). *Evolving cities. Geocomputation in Territorial Planning*.

Dinter, O. V. & Roşu, L. (2021). Evaluating the potential conflicts of collective housing development in the suburbs of Iasi, Romania. *Journal of Settlements & Spatial Planning*.

Draghici, A., Vaduva, R., Capotescu, S., Banaduc, G., & Robescu, D. (2022). Innovations for tackling post-pandemic related challenges–a collaborative research to discover new solutions for hybrid work in the context of 15-minute cities. *Acta Technica Napocensis, Series Applied Mathematics, Mechanics, and Engineering, 65. Issue Special I*, February 2022.

Drobniak, A., Polko, A., & Sucháček, J. (2017). Transition and resilience in central and eastern European regions. In *The Routledge handbook to regional development in central and Eastern Europe*. Routledge.

Drobniak, A., & Stangel, M. (2024). Fostering local urban centers in relation to district train stations as part of the transition from a post-industrial to a sustainable city: A case study of Katowice. *Bulletin of Geography. Socio-Economic Series, 64*, 103–116. https://doi.org/10.12775/bgss-2024-0016

Ferrer-Ortiz, C., Marquet, O., Mojica, L., & Vich, G. (2022). Barcelona under the 15-Minute City Lens: Mapping the accessibility and proximity potential based on pedestrian travel times. *Smart Cities, 5*, 146–161. https://doi.org/10.3390/smartcities5010010

Földényi, K. (2023). *The 15-minute city in Budapest, Jakobsleiter–reconstructions,* Milestone Institute.

Foşalău, C. M. (2023). *Iaşi, oraş al celor 15 minute. Implicații privind sustenabilitatea urbană.* Master thesis,. Alexandru Ioan Cuza University, Iaşi.

French, R. A., & Hamilton, F. E. I. (Eds.). (1979). *The socialist city: Spatial structure and urban policy*. Wiley, Chichester.

Gaglione, F., Gargiulo, C., Zucaro, F., & Cottrill, C. (2022). Urban accessibility in a 15-minute city: A measure in the city of Naples, Italy. *Transportation Research Procedia, 60*, 378–385. https://doi.org/10.1016/j.trpro.2021.12.049

Gehl, J. (2001). *Life between buildings: Using public space*. Arkitektens Forlag.

Gogishvili, D., Aghasyan, V., Asabashvili, L., & Blaauw, B. (2017). *Micro-rayon living - everyday life strategies and DIY practices in the Post-Soviet micro-rayon*. In Project: genealogy of Soviet mass housing.

Grabkowska, M. (2022). *Post-socialist cities and the urban common good: Transformations in central and Eastern Europe*. Routledge. https://doi.org/10.4324/9781003089766

Guzman, L. A., Arellana, J., Oviedo, D., & Aristizabal, C. A. M. (2021). COVID-19, activity and mobility patterns in Bogota. Are we ready for a '15-minute city'? *Travel Behaviour and Society, 24*, 245–256. https://doi.org/10.1016/j.tbs.2021.04.008

Hirt, S. (2013). Whatever happened to the (post)socialist city? *Cities, 32*, S29–S38. https://doi.org/10.1016/j.cities.2013.04.010

Horton, A. (2024). *Citizen Centred Decarbonization of urban mobility in post-socialist cities: A comparative analysis of Tallin and Skopje. Centre international de formation europeenne, Master thesis in Global Energy Transition and Governance*, 2023–2024.

Ibănescu, B.-C., Bănică, A., Eva, M., & Cehan, A. (2020). The puzzling concept of Smart City in central and Eastern Europe: A literature review designed for policy development. *Transylvanian Review of Administrative Sciences, 70–87.* https://doi.org/10.24193/tras.61E.4

Ilchenko, M., & Dushkova, D. (2018). Editorial: In search of the post-socialist urban geography. How do we see the post-socialist city today? *Belgeo, 4.* https://doi.org/10.4000/belgeo.31467

Khavarian-Garmsir, A. R., Sharifi, A., Abadi, H. H., & M. & Moradi, Z. (2023). From Garden City to 15-Minute City: A historical perspective and critical assessment. *Land, 12*, 512.

Kissfazekas, K. (2022). Circle of paradigms? Or '15-minute' neighbourhoods from the 1950s. *Cities, 123*, 103587. https://doi.org/10.1016/j.cities.2022.103587

Kola-Bezka, M., Czupich, M., & Ignasiak-Szulc, A. (2016). Smart cities in central and Eastern Europe: Viable future or unfulfilled dream? *Journal of International Students, 9*, 76–87. https://doi.org/10.14254/2071-8330.2016/9-1/6

Kovács, Z., Farkas, Z. J., Egedy, T., Kondor, A. C., Szabó, B., Lennert, J., Baka, D., & Kohán, B. (2019). Urban sprawl and land conversion in post-socialist cities: The case of metropolitan Budapest. *Cities, 92*, 71–81. https://doi.org/10.1016/j.cities.2019.03.018

Logan, T., Hobbs, M., Conrow, L., Reid, N., Young, R., & Anderson, M. (2022). The x-minute city: Measuring the 10, 15, 20-minute city and an evaluation of its use for sustainable urban design. *Cities, 131*, 103924.

Milanovic, N., Gutry-Korycka, M., & Rink, D. (2007). Sprawl in the post-Socialist City: The changing economic and institutional context of central and eastern European cities. In C. Couch, L. Leontidou, & G. Petschel-Held (Eds.), *Urban Sprawl in Europe* (pp. 102–135). Blackwell Publishing Ltd. https://doi.org/10.1002/9780470692066.ch4

Miljanović, D., Vuksanović-Macura, Z., & Doljak, D. (2023). Rethinking the spatial transformation of postsocialist cities: Shrinking, sprawling or densifying. *Cities, 140*, 104443. https://doi.org/10.1016/j.cities.2023.104443

Milovanovic, S., Cvitković, I., Stojanović, K., & Klečina, A. (2024). *Belgrade's sustainable transition: Reflections by urban planners on change in planning urban green space.* https://doi.org/10.20944/preprints202407.1155.v1

Mocák, P., Kvetoslava, M., René, M., János, P., Piotr, P., Mishra, P. K., Katarína, K., & Michaela, D. (2022). 15-minute city concept as a sustainable urban development alternative: A brief outline of conceptual frameworks and Slovak cities as a case. *Folia Geogr., 64*, 69.

Moreno, C. (2016). *La ville du quart d'heure: pour un nouveau chrono-urbanisme*. La Tribune.

Moreno, C., Allam, Z., Chabaud, D., Gall, C., & Pratlong, F. (2021). Introducing the "15-minute. City": Sustainability, resilience and place identity in future post-pandemic cities. *Smart Cities, 4*, 93–111. https://doi.org/10.3390/smartcities4010006

Muntele, I., Sîrbu, C.-C., & Ostopocivi, I.-V. (2019). Historical Moldavia–from demographic expansion to a shrinking region. *Revue Roumaine de Géographie/Romanian Journal of Geography, 63(1)*, 3–17.

Noworól, A., Kopycinski, P., Hałat, P., Salamon, J., & Hołuj, A. (2022). The 15-Minute City—The geographical proximity of Services in Krakow. *Sustainability, 14*, 7103. https://doi.org/10.3390/su14127103

Rišová, K. (2020). Walkability research: Concept, methods and a critical review of post-socialist studies. *Geografický Časopis–Geographical Journal, 72(3)*, 219–242. https://doi.org/10.31577/geogrcas.2020.72.3.11

Roșu, L., Istrate, M., & Bănică, A. (2018). Passenger car dependency and consequent air pollutants emissions in Iasi metropolitan area (Romania). *Environmental Engineering & Management Journal (EEMJ), 17(4)*.

Rufat, S. (2012). Existe-t-il une « mauvaise » résilience ? In *La résilience urbaine* (pp. 195–241). *Les villes face aux catastrophes.*

Sailer-Fliege, U. (1999). Characteristics of post-socialist urban transformation in East Central Europe. *GeoJournal, 49*, 7–16. https://doi.org/10.1023/A:1006905405818

Sandu, A. (2023). The post-socialist cities from central and Eastern Europe: Between spatial growth and demographic decline. *Urban Studies*. https://doi.org/10.1177/00420980231189261

Sandu, A., Bănică, A., & Muntele, I. (2021). Urban resilience: An instrument to decode the post-socialist socio-economic and spatial transformations of cities from central and Eastern Europe. *East. J. Eur. Stud., 12*, 170–195. https://doi.org/10.47743/ejes-2021-SI08

Sfîcă, L., Ichim, P., Apostol, L., & Ursu, A. (2018). The extent and intensity of the urban heat island in Iaşi city, Romania. *Theoretical and Applied Climatology, 134*, 777–791.

Schmidt, S., Fina, S., & Siedentop, S. (2015). Post-socialist sprawl: A cross-country comparison. *European Planning Studies, 23*, 1357–1380. https://doi.org/10.1080/09654313.2014.933178

Smith, D. M. (1996). The Socialist City. In G. Andrusz, M. Harloe, & I. Szelenyi (Eds.), *Cities after socialism* (pp. 70–99). Blackwell Publishers Ltd. https://doi.org/10.1002/9780470712733.ch3

Song, G., He, X., Kong, Y., Li, K., Song, H., Zhai, S., & Luo, J. (2022). Improving the spatial accessibility of community-level healthcare service toward the '15-Minute City' goal in China. *ISPRS International Journal of Geo-Information, 11*, 436. https://doi.org/10.3390/ijgi11080436

Stanilov, K. (Ed.) (2007). The post-socialist city: Urban form and space transformations. *In Central and Eastern Europe after socialism, GeoJournal Library*. Springer, Dordrecht.

Sun, Y., & Luo, S. (2024). A study on the current situation of public service facilities' layout from the perspective of 15-minute communities—Taking Chengdu of Sichuan Province as an example. *Land, 13*, 1110. https://doi.org/10.3390/land13071110

Sykora, L. (1998). Commercial property development in Budapest, Prague and Warsaw. In G. Enyedi (Ed.), *Social change and urban restructuring in Central Europe*. Akadémiai Kiadó.

Sykora, L., & Bouzarovski, S. (2012). Multiple transformations: Conceptualising the post-communist urban transition. *Urban Studies, 49*, 43–60. https://doi.org/10.1177/0042098010397402

Tarasevych, O., Akhromkin, Y., Volkova, A., Zablodska, D., & Akhromkin, A. (2023). Resilient openness of eastern European cities in the conditions of sustainable development. *Probl. Ekorozwoju, 18*, 111–119. https://doi.org/10.35784/pe.2023.1.11

Teixeira, J. F., Silva, C., Seisenberger, S., Büttner, B., McCormick, B., Papa, E., & Cao, M. (2024). Classifying 15-minute cities: A review of worldwide practices. *Transportation Research Part A: Policy and Practice, 189*, 104234. https://doi.org/10.1016/j.tra.2024.104234

Tsenkova, S. (2006). Beyond transitions: Understanding urban change in post-socialist cities. In Tsenkova, S., Budic, Z.N. (Eds.), *The urban mosaic of post-socialist Europe: Space, institutions and policy, contributions to economics*. Physica-Verl,, pp. 21–50.

Tuvikene, T., Sgibnev, W., Zupan, D., Jovanović, D., & Neugebauer, C. S. (2020). Post-socialist infrastructuring. *Area, 52(3)*, 575–582. https://doi.org/10.1111/area.12590

United Nations. (2018). *Tracking progress towards inclusive, safe, resilient and sustainable cities and human settlements: SDG 11 synthesis report–high level political forum 2018*. United Nations.

Urban Accessibility

Assessing Spatially the "15-Minute City" in Spanish Cities: The Case of Granada

Carme Miralles-Guash, Laia Mojica, Carmen Lizárraga, and Alejandro Luis Grindlay

1 Introduction

In recent years, the concept of urban proximity, epitomised by the "15-minute city" (Moreno et al., 2021), has emerged as a key tool to support sustainable mobility and the reorganisation of urban life around accessibility rather than speed. This concept has spread beyond academic literature to the point of representing one of the most prominent ideas in the field of urban policies. Compact, diverse, vital, and accessible cities promote walking. Such urban settings encourage a healthy lifestyle, reducing car dependency and benefitting the environment. In addition, they foster social interaction, equal opportunities, and provide economic advantages for residents and the public sector. The relevance of this model, reducing mobility and urban transport needs, was evidenced during the recent global pandemic. It emphasises the need for residents to access essential services—such as schools, health care, shops, parks and transport—within a short walking distance. Its importance lies in addressing multiple urban challenges: Mitigating climate change through reduced car dependency, enhancing public health by promoting active mobility, and strengthening local economies and social cohesion. However, it also raises key challenges, including governance coordination, data availability, and the risk of spatial inequality if proximity benefits remain concentrated in central districts.

Access inequality represents one of the main barriers to the implementation of the "15-minute city." Unequal distribution of urban services can perpetuate socio-spatial segregation, excluding peripheral or disadvantaged neighbourhoods from the benefits of proximity. Addressing this inequality is therefore essential to ensure that

C. Miralles-Guash · L. Mojica
Autonomous University of Barcelona, Barcelona, Spain

C. Lizárraga · A. L. Grindlay (✉)
University of Granada, Granada, Spain
e-mail: grindlay@ugr.es

L. Bouzouina et al. (eds.), *15-Minute Cities*, Footprints of Regional Science,
https://doi.org/10.1007/978-3-032-20840-8_13

proximity policies are socially just and inclusive, providing evidence-based guidance for governments seeking to align urban planning with sustainability and equity goals (Pozoukidou & Chatziyiannaki, 2021; Khavarian-Garmsir et al., 2023).

In Spain, the "15-minute city" has gained increasing relevance within the National Urban Agenda (Ministerio de Fomento, 2019) and municipal sustainable mobility plans. Spanish cities, characterised by compact morphology, mixed uses, and a tradition of walkable public spaces, offer favourable conditions for testing this model. At the same time, recent trends—such as urban sprawl, rising housing costs, and mobility inequalities—highlight the urgency of assessing how equitable access to basic services truly is (Ministerio de Transportes, Movilidad y Agenda Urbana, 2023).

This study analyses spatial proximity across the districts of four Spanish cities, examining the extent to which basic urban services and everyday destinations are equitably distributed. Through a spatially explicit approach, it contributes to the understanding of intra-urban disparities and the implications of proximity planning in the Spanish context. The aim is to identify patterns that can inform a better application of proximity policies and to support the transition toward more equitable and sustainable urban models (GEMOTT, 2024b).

The analysis has been carried out in four compact and diverse Mediterranean cities, showing that their historical urban development fits perfectly within the concept explored in this book. Their exceptional suitability stems from their long-standing compactness, mixed-use structure, and dense service networks, which naturally align with proximity principles. For instance, the average urban density of these cities (between 10,000 and 16,000 inhabitants per km^2) and their traditional street patterns facilitate short-distance travel and active modes of mobility (INE, 2024).

2 Methodology

The concept of proximity, typified in the "15-minute city," refers to the possibility of accessing everyday destinations and services on foot within a given time. Proximity is calculated as the time needed to reach these destinations via pedestrian routes, such as pavements, pedestrian crossings, parks, and squares (Moreno et al., 2021; Ferrer-Ortiz et al., 2022).

These destinations and services are grouped in five basic functions that any city should offer so that residents can enjoy a quality urban life: Care, education, provisioning, leisure and culture, and public and active transport. This functional grouping is directly based on the classification proposed by Ferrer-Ortiz et al. (2022), which identifies these five categories as fundamental to the daily accessibility of residents at the neighbourhood scale. It is consistent with other frameworks in the literature addressing proximity and walkable accessibility (Marquet & Miralles-Guasch, 2015; Moreno et al., 2021). Each category has been assigned equal weight, assuming that all contribute similarly to fulfilling residents' everyday needs within walkable proximity. Future research could refine this approach by applying differentiated weights to reflect the relative importance of specific services for different social groups (e.g., children, elderly, or workers).

The urban functions and facilities considered are displayed in Table 1.

Based on previous studies (Marquet & Miralles-Guasch, 2015; Ferrer-Ortiz et al., 2022), a spatial analysis to determine travel times to these facilities through ArcGIS has been carried out. They have been calculated based on the walkable urban layout, which includes pavements, pedestrian crossings and paths in parks and gardens, excluding roads, bicycle lanes, and other routes not suitable for pedestrians. A speed of 5 km/h is used, which is the average walking speed of a standard adult in the urban environment. The shortest route is chosen, calibrated with Google Maps. With this result, the minimum time needed to reach each service is obtained, as well as the provision of facilities available within a 15-min range. In order to be comparable, the results have been adjusted to an orthogonal grid of 100 × 100 m for all cities. By relating the result acquired with population data for each district, the percentage of the population that can access basic urban functions on foot in less than 15 min is obtained.

Proximity values are therefore derived from the minimum walking time to a set of 27 facilities considered fundamental at the neighbourhood scale (Ferrer-Ortiz et al., 2022). This framework justifies the inclusion of facilities such as nurseries and schools, while excluding city-scale services such as hospitals or universities, as these serve wider catchment areas and correspond to non-daily urban functions.

These proximity areas have been calculated at the municipal level for the four cities considered. In each municipality, the resulting maps have been restricted exclusively to the urban area. To define this delimitation, the cadastral urban limits have been used, excluding all rural areas. Proximity of facilities and services at the neighbourhood level have been taken into account. This explains the exclusion of facilities at the city level, such as hospitals and universities. Large transport infrastructures, such as ports and airports in these cities, have also been excluded.

A wide variety of data sources have been used. For each city, the location of facilities has been obtained from a range of sources, such as the databases of city

Table 1 Urban functions and facilities considered in the spatial analysis

Care	Education	Provisioning	Leisure and Culture	Public and Active Transport
– Health centres. – Social services. – Day centres.	– Nurseries. – Schools. – Secondary schools,	– Supermarkets. – Markets. – Fresh food. – Other daily shopping. – Catering. – Miscellaneous services.	– Civic centres. – Libraries. – Theatres and cinemas. – Gyms. – Sports facilities. – Children's playgrounds. – Small parks. – Large parks.	– Daytime bus stops. – Night bus stops. – Train stations. – Bike sharing stations. – Cycling infrastructure.

Source: GEMOTT (2024b)

councils, transport entities, or chambers of commerce for facilities related to provisioning. To determine proximity to these provisions, the categories established by the Census of Economic Activities to identify food stores and other non-food daily purchases have been taken into account.

Activities primarily serving non-residential users (e.g. business services, wholesalers, construction) and those of occasional or non-daily use (e.g. communication centres, souvenir shops, antique shops) have been excluded, as the analysis focuses exclusively on basic, day-to-day services relevant to residents' local lives. Including such activities would distort the analysis, since their spatial distribution responds to commercial or tourist logics rather than everyday urban accessibility.

Walking times were calculated using the Open Street Map (OSM) road map, version 2022. OSM was selected because it provides a comprehensive and up-to-date representation of the walkable urban fabric, including sidewalks, pedestrian crossings, paths in parks and gardens, and public transport stops. In contrast, official municipal datasets typically represent only the vehicular network, without differentiating pedestrian areas from motorised traffic lanes. Therefore, OSM data enable a more accurate estimation of pedestrian routes and walking times for the purposes of proximity analysis.

As indicated, an average walking speed of 5 km/h—representing the mean pace of an adult in urban conditions—was applied to calculate travel times. The shortest routes were identified through network analysis and calibrated using Google Maps to ensure empirical accuracy.

Since proximity is conceived here as a fundamental condition of everyday urban life, the analysis focuses on inhabited areas where proximity to education, daily supplies, health services, and cultural or leisure amenities is most relevant. Consequently, large mono-functional zones devoid of residential population—such as forests, industrial estates, beaches, or port infrastructures—were excluded from the analysis.

For each district in the cities analysed, the following information has been determined and articulated as the core objectives of the study:

- First, to quantify the proportion of the population able to reach essential urban services and facilities on foot within 15 min, thereby evaluating the general level of spatial proximity achieved in each district.
- Second, to measure the diversity and concentration of destinations accessible within that time threshold, producing a comparative indicator of the richness of local opportunities.
- Third, to identify and interpret intra-urban disparities in accessibility, comparing the relative performance of districts to reveal spatial inequalities and to inform more balanced, inclusive, and evidence-based proximity policies in each city.

3 Overall Results

Our study examined the population's access to essential urban services, such as care, education, provisioning, leisure and culture, and public and active transport, in four Spanish cities: Barcelona, Palma de Mallorca, Valencia, and Granada. These cities were selected because they represent four compact and diverse Mediterranean urban models that combine high population density, mixed land use, and strong municipal commitment to sustainable mobility policies (GEMOTT, 2024a; MITMA, 2023). Their urban form and policy orientation make them particularly suitable for assessing the "15-minute city" approach within the Spanish context.

The analysis relied on the same harmonised geospatial database for all cities, in which facility locations were compiled, validated, and processed following a common protocol. The data were integrated and handled in ArcGIS, where the pedestrian network—derived from OpenStreetMap (2022)—was used to calculate walking times to each facility type. The resulting accessibility indicators were subsequently aggregated at the district level to allow comparison across cities.

The proximity results of districts in these four cities are summarised, and the case of the latter will be examined in more detail in the next section. Granada was chosen for detailed discussion because it represents a medium-sized Spanish city whose compact morphology and recent transport transformations provide a relevant contrast to the larger metropolitan cases of Barcelona and Valencia. Its analysis illustrates how the "15-minute city" model can be adapted to smaller urban scales and to different socio-spatial contexts within Spain.

The results confirm the general high proximity level existing in all the cities studied, but they also demonstrate the relevant urban inequality existing among districts reflected in the differing average number of facilities available in them. These variations are derived from the actual distribution of services obtained from the compiled municipal and OSM-based datasets rather than from any predefined assumptions. The frequency of facilities was determined empirically from these georeferenced data, and the percentage of population with access within 15 min was calculated considering all service categories included in the study (care, education, provisioning, leisure and culture, and public and active transport).

In the case of Barcelona particularly, the percentage of the population living in areas of its ten districts meeting the "15-minute city" accessibility requirements range from Nou Barris (72%), in the north, the most peripheral district, to the Eixample (92%), the most central and well-supplied sector. These results show how the level of urban facilities distribution follows a naturally decreasing law towards the peripheral areas, as the other two most central districts—Gracia (89%) and Ciutat Vella (85%)—show a high level of provision (Fig. 1).

In Palma de Mallorca, the capital city of the Balearic Islands, there is more disparity among its five districts. All of the population of the most central sector, Centre (100%), meet these accessibility requirements giving the highest level of distribution, whereas the lowest district, Platja de Palma (65%), the touristic beach area full of hotels and apartments, has limited urban facilities (Fig. 2).

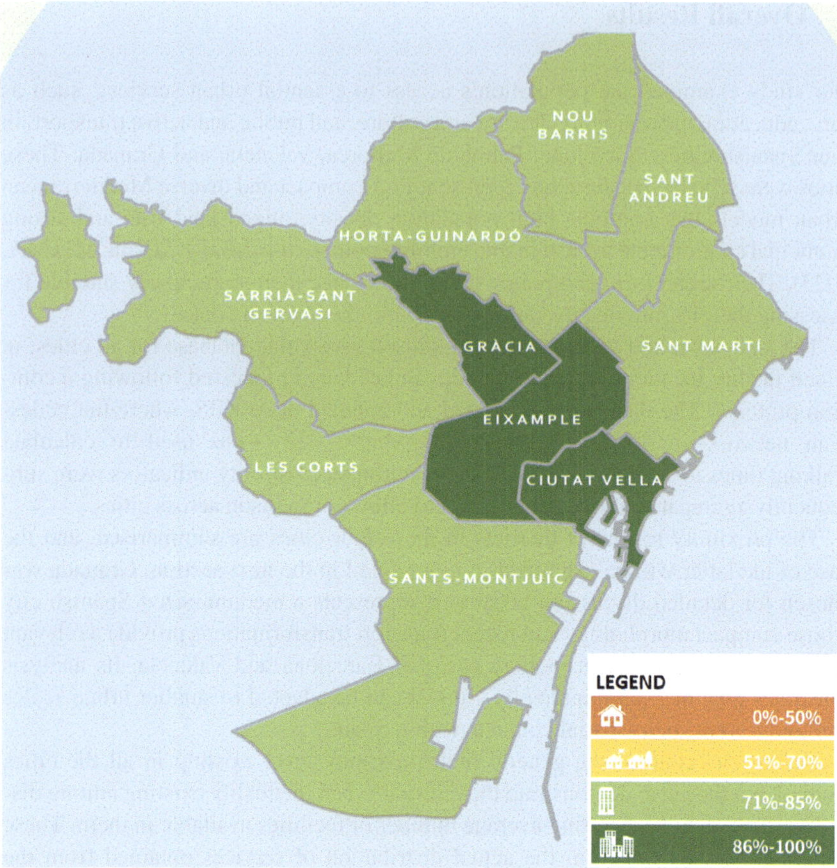

Fig. 1 The districts of Barcelona. Source: GEMOTT (2024a)

The Valencia case is even more diverse, as it has 19 districts and the disparity between them is even more acute. The 16 districts around the city centre, which have all or almost all of the population meeting the accessibility requisites, each of them with more than 90%—such as Extramurs (100%), Ciutat Vella and Camins al Grau (99%), to Quatre Carreres (91%)—in contrast to the three peripheral districts in the west, Poblats de L'Oest (87%), south, Poblats del Sud (63%), and in the north, Poblats del Nord (38%), respectively, with the lowest proximity level of all the districts studied, as the number of available facilities in these areas is very low (Fig. 3).

It is also worth noting the internal variation of accessibility among different service categories: While educational and care facilities tend to be more evenly distributed, leisure and cultural amenities or certain provisioning functions often concentrate in central districts, reinforcing spatial inequalities in daily accessibility.

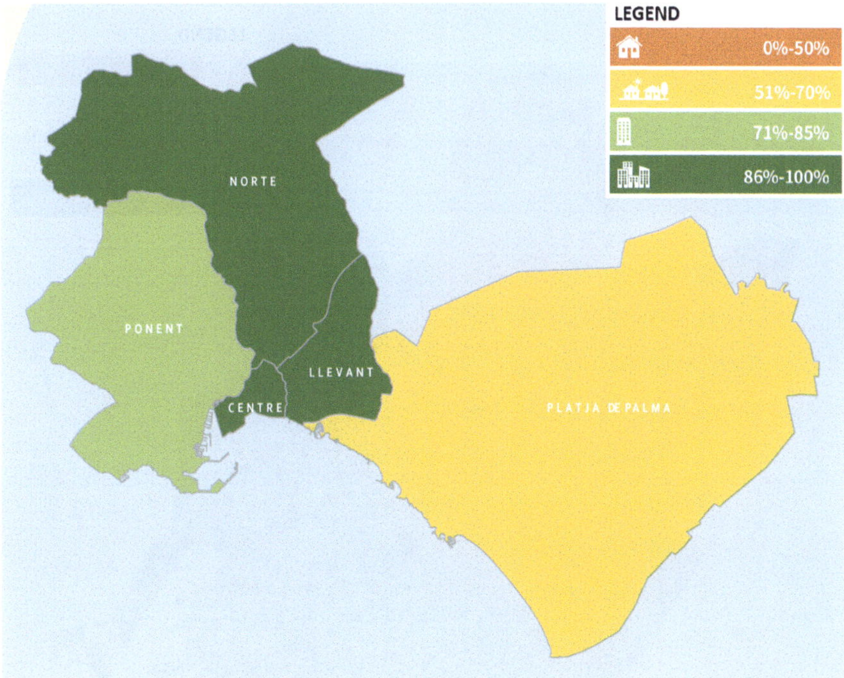

Fig. 2 The districts of Palma de Mallorca. Source: GEMOTT (2024a)

4 The Case of Granada

4.1 Location and Evolution

Granada city, located in the southeast of Spain, is well known for its most famous monument: the Alhambra, which is a magnificent medieval Arab palace and fortress complex, a UNESCO World heritage site since 1984, and one of the most popular tourist destinations in the country.

This city lies in the heart of the Baetic mountain ranges on the northwestern border of the Sierra Nevada and is situated on the fertile plain fed by three rivers, the Genil River and its two tributaries, the Darro and Beiro Rivers, which create a natural crossroads (Fig. 4).

Its historical evolution dates from before the Romans, who had a settlement on a hill on the western bank of the Darro River. However, the present city was founded in the eighth century, after the Muslim invasion of the Iberian Peninsula, on the river's eastern side. The city experienced major growth during the Middle Ages, although Granada's urban growth suffered stagnation after the Christian recovery of the Nasrid kingdom. The urban area remained roughly within the limits of the

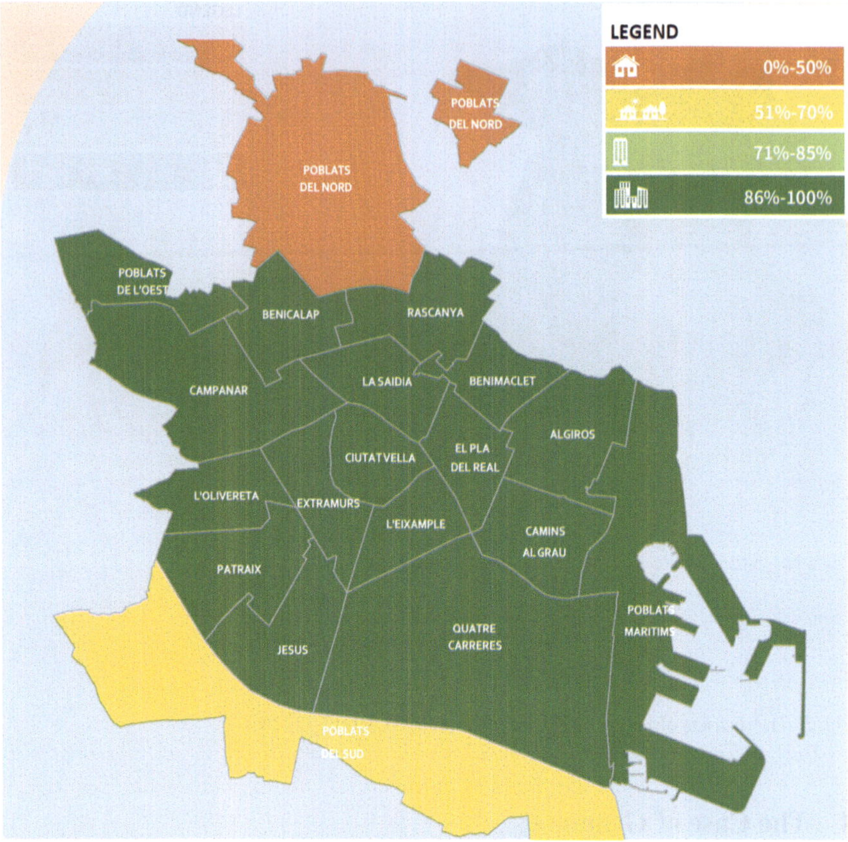

Fig. 3 The districts of Valencia. Source: GEMOTT (2024a)

Fig. 4 Siting of Granada. Source: Own elaboration over a *Google Maps* image

historical walls until the mid-twentieth century but it experienced a relevant growth in the last four decades of the last century. The defences were mostly demolished during the mid-nineteenth century, and the urban advance was densely concentrated along the main road axes, with the city reaching a population of approximately 250,000 inhabitants (Barrios Rozúa, 2025).

The city is surrounded by more than 30 traditional farming villages, which were connected in the 1920s by a historical tram system used for the transport of both people and the plain's agricultural products, leading to a metropolitan configuration. This transportation network disappeared during the 1970s due to the growth of motorisation and an obsolescence process, and since then the public transport system relied solely on buses, until a new metropolitan Light Rail Transit (LRT) was completed in 2017, with a single line connecting some of the north/south municipalities (Grindlay et al., 2021).

During the first decades of the present century urban growth was higher in the surrounding municipalities around Granada than in the capital city, which stagnated in terms of population growth. This was supported by an escalating rate of motorisation, and by the formation of a low-density suburban sprawl, which created a small metropolis of a quarter of a million inhabitants. Its daily mobility mainly consists of private motorised vehicles and the use of public transport scarcely makes up 10%, resulting in high levels of congestion and severe pollution problems, exacerbated by its geographical and climatic conditions, as was evidenced during the recent pandemic (Grindlay et al., 2021). However, the new tram development is reinforcing the centrality of the secondary populations around Granada and increasing the public transport percentage in the modal share (Grindlay et al., 2024).

These historical and morphological dynamics have direct implications for the current spatial distribution of services and accessibility levels. The compact urban core, with its dense and mixed-use fabric, tends to concentrate public facilities, retail and cultural amenities, whereas the peripheral and suburban municipalities, shaped by more recent and lower-density development, exhibit more limited-service provision. Consequently, the population density and the concentration of everyday destinations vary substantially across districts, influencing their proximity performance in the 15-minute city assessment.

Granada also possesses a truly unique cluster of academic institutions, with its 5 campuses serving more than 60,000 students spread throughout the city. In addition, its historical central district has experienced a pedestrianisation process in the last four decades establishing car-restricted areas.

The strong educational presence and pedestrian-friendly central areas reinforce the city's potential for proximity, while also generating specific accessibility challenges linked to tourist pressure and uneven housing availability. These features are crucial for understanding the observed results in the following analysis.

4.2 Present Situation

Despite its compact and dense urban development, the highest percentages of the "fifteen-minute city," as can be seen in Fig. 5, are concentrated in the central and southern parts of the city, due to the disparity in the levels of facility provision among districts. Conversely the peripheral areas are more urbanistically dispersed and have lower numbers of facilities.

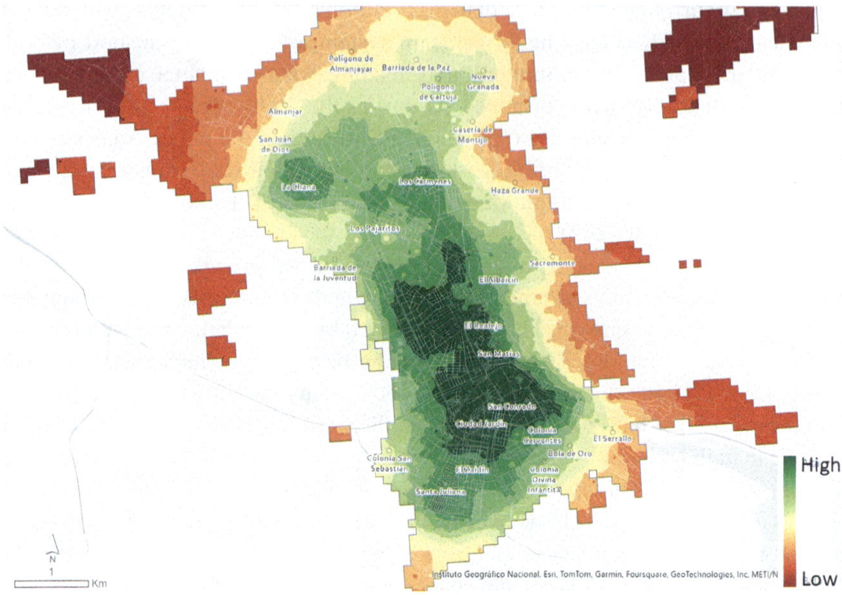

Fig. 5 The "15-minute city" Granada. Source: GEMOTT (2024a)

The city has been divided into eight districts, and all of them have over 80% of the population meeting the proximity requirements (Fig. 6). Table 2 shows the detailed results for each district in the five basic functions considered: Care, education, provisioning, leisure and culture, and public and active transport.

Table 2 presents two complementary indicators for each district. The first group of columns (POP_variables) indicates the percentage of the resident population that can access each functional category within 15 min on foot. The second group (FAC_variables) refers to the relative availability of facilities per district. This latter indicator is expressed as a percentage of the total number of facilities accessible in less than 15 min in each district compared to the district with the highest provision in the city. In other words, it allows a comparison of facility concentration among districts, with 100% representing the best-served district (Centro) and lower percentages indicating relatively fewer accessible destinations.

The district with the lowest proximity level (81%) is the most historical and touristic: The Albayzín. It is suffering a gentrification process, losing its traditional inhabitants for the new occasional residents, and having a reduced number of facilities for the local population—for example health centres (40%), nurseries (33%), supermarkets (38%), and sport facilities (17%). Given its old and limited street pattern, and its pedestrian-oriented character, it was the first quarter to become a car-restricted area—except for residents—in the late 90s (Fig. 7).

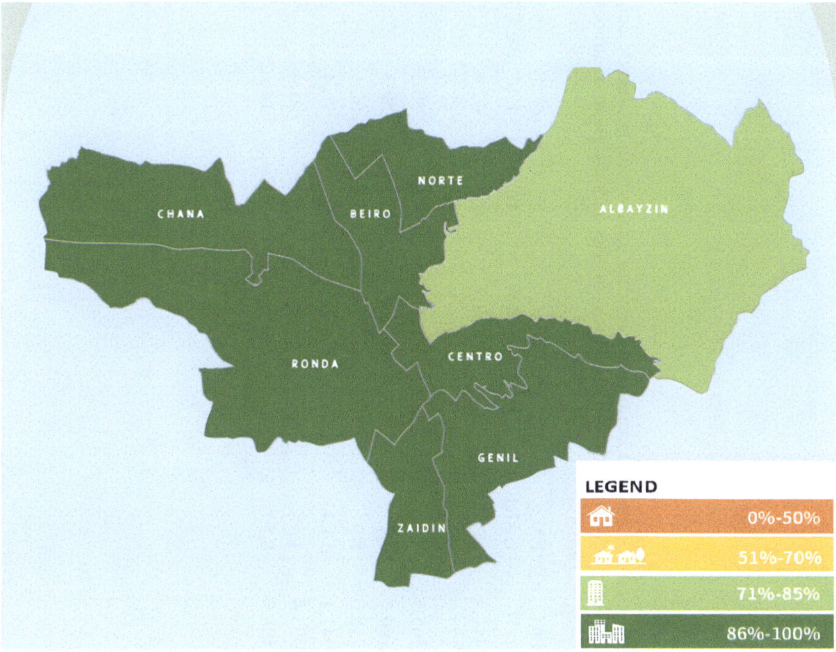

Fig. 6 The districts of Granada Source: GEMOTT (2024b)

The area with the next lowest proximity level is the Genil district (88%), a more modern locality in the southeast of the city, developed in the second half of the last century. It also has a relatively low level of facilities, for example supermarkets (41%) and libraries (33%).

The remaining six districts have more than 90% of the population living in areas meeting the 15-minute city accessibility requirements, Beiro and Norte (92%), both in the northern and traditionally poorer area of the city, Chana (95%), and Centro, Ronda, and Zaidín with the same high proximity level (96%).

Centro and Ronda are the most central districts and are those with the highest level of facilities, thus the highest percentage of population meeting these accessibility requirements. These are followed by Chana and Zaidín, both popular and vibrant working-class quarters formed in the fifties and sixties, with larger developments of public housing at opposite ends -eastern and western- of what was then the existing city and is now its central area.

Table 2 Percentage of population, which can access the different urban functions and facilities within 15 min (care, education, provisioning, leisure and culture, and public and active transport), and percentage of facilities accessible in less than 15 min in a district compared to the district with the greatest number of accessible destinations in the city

District	POP_ TOT	POP_ CARE	POB_ EDU	POP_ PROV	POP_L&C	POP_P&AT	FAC_ TOT	FAC_ CARE	FAC_ EDU	FAC_ PROV	FAC_L&C	FAC_P&AT
Albayzín	81%	94%	80%	80%	85%	64%	441.9	8	23.7	258.3	109.2	42.6
Genil	88%	92%	93%	83%	86%	87%	377	8.4	27	169.8	83	88.8
Beiro	92%	99%	94%	74%	93%	100%	365.5	11	24.1	169	71.2	90.2
Norte	92%	90%	94%	100%	89%	86%	160.8	5.6	19.9	25.1	50.9	59.3
Chana	95%	96%	96%	97%	94%	95%	218.6	8.6	20.5	86.4	41.1	62
Centro	96%	100%	92%	100%	94%	93%	941.1	14.4	31.3	664.8	154.2	76.4
Ronda	96%	100%	88%	94%	96%	100%	903.4	11.6	30.3	606.3	163.8	91.4
Zaidín	96%	99%	100%	83%	97%	100%	498.9	8.5	34.7	215.2	124.2	116.2

Source: GEMOTT (2024b)

Fig. 7 View of the Albayzín quarter from the tower of the Alhambra fortress. Photo: Grindlay Moreno et al., 2024

5 Closing Remarks

The urban characteristics of a city play a fundamental role in stimulating and encouraging active mobility and connectivity, as has been shown in the cases studied. Compact, diverse, vibrant and liveable "proximity cities" promote walking, providing substantial and equitable benefits for health, the environment and the economy. In the four Spanish cities analysed, the morphological compactness and the concentration of facilities within mixed-use neighbourhoods have been key for supporting short-distance travel and to reduce the reliance on motorised transport. These structural features confirm the potential of Mediterranean urban forms to operationalise the 15-minute city concept.

The results have confirmed the general high proximity level existing in each of the historical and Mediterranean cities analysed, but have also highlighted the relevant inequality existing between districts, shown in the different average number of facilities available in them. These findings provide an empirical basis for recommending specific policy actions aimed at improving the spatial distribution of essential services. First, urban planning authorities should prioritise the location of underrepresented facilities—particularly care, leisure, and cultural amenities—in peripheral districts where accessibility gaps are most evident. Second, investments in public and active transport infrastructure could reinforce connectivity between districts, mitigating inequality while maintaining a low-carbon mobility model. Third, the integration of accessibility indicators into local planning instruments would allow municipalities to continuously monitor proximity levels and evaluate the social equity of urban interventions.

However, as these results are based on an average adult walking speed, future research can be more precise with respect to other social sectors such as the elderly and children. Further methodological refinements could also include differentiated weighting of services according to user needs and temporal variations in accessibility (e.g., service opening hours or topographical constraints). Expanding the analysis to the metropolitan scale would also help to understand how suburban areas relate to the proximity principles observed within the compact city.

This analysis constitutes a useful tool for public administrations to promote proximity policies. There is a necessity to balance the level of provision of amenities and services among urban districts to reduce the existing wide disparity and to ensure that all areas of our cities adhere more closely to the concept of the "fifteen-minute city" model, creating a human-centred urban environment.

In summary, the chapter demonstrates that proximity-based planning is not only a theoretical model but a measurable spatial condition that can guide equitable and sustainable urban policies. Strengthening everyday accessibility at the neighbourhood level is essential for building cities that are socially cohesive, resilient, and aligned with the principles of the 15-minute city.

References

Barrios Rozúa, J. M. (2025). *Granada, ciudad milenaria: historia urbana y arquitectónica*. Comares.

Ferrer-Ortiz, C., Marquet, O., Mojica, L., & Vich, G. (2022). Barcelona under the 15–Minute City lens: Mapping the accessibility and proximity potential based on pedestrian travel times. *Smart Cities, 5*, 146–161. https://doi.org/10.3390/smartcities5010010

GEMOTT, Grup d'Estudis en Mobilitat, Transport i Territori. (2024a). *Atlas de Ciudades para la Movilidad Activa en España*. UAB. Available at: https://www.movactiva.es/

GEMOTT, Grup d'Estudis en Mobilitat, Transport i Territori. (2024b). La proximidad en las ciudades españoles. In *La ciudad de los 15 minutos: Barcelona, Granada, Palma de Mallorca y Valencia*. UAB. Available at: https://gemott-uab.cat/wp-content/uploads/2024/02/GEMOTT_Informe_Final-1.pdf

Grindlay, A. L., Ochoa-Covarrubias, G., & Lizárraga, C. (2021). Sustainable mobility and urban space quality: The case of Granada, Spain. *International Journal of Transport Development and Integration, 5*(4), 309–326. https://doi.org/10.2495/TDI-V5-N4-309-326

Grindlay Moreno, A. L., Molero Melgarejo, E. & Lizárraga Mollinedo, C. (2024). Metropolitan centrality analysis and development potential for Granada's tram stations. *Book of Abstract. Proximity Planning International Congress*. UPM, 79-80. https://doi.org/10.20868/UPM.book.82998

Instituto Nacional de Estadística (INE). (2024). *Padrón municipal y densidad urbana de las principales ciudades españolas*. INE. Available at: https://www.ine.es

Khavarian-Garmsir, A. R., Sharifi, A., & Sadeghi, A. (2023). The 15-minute city: Urban planning and design efforts toward creating sustainable neighborhoods. *Cities, 132*, 104101. https://doi.org/10.1016/j.cities.2022.104101

Marquet, O., & Miralles-Guasch, C. (2015). The walkable city and the importance of the proximity environments for Barcelona's everyday mobility. *Cities, 42*, 258–266. https://doi.org/10.1016/j.cities.2014.10.012

Ministerio de Fomento. (2019). *Agenda Urbana Española*. MFOM. Available at: https://cdn.mivau.gob.es/portal-web-mivau/aue/doc/AUE-Doc-Completo_compressed.pdf

Ministerio de Transportes, Movilidad y Agenda Urbana. (2023). *Estrategia de Movilidad Segura, Sostenible y Conectada 2030*. MITMA. Available at: https://www.transportes.gob.es/movilidad-sostenible/estrategia-de-movilidad-segura-sostenible-y-conectada-2030

Moreno, C., Allam, Z., Chabaud, D., Gall, C., & Pratlong, F. (2021). Introducing the "15-minute city": Sustainability, resilience and place identity in future post-pandemic cities. *Smart Cities, 4*(1), 93–111. https://doi.org/10.3390/smartcities4010006

Pozoukidou, G., & Chatziyiannaki, Z. (2021). 15-minute city: Decomposing the new urban planning Eutopia. *Sustainability, 13*(2), 928. https://doi.org/10.3390/su13020928

Enhancing Urban Mobility: Analysis of the 15-Minute City for an Accessible and Eco-Friendly Urban Environment in Utrecht

Nolwenn Sourimant, Peter Nijkamp, Karima Kourtit, and Louafi Bouzouina

1 Introduction

Today, the concept of spatial mobility and accessibility is becoming increasingly challenging: the Covid-19 pandemic, the geopolitical tensions and wars, and the different natural disasters are all events that need to be considered when crafting public policies for sustainable transportation, especially in an urbanised world. In fact, more than half of the world's population lives in cities, and the World Bank's trends estimate that this figure will rise to 70 percent by 2050. Managing the city is one of today's key issues in both developed and developing countries. Various public policies are being implemented in large cities, including the 15-minute city idea designed by Carlos Moreno (Moreno, 2024; Moreno et al., 2021; Pozoukidou & Chatziyiannaki, 2021). Its aim is to reshape the city so that all its inhabitants can

N. Sourimant
Regional and Interdepartmental Directorate for the Environment, Planning and Transport, Île-de-France Roads Authority, Paris, France

P. Nijkamp (✉)
EHERO–Erasmus University, Erasmus University, The Netherlands, Rotterdam, Netherlands

University College Wrocław (WSKZ), Wrocław, Poland

SPPM Tsinghua University, Beijing, China, Tsinghua University, Beijing, China, Beijing, China

K. Kourtit
Faculty of Management, Open University of the Netherlands, Heerlen, The Netherlands

EHERO – Erasmus University Rotterdam (EUR), Rotterdam, The Netherlands

University College Wrocław (WSKZ), Wrocław, Poland

L. Bouzouina
LAET, ENTPE, University of Lyon, Vaulx-en-Velin, France
e-mail: louafi.bouzouina@entpe.fr

L. Bouzouina et al. (eds.), *15-Minute Cities*, Footprints of Regional Science,
https://doi.org/10.1007/978-3-032-20840-8_14

access essential services in less than 15 minutes by active modes such as walking or cycling. This is a key element in facilitating travel, reducing traffic congestion, and improving air quality, which are objectives of the 15-minute city. Indeed, the concept is in line with the UN (United Nations) Sustainable Development Goals, specifically Sustainable Cities and Communities, Goal 11 (Moreno et al., 2021). By reducing the role of motorised transport in the city (particularly the car), we will be able to reduce the share of GHG (Greenhouse Gas) emissions (31% of CO_2 emissions for the transport sector) and reduce the number of deaths and diseases due to air pollution (Albayati et al., 2021), which is currently a major public health problem. The aim of the study is to examine how the current car-dominated model of the city may be influenced by the growing use of new transport modes such as cycling. In fact, this study aims to investigate and shed light on various aspects of the 15-minute city concept in the metropolitan area of Utrecht by studying the quality and accessibility of cycling infrastructures.

The 15-minute city concept has been tested in cities around the world, including Paris, Melbourne, and Barcelona. These experiments have demonstrated the benefits of urban planning based on the proximity of services and activities by active modes. For example, in the city of Utrecht in the Netherlands, accessibility to essential services is possible within 15 minutes by bicycle (Knap et al., 2023) for 100% of the inhabitants. However, this study also shows the large differences in accessibility between Utrecht city centre and the outskirts, indicating a problem of equity. This issue is also highlighted in the study of Logan et al. (2022). This article shows in particular that the way in which accessibility is calculated may hide certain inequalities that actually exist. This study is therefore confronted with intertwining differences between theoretical and practical results. The Netherlands is one of the countries that has experimented with the 15-minute city model. This country has embraced cycling as a mode of transport, with more than 27 percent of all trips made by bicycle. Utrecht is a city that has invested heavily in cycling infrastructure and public transport. The quality of transport infrastructure plays an important role in the 15-minute city, as this concept includes a 'quality of life' component in relation to accessibility.

This article aims to investigate the quality of the cycling network in Utrecht in terms of the key points of the 15-minute city concept. First, the accessibility of cycle paths to services (leisure, living, and work) is analysed. In fact, Moreno (2024) advocated returning to short travel routes accessible by active modes. Next, the quality of cycling (infrastructure, physio-chemical parameters) will be studied, which corresponds to the objective of a high quality of life. To what extent do the cycle paths in Utrecht fulfil two key goals of the 15-minute city concept: accessibility to essential services (leisure, housing, work) and the provision of high-quality living conditions? By answering this research question, this paper aims to provide evidence-based information on the effectiveness of the 15-minute city approach in different urban contexts and to support decision-making in the implementation of public policies regarding urban sustainability.

2 Literature Review

Against the backdrop of climate urgency, a change in the way people live in cities is becoming a necessity. In his interview at the ESPON conference—European Observation Network for Territorial Development and Cohesion (ESPON, 2022), Carlos Moreno announced that the concept of the 15-minute or 30-minute city in low-density areas must incorporate six components. Two of these are the desire to live in cities with good conditions (air quality, temperatures, noise) and to access a small-scale economy quickly by active modes such as walking or cycling.

Walking as the main mode of transport has been studied on numerous occasions in order to examine the various consequences for human health and the improvement in quality of life in city centres. In Weng et al. (2019), the authors start from the hypothesis that the possibility of getting around a neighbourhood on foot is a key factor in promoting the health and well-being of residents. Good pedestrian accessibility enables residents to reach services, green spaces, and other amenities quickly and easily, thereby promoting an active lifestyle and a better quality of life. Similar conclusions are also drawn from Riggs (2016), who shows that the areas least accessible by foot are the areas with the highest percentage of obesity. The issue of access to green spaces to improve people's health is also addressed by Lin and Chen (2023). The aim of this article is to study whether the city of Qingdao in China complies with the 2018 accessibility standards. In particular, they mention access to parks larger than 4000 square metres, which must be accessible within 300 m of people's homes. People living near green areas are more likely to be physically active (Mytton et al., 2012).

The services that must be accessible to all are defined using a basket of goods. This basket is presented by Caubel (2006) in his study. It includes leisure activities, purchases necessary for survival, human health, and administrative procedures. Access to these services by active modes of transport would be possible in less than 15 minutes (Steuteville, 2023). Within the perimeter defined in their study, a 15-minute cycling journey would provide access to the main cultural, medical, and higher education establishments, as well as to parks and the main employers. Access to intercity public transport may also be a possible sustainable option. Accessibility by bicycle, therefore, includes a connectivity component to intercity public transport.

Both empirical and theoretical approaches postulate that infrastructure dedicated to bicycles makes it possible to improve accessibility to services belonging to a basket of goods. In the study conducted by Mora et al. (2021), the authors hypothesise that cycling infrastructure is a determining factor in accessibility: they postulate that the presence of a well-developed cycling infrastructure, in particular safe and well-connected cycling paths, facilitates accessibility for cyclists in a given city. This article proposes public policy recommendations to improve the cycling network and enable equitable accessibility by bicycle, including disadvantaged urban areas. The quality of the cycling infrastructure is frequently neglected. However, it would be interesting to study it by proposing a travel quality indicator, including both air quality and infrastructure quality.

The methods and measurement tools used to study travel in different European and American cities are diverse. They are based in particular on the use of GIS (Geographic Information System) data, mobility data, and population surveys (Knap et al., 2023; Mora et al., 2021; Riggs, 2016). These different studies take approaches at different scales, ranging from the microscopic (individuals) to the larger scale of neighbourhoods and districts in a province. The various articles look at active travel from the angle of the 15-minute city concept introduced by Moreno. In Knap et al. (2023), the authors suggested creating an accessibility indicator for cyclists in Utrecht and linking it to the city's spatio-socio-economic indicators. They use data that correspond to recorded trips from individual panel members. The results for Utrecht appear to be good, as 100% of the areas have the potential for 15-minute city accessibility to core services. However, the article focuses only on the accessibility component and not on the journey quality component, which may impact the experienced travel time. Most research on the 15-minute city focuses on accessibility. Consequently, many articles investigate the issue of cycling in the city and measure it using indicators of cyclability. However, only a few articles addressed the quality of travel, despite its importance.

3 Study Area and Data

3.1 Data Source

Located in the centre of the Netherlands, the city of Utrecht has frequently been the subject of studies on the 15-minute city concept. The city is structured in the form of several centres, making it a good example of a 15-minute city. The area of study in our article focuses on the city of Utrecht as well as the surrounding villages (Vleuten-De Meern). The data used in this article come from two sources. First, there is the OpenStreet Map data, which lists housing, services, and cycle paths in the city of Utrecht. This data is continually updated by contributors to the OSM community around the world. The Utrecht data is relatively recent (between 2019 and 2023), but does not appear to be complete for all areas. The results obtained will therefore have to take into account this missing data. To extract this data from Open Street map, QuickOSM plugin was integrated into QGIS. The Open Street map housing data was extracted by listing all buildings with the key apartment or house. For the data related to cycle paths, roads, cycle paths and pedestrian zones where bikes are allowed have been identified. These metrics characterise the extent and connectivity of Utrecht's cycling network. To this Open Street Map cycle network, data from Gezond Stedelijk Leven ('Healthy urban life') has been integrated to identify the regional cycle network. Finally, a similar procedure was followed to identify the services that are relevant to map out. Certain services, cycle paths, or housing information may be missing or inaccurately represented in the dataset. This limitation could have implications for the research outcomes, as it may lead to an underestimation or overestimation of accessibility to essential services or the quality of the cycling network in some areas. To address this limitation, a cautious approach was taken during the data analysis and interpretation. Sensitivity analyses have been conducted to assess the potential impact of missing data on the overall findings. Additionally, discussions and conclusions drawn from this study take into account the potential uncertainties arising from incomplete data.

3.2 Choice of Goods Basket

The choice of services included in the accessibility measures is based on careful consideration of their frequency of use. It is important to consider that some services are used more frequently by the population than others. For example, a local grocery shop is often visited by local residents to meet their daily food needs, while an administrative centre may be less frequented, as it is mainly used for specific administrative formalities. Another crucial factor in the selection of services is the difference in priority given according to people's needs and constraints. In the case of childcare facilities such as kindergarten, it is essential to make them quickly accessible. Transporting young children can be difficult, so it is important to reduce the distance between home and kindergarten. Proximity to kindergarten is therefore crucial in facilitating the daily lives of parents and ensuring the well-being of children. In addition, the importance of rapid access to green spaces such as parks should be emphasised in accessibility measures. The availability of a park within a 5-minute walk can promote an active lifestyle and encourage people to take part in outdoor leisure activities. It will promote better health and general wellbeing for individuals, helping to create healthier, more balanced urban environments. In short, the selection of services summarised in Table 1 takes into account the frequency of use by the population, but also the specific needs and constraints faced by individuals. This approach aims to reduce the cognitive and psychological cost to the population by making essential services more easily accessible.

Table 1 Summary table of services studied

Services		
Destination	X-minute bike service	Number of points
Kindergarten	5	46
Convenience store	5	22
Green open space	5	78
Primary school	10	Unknown
Open market	10	Unknown
Bus stop	10	3500
Middle school	15	Unknown
Supermarket	15	85
Public service	15	75
Sport facility	15	123

The other data comes from Gezond Stedelijk Leven ('Healthy urban life') and is 'Green Lines' data. This is a project in which data on the quality of cycling paths and air quality is collected by placing sensors on bicycles. This data from 2018 is owned by and sourced from the municipality of Utrecht.

4 Methodology for Measuring Indicators

An important aim of our study is to determine an overall indicator of the quality and accessibility of cycle paths in Utrecht. The first step in assessing the quality of cycle paths is to calculate their accessibility. The method used is to obtain an Origin–Destination matrix between dwellings and services using the cycling network. This matrix is used to determine the journey time between each dwelling and all the related services in the city.

4.1 Creating the Cycle Network

Using OpenStreetMap map data, a cycle network was created in QGIS in Fig. 1. This network was created by grouping together all the roads that can be used by bicycles:

- For pedestrian routes, the aim is to see whether they are shared by pedestrians and cyclists.
- For roads with cars, Open Street Map identifies whether bikes are allowed. We therefore only keep roads where bikes are allowed.

4.2 Origin–Destination Matrix

The approach used to measure accessibility to services in the city of Utrecht employs the generation of origin–destination matrices. These matrices are created by considering the cycling network, where a travel speed of 20 km/h maximum is assumed. The calculated journeys from each dwelling to a service aim to determine the shortest route. However, it is important to acknowledge that individuals do not always opt for the shortest or fastest path when traveling. In fact, according to various studies (Harms et al., 2021; Manley et al., 2015), people do not systematically take the shortest route and prefer the best-known routes. This limit is hard to take into account because at this stage, actual GPS data of the individuals is not available. In order to generate the origin-destination matrices, the algorithm implements OD-Matrix analysis. This analysis results in a matrix consisting of origin-destination pairs represented as lines, which represent the network-based costs associated with

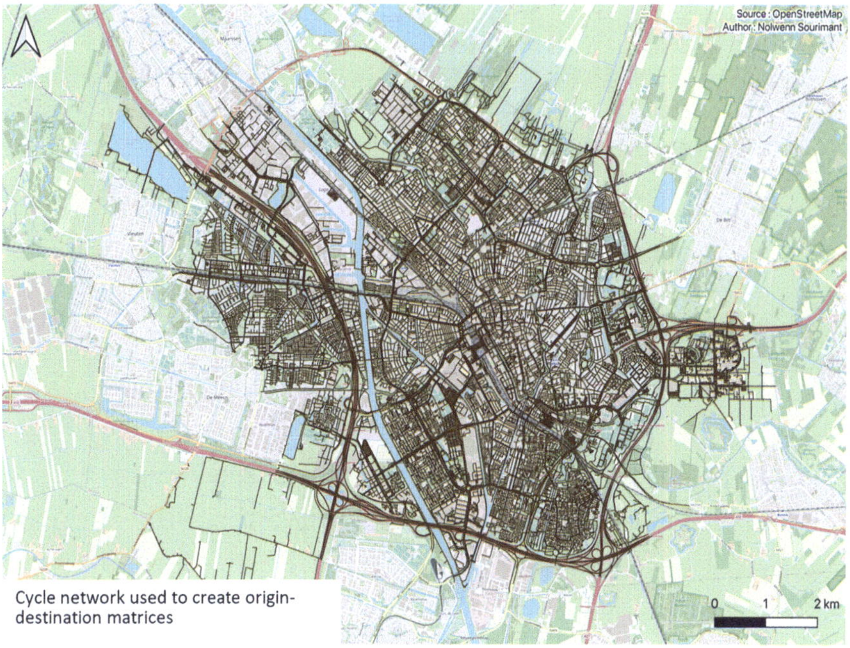

Fig. 1 Cycle network in Utrecht

traveling between two layers of points in the dataset (m:n). These cost values are listed in the attributes in the table of the resulting matrix (see Table 2).

4.3 Converting OSM Data into GPS Data

To geocode the results, OSM (OpenStreetMap) identifiers were converted into GPS coordinates using OpenStreetMap API. The OSM identifiers were extracted from the source and destination data, corresponding to entities such as dwellings and services. These identifiers were then divided into batches to facilitate easier processing. For each batch of OSM identifiers, OpenStreetMap API was used to retrieve the corresponding geolocation information by automating the requests using Python code. This approach allowed to efficiently convert OSM identifiers into GPS coordinates, which was essential for analysing and visualising the results using mapping tools such as QGIS.

Table 2 Example of origin–destination matrix

	Origin_id	Destination	Entry_cost	Network_cost	Exit_cost	Total_cost	Minute
1	w991087827	w35142657	29,9,073,754	1087,6,917,651	28,5,911,608	1146,1,903,014	3
2	w991087827	w35142657	29,9,073,754	1087,6,917,651	28,5,911,608	1146,1,903,014	3
3	w991087826	w35142657	22,8,348,680	1110,0461405	28,5,911,608	1161,4,721,693	3
4	w991087826	w35142657	22,8,348,680	1110,0461405	28,5,911,608	1161,4,721,693	3
5	w991087825	w35142657	48,4,875,717	1192,6,978,541	28,5,911,608	1269,7,765,856	4
6	w991087825	w35142657	48,4,875,717	1192,6,978,541	28,5,911,608	1269,7,765,856	4
7	w991087824	w35142657	16,0346153	1072,8,879,455	28,5,911,608	1117,5,137,216	3
8	w991087824	w35142657	16,0346153	1072,8,879,455	28,5,911,608	1117,5,137,216	3
9	w976399880	w306433879	50,5,211,425	399,7,037,491	47,3,588,507	497,5,837,423	1
10	w976399880	w306433879	50,5,211,425	399,7,037,491	47,3,588,507	497,5,837,423	1
11	w976399879	w306433879	26,3,385,909	494,8,507,504	47,3,588,507	568,5,481,920	2
12	w976399879	w306433879	26,3,385,909	494,8,507,504	47,3,588,507	568,5,481,920	2
13	w975767255	w583356110	25,1,931,036	346,8,238,950	33,4,688,299	405,4,858,286	1

4.4 Method for Assessing Air Quality

To record the air quality, we started with the Groene Fietroutes ('Free Bicycle Routes') database, which lists the cycle paths in Utrecht and provides data on pollutant concentration values. We encountered several problems with this database. First, not all the cycle paths are represented and not all the districts of Utrecht are covered. It therefore seems difficult to obtain representative air quality scores. In addition, the values were only taken at one point in time in 2018 and may not be consistent with annual averages. Therefore, two additional databases were included from the above-mentioned database on Gezond Stedelijk Leven. These data consist of layers covering the entire city of Utrecht and giving the pollution classification for particulate matter PM2.5 and PM10. These classes are divided according to the GES (Gezondheids Effect Screening) (Health Effect Screening) scale. A given concentration interval is assigned for each pollution class, ranging from 1 to 8 (Fast et al., 2017), as in the example of Table 3.

It is possible to find the share of each class for each district. Each class is given a score (e.g., class 6a has a score of 3.7/10). The final score associated with each district is calculated using the formula below:

$$s = p1.s1 + p2.s2 + \ldots + pn.sn$$

with pn, the proportion of each class in the total area of the district and sn, the score assigned to the class. These data are then used as the basis for empirical experiments.

Table 3 Example of GES-scores assigned based on PM10 concentration for a year, from Fast et al. (2017)

PM^{10} (annual average)	GES score		Environmental health quality	Colour		RGB-code
<2		1a				
2-4	1	1b	Good		Green	000-153-000
4-6		2a				
6-8	2	2b	Reasonable		Green-yellow	220-248-010
8-10		3a				
10-12	3	3b	Fairly moderate		Yellow	255-255-000
12-14		4a				
14-16	4	4b	Moderate		Yellow-orange	255-210-000
16-18		5a				
18-20	5	5b	Very moderate		Orange	255-165-000
20-22,5		6a				
22,5-25	6	6b	Unsatisfactory		Red	255-000-000
25-27,5		7a				
27,5-30	7	7b	Well unsatisfactory		Red-purple	192-000-000
>=30	8		Very unsatisfactory		Dark purple	128-000-128

5 Results

The results of processing the various accommodation-cycling services matrices reveal some information gaps. The majority of residential properties in Utrecht are well served within the times selected in the data section. There are still areas, particularly in the north of Utrecht and in the east, where services are less easily accessible. However, 90% of them can be reached in less than 15 minutes by bike, which confirms the results of Knap et al. (2023).

5.1 Accessibility Maps

Services that might be accessible in less than 5 minutes by bicycle (kindergartens, convenience stores, and parks of more than 4000 square metres) were studied using origin–destination matrices. The total time taken to get from homes to study services is shown on a map for all the dwellings (see Fig. 2).

It can be seen that the northern, western, and eastern districts are less accessible. Indeed, when we look at the journey times, they are more than 5 minutes by bicycle. Nevertheless, the initial results for 5-minute services are consistent with the results of the case study by Knap et al. (2023). In fact, the conclusions show that the entire study area is a 15-minute city for almost 100% of the population. Similar results were found for other services studied (kindergarten, parks, and grocery shops), which may help to validate the model presented. Clearly, these initial maps show that the majority of services meet the 5-minute accessibility constraint. We may also

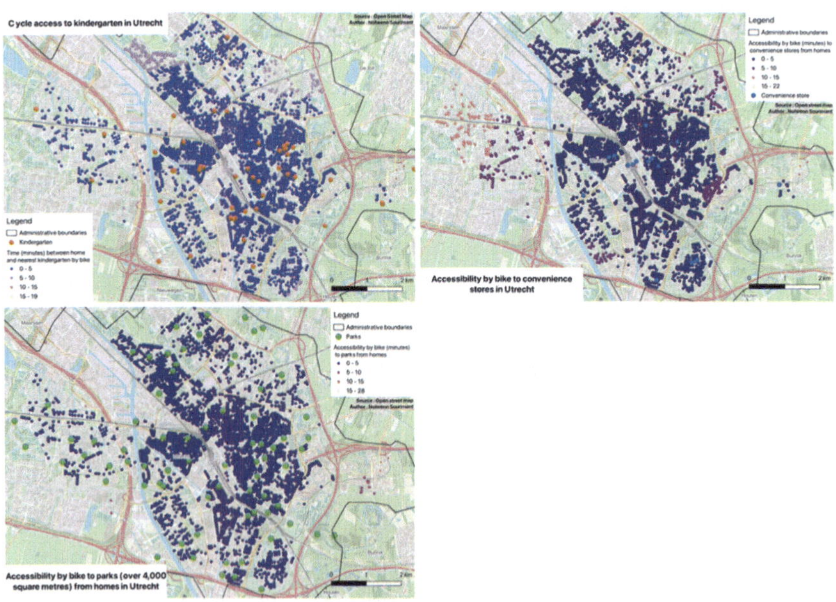

Fig. 2 Accessibility of services (kindergarten, stores, and parks) in less than 5 minutes by bike

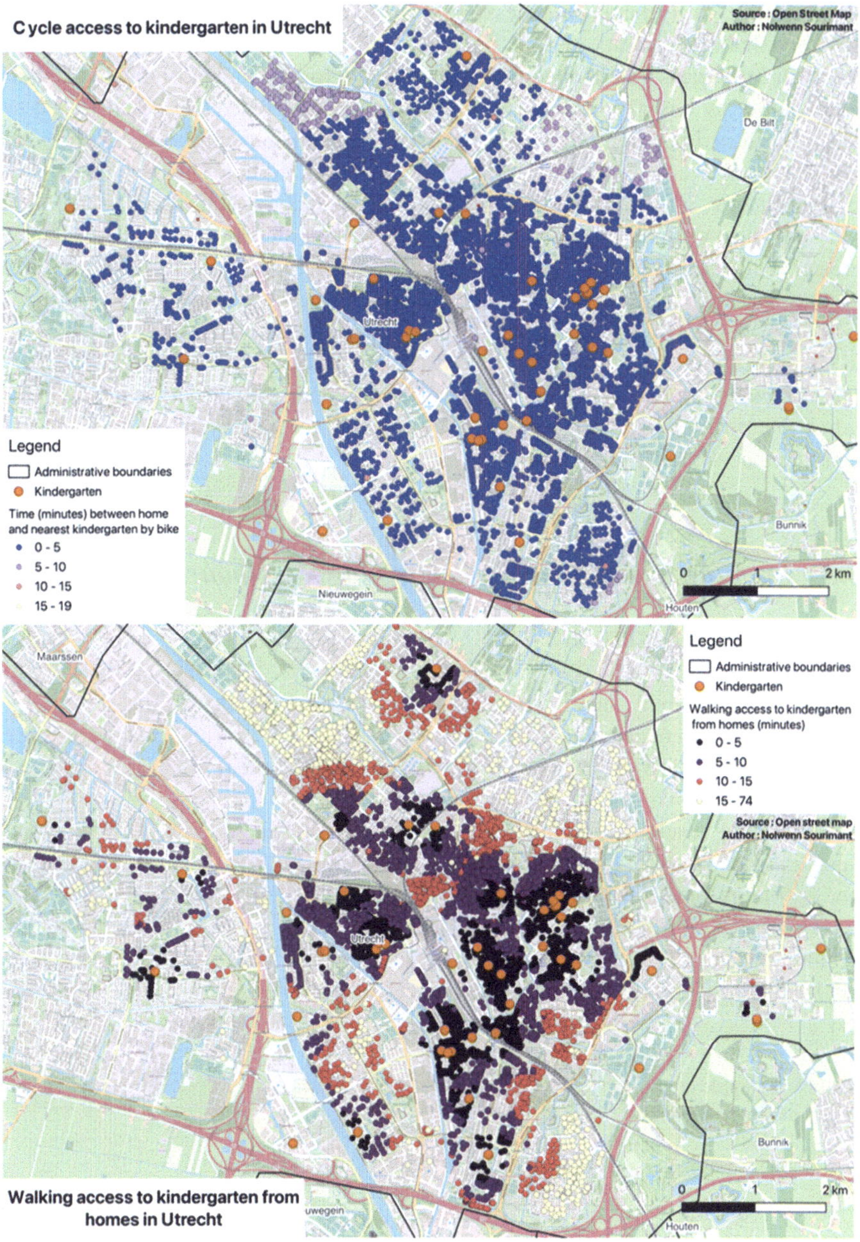

Fig. 3 Comparison of accessibility to kindergartens by bike and on foot

look at accessibility to these same services on foot to analyse the difference between the two active modes. To do this, the same origin–distance matrix method was used and adapted to a walking speed of 5 km/h. The results obtained for kindergartens are shown in Fig. 3. Analysis of accessibility to kindergartens from dwellings reveals a significant disparity between accessibility on foot and by bicycle. In fact, it was observed that accessibility on foot, i.e., the possibility of reaching a kindergarten in less than 5 minutes' walk, concerned a significantly lower number of dwellings than accessibility by bicycle over the same period. This difference can be attributed to several factors. First, walking is considerably slower than cycling, which limits the distance people can cover in 5 minutes. In addition, certain environmental constraints can also affect accessibility on foot, such as inconvenient roads or intersections, physical obstacles or less direct routes. Cycling, on the other hand, can cover greater distances in less time. In addition, well-developed cycling infrastructure, such as cycle paths, can encourage accessibility by bicycle by providing dedicated, safe routes for travel. In conclusion, the analysis shows that accessibility by bicycle in less than 5 minutes is clearly superior to accessibility on foot in the specific context of kindergartens. This highlights the need to improve pedestrian and cycling infrastructure to ensure equitable accessibility for all residents, independently of the chosen mode of transport.

5.2 *Creating Accessibility Indicators for Neighbourhoods*

5.2.1 Statistics by District

Using the accessibility maps above, statistics on accessibility to the service by neighbourhood were calculated. The city of Utrecht has 10 districts, and the number of data items per district makes it possible to see whether the results are representative or not, as shown in Table 4.

For further results from our study, the Vleuten-De Meern districts will not be taken into account because the study area is not complete and is therefore probably not well-representative of the district.

Figure 4 shows the uneven distribution of cycling journey times to access 5-minute services (kindergarten, parks, and convenience stores). As for the study of kindergartens (Fig. 4a and b), it can be seen that kindergartens in the centre of

Table 4 Example of origin–destination matrix

ID	Neighbourhood	NUMPOINTS
1	Noordwest	4580
2	Overvecht	727
3	West	4414
4	Leidsche Rijn	2853
5	Zuidwest	2218
6	Binnenstad	2317
7	Noordoost	4411
8	Oost	3193
9	Zuidwest	1628
10	Vleuten-De Meern	128

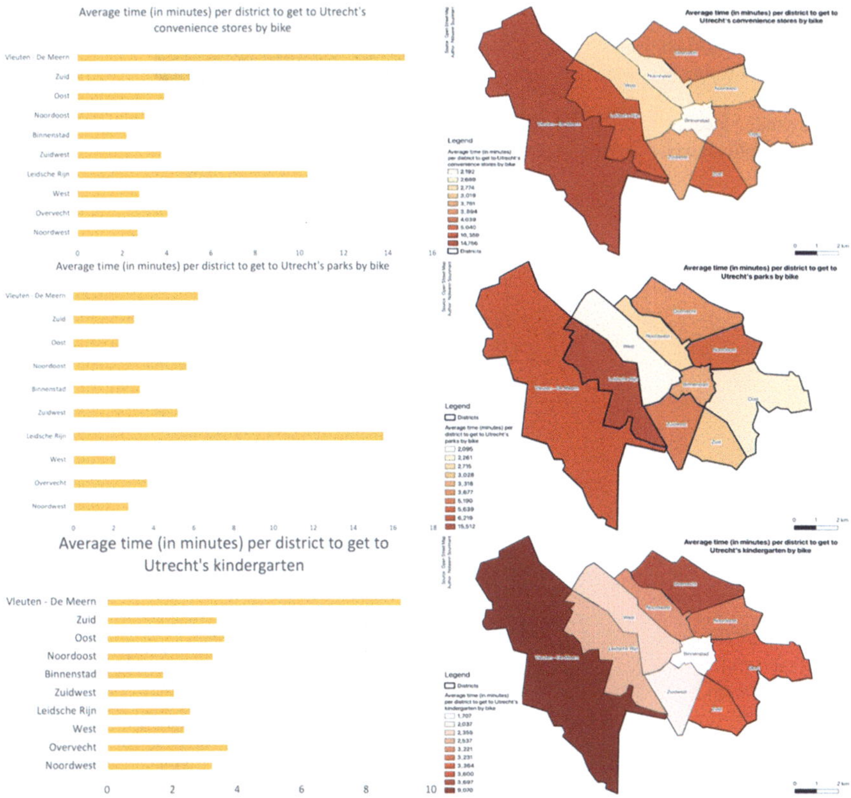

Fig. 4 Comparison of 5-minute service accessibility

Utrecht are very quickly accessible by bicycle. Overall, kindergartens throughout the municipality can be reached in less than 5 minutes by bicycle, except for those in the Vleuten-De Meern district, for which data are missing. The same applies to convenience stores (Fig. 4c and d), with the exception of the Leidsche Rijn district, where the average journey time exceeds 10 min. The results for parks (Fig. 4e and f) show similar trends, i.e., the average travel time for the neighbourhoods is lower in the central areas, but gradually increases on the outskirts.

The assessment of accessibility to parks can be correlated with the health of residents. In fact, a US study conducted by the NIH (National Institute of Health) shows that access to free services such as parks and sports facilities increases sports activity (De Andrade et al., 2021). The Dutch government recommends at least 30 minutes of physical activity a day. Statistics carried out by RIVM for the municipality of Utrecht show the percentage of people who reach the government's daily recommendations. The districts of Overvecht, Leidsche Rijn, and Vleuten-de Meern are below the provincial average (64%) (Fig. 5). When we look at the accessibility results, we see that these are the districts where parks are least easily accessible by bike. In contrast, some neighbourhoods with above-average levels of physical

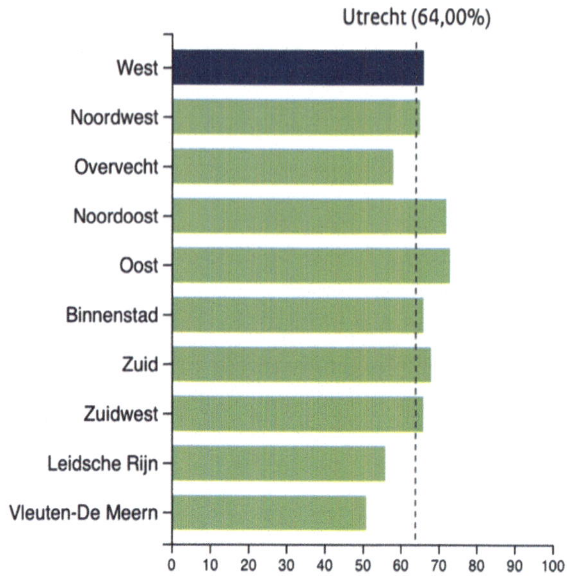

Fig. 5 Statistics in the province of Utrecht on the percentage of residents achieving the recommended level of sporting activity, taken from RIVM

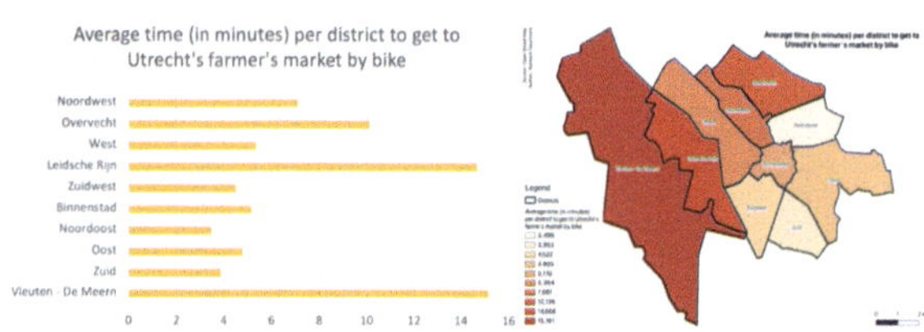

Fig. 6 Example of a 10-minute service accessibility map

activity per day, such as Zuid, Oost, and West, have parks that are easily accessible. These statistics are interesting because it confirms the correlation between good accessibility to parks and compliance with physical activity guidelines.

Figure 6 shows the uneven distribution of cycling journey times to access 10-minute services (e.g., open market). The figure reveals that open-air markets in the North–East district are very accessible by bicycle. Overall, open markets throughout the municipality can be reached in less than 10 minutes by bicycle, except for those in the Leidsche Rijn and Overvecht districts and the Vleuten-De Meern district, for which data are missing.

5.2.2 Creating Accessibility Scores

To establish an accessibility indicator, our study grouped the residential properties by neighbourhood in order to obtain the average time taken to access the various services by bicycle. Accessibility scores by neighbourhood were assigned as follows (see Fig. 7):

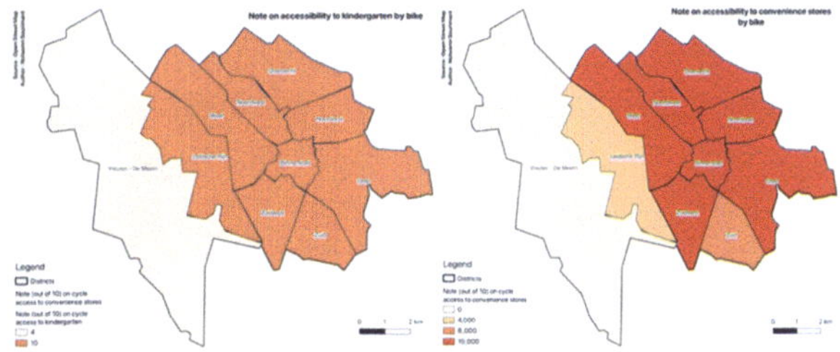

Fig. 7 Accessibility score of 5-minute services by bike

- The shorter the journey time, the higher the score, and vice versa
- When the time taken to access the service is the same or less than the desired time, the score is 10/10
- When the time is slightly longer (more than 2 min), the score is 8/10
- When the time is longer (more than 4 min), the score is 6/10
- When the time is much longer (more than 6 min), the score is 4/10
- When the time is much longer (more than 8 min), the score is 2/10

5.2.3 Pollution Indicators

To assess the concentrations of PM10 and PM2.5 particles in the districts of Utrecht and to identify polluted and non-polluted areas, an indicator has been created. Fine particles, such as PM10 (particles with a diameter of less than 10 micrometres) and PM2.5 (particles with a diameter of less than 2.5 micrometres), are known to have a significant impact on air quality and human health. By analysing the data collected on PM10 and PM2.5 concentrations in each district, an indicator has been defined to assess the level of pollution in each zone. This indicator takes into account the concentration levels of particles in the air and makes it possible to distinguish between areas with low pollution and areas with high pollution. From the two maps in Figs. 8 and 9, the least polluted areas are located in the neighbourhoods furthest from the centre of Utrecht.

These results can be validated using RIVM statistics, which evaluate the share of people exposed to pollution in each district in Fig. 10. It can be seen that the districts with the lowest scores correspond to districts where the proportion of people exposed to pollution is highest, and vice versa.

5.2.4 Global Quality of Life Indicator

The overall quality of life indicator was calculated by averaging all the scores (accessibility and air quality). The map was made using the determined scores. The results are shown in Fig. 11.

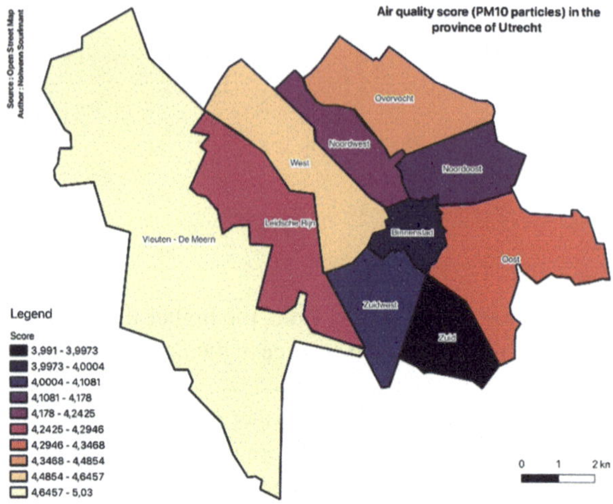

Fig. 8 Score PM10

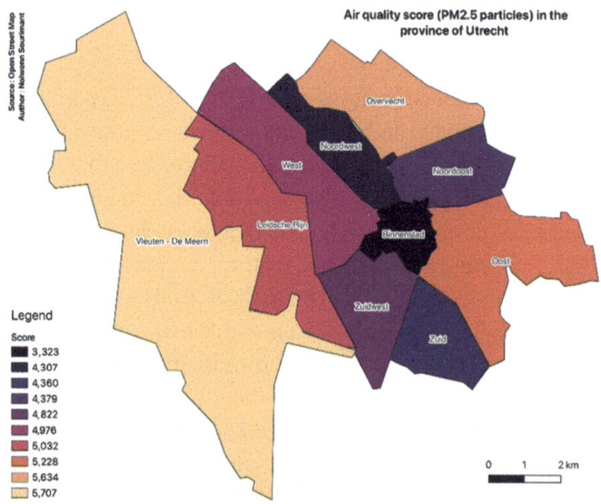

Fig. 9 Score PM2.5

6 Discussion

6.1 Interpretation

Accessibility of services by bicycle is an important issue to consider in order to ensure the mobility and quality of life of the residents of a city like Utrecht. From the study carried out, the outlying districts of the city of Utrecht are at a disadvantage in terms of accessibility by bike compared with the central districts. On the other hand, air quality counterbalances this lack of accessibility, as the outlying districts seem to have better air quality.

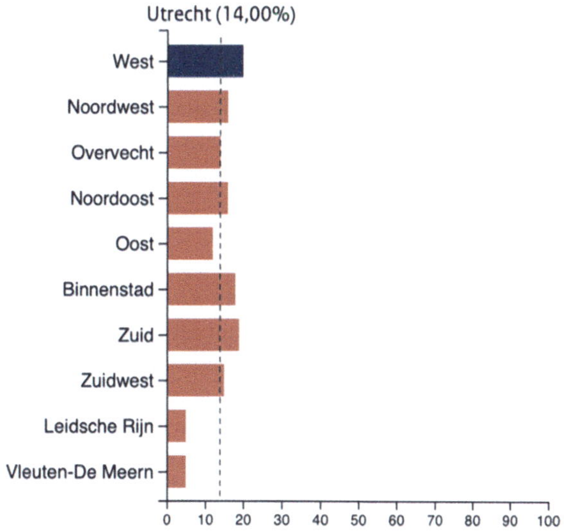

Fig. 10 Proportion of residents exposed to pollution according to RIVM data

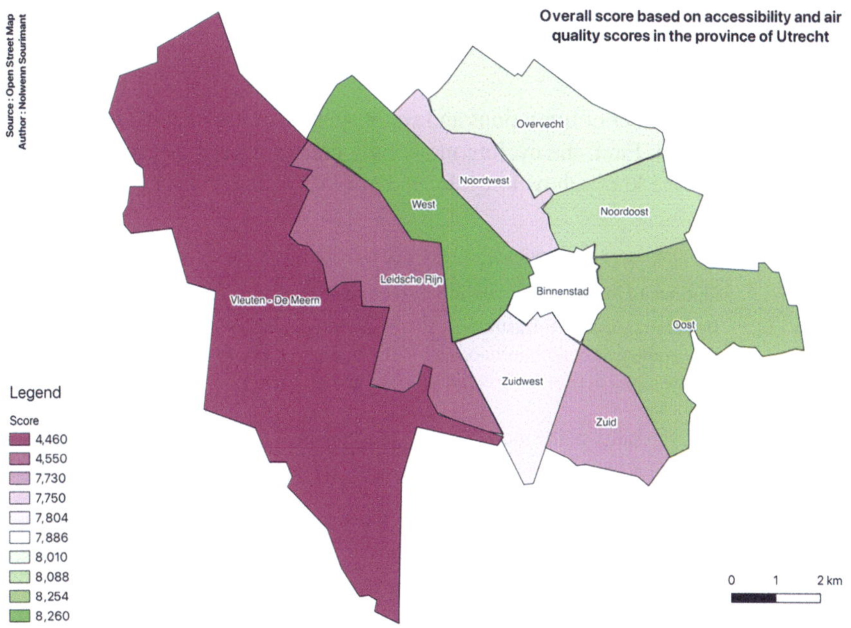

Fig. 11 Overall score for quality of life

Neighbourhoods on the outskirts of Utrecht can be at a disadvantage in terms of cycling accessibility for a number of reasons. First, the density of essential services and facilities (such as schools, local shops, etc.) may be lower in these neighbourhoods far from the city centre. This means that residents in these areas have to cycle longer distances to access these services. On the other hand, peripheral neighbourhoods can benefit from better air quality than more central and densely populated areas. In general, city centres tend to have heavier traffic and higher building density, which can lead to higher concentrations of fine particles (PM2.5 and PM10).

The overall score obtained for all neighbourhoods represents a compromise between good accessibility and good air quality. Good accessibility in the city centre comes at the expense of poorer air quality, which affects people's health. The duality of accessibility and air quality is an interesting concept to study when drawing up public policies. To gain a better understanding of how public policy can improve both accessibility and air quality, a decision-support grid (Table 5) has been drawn based on three relevant policy areas: equity, impact on quality of life, and impact on the economy. It provides potential insights into improving the overall score for the city of Utrecht.

6.2 *Limitations and Bottlenecks of the Study*

This study has a number of limitations and simplifying assumptions that may distort the results obtained. First, the cycling mode only concerns people who are able to cycle. Not everyone knows how to cycle, and not everyone can cycle because of a

Table 5 Decision-making grid for public policies between air quality and accessibility

Criteria	Options to improve accessibility	Options to improve air quality
Equity	Invest in transport infrastructure in disadvantaged neighbourhoods to ensure equal access to essential services	Develop public policies to limit motorised transport in cities, such as low emission zones
	Develop safe cycling routes to encourage bicycle use as a means of transportation	Vegetation around cycle routes
Impact on quality of life	Redesign urban spaces to prioritise walking and cycling, creating more pleasant environments for residents	Raise awareness among residents about the health effects of air pollution and promote environmentally friendly behaviours
	Promote an active and healthy lifestyle by encouraging cycling and walking	
	Strengthen sustainable urban planning policies to create parks	
Economic impact	Encourage walking and cycling can foster local businesses and neighbourhood commerce	Promote the transition to a green economy by encouraging the adoption of eco-friendly transportation solutions, attracting sustainable investments

physical disability. The 20 km/h speed limit for cycle journeys could therefore be discussed and revised downwards to take account of the entire population, including young children and older adults. In addition, when the origin–destination matrices are drawn up, the quickest route between a dwelling and a service is taken into account. However, based on existing literature, people do not systematically take the shortest or fastest route. External factors such as the quality of the journey (taking a cycle lane rather than a lane shared with cars) may explain this choice. The decision support grid between accessibility and air quality is therefore interesting because it gives a better understanding of the complexities of people's travel choices. By focusing solely on the quickest route in deriving the origin-destination matrices, there is a risk of overlooking crucial factors that influence users' urban mobility decisions. Data validation, however, is complex to establish as we did not have access to official databases to compare with my OpenStreetMap data. It is essential to consider that the data used may be inaccurate and could potentially impact the current model obtained. For future research, the use of more reliable but non-open-access data would be necessary to conduct sensitivity analysis and evaluate the quality of the obtained model.

6.3 Implications for Urban Planning

The findings of this study reveal the complexity of the choice made by political decision-makers between prioritising air quality or improving accessibility to services. Disparities in cycling accessibility to essential services between central and outlying districts indicate potential challenges in ensuring equitable access to amenities for all residents. Furthermore, the observed trade-off between good accessibility and air quality in the city centre presents a critical dilemma for policymakers. While the central districts have better access to services, they also experience higher levels of air pollution due to increased traffic congestion and urban density. This trade-off highlights the importance of adopting sustainable transportation policies that promote active mobility options, public transportation, and eco-friendly initiatives to improve air quality while maintaining acceptable levels of accessibility in densely populated areas. To address these challenges, urban planners can prioritise the creation of pedestrian and cycling—friendly zones in the city centre. Simultaneously, efforts should be made to enhance public transportation networks, providing efficient alternatives to private car use. Such measures can alleviate traffic congestion, mitigate pollution, and improve overall air quality.

7 Conclusion

The main objective of this study was to analyse the cycling infrastructure in the city of Utrecht, taking air quality into account. The results show that accessibility is very good in the city centre and less so on the outskirts. However, when it comes to air quality, the results are the opposite. This duality between accessibility and air quality raises important questions for public policy on urban mobility and environmental sustainability. The aim of the article was to experiment with a new approach to calculating accessibility radii using housing databases. The results, while consistent with those of other studies, are a first outline for the analysis of accessibility radii to services in the 15-minute city. These results can be improved by using official databases and including additional services (by optimising the calculation of origin-destination matrices, which was not possible for services such as sports centres, colleges, etc.). More realistic results could support policy-making for the creation of a sustainable city. Utrecht is a 15-minute city and is known to be a model for other cities. So, if obtained results are consistent with those of existing studies, i.e., that accessibility by bike is within the 15 minutes constraint, then it is possible to transpose the used model to other cities that are not currently 15-minute cities. The model could be adapted, for example, for cities with a high spatial gradient, such as the major Portuguese cities of Porto and Lisbon. The study of accessibility, taking into account external environmental factors that can influence citizens' journeys (air quality) on real mobility databases, could enable the refinement of existing models and thus help urban transport policies. This is an important problem to solve to move towards (Moreno et al., 2021; Moreno, 2024) 15-minute city model.

References

Albayati, N., Waisi, B. I., Al-Furaiji, M., Kadhom, M., & Alalwan, H. A. (2021). Effect of COVID-19 on air quality and pollution in different countries. *Journal of Transport & Health, 21*, 101061. https://doi.org/10.1016/j.jth.2021.101061

Caubel, D. (2006). *Politique de transports et accès à la ville pour tous?* Une méthode d'évaluation appliquée à l'agglomération lyonnaise [Doctoral dissertation, Université Lumière Lyon II].

De Andrade, L. S., Geffin, R., Maguire, M., Rodriguez, P., Castro, G., Alkhatib, A., & Barengo, N. C. (2021). The associations between access to recreational facilities and adherence to the American Heart Association's physical activity guidelines in US adults. *Frontiers in Public Health, 9*, 666. https://doi.org/10.3389/fpubh.2021.666

ESPON. (2022). *Carlos Moreno–The full interview*. https://www.espon.eu

Fast, T., Hazel, P., Hans, H., & Van De Weerdt, D. (2017). *Gezondheidseffectscreening*.

Harms, I. M., Burdett, B. R., & Charlton, S. G. (2021). The role of route familiarity in traffic participants' behaviour and transport psychology research: A systematic review. *Transportation Research Interdisciplinary Perspectives, 9*, 100331. https://doi.org/10.1016/j.trip.2021.100331

Knap, E., Ulak, M. B., Geurs, K. T., Mulders, A., & Van Der Drift, S. (2023). A composite X-minute city cycling accessibility metric and its role in assessing spatial and socioeconomic inequalities—A case study in Utrecht, The Netherlands. *Journal of Urban Mobility, 3*, 100043. https://doi.org/10.1016/j.urbmob.2023.100043

Lin, L., & Chen, T. (2023). Activity space and the 15-minute neighborhood: An empirical study using big data in Qingdao, China. *Journal of Transport and Land Use, 16*(1), 1–17. https://doi.org/10.5198/jtlu.2023.2252

Logan, T. M., Hobbs, M., Conrow, L., Reid, N., Young, R., & Anderson, M. W. (2022). The X-minute city: Measuring the 10-, 15-, and 20-minute city and an evaluation of its use for sustainable urban design. *Cities, 131*, 103924. https://doi.org/10.1016/j.cities.2022.103924

Manley, E., Addison, J. D., & Cheng, T. (2015). Shortest path or anchor-based route choice: A large-scale empirical analysis of minicab routing in London. *Journal of Transport Geography, 43*, 123–139. https://doi.org/10.1016/j.jtrangeo.2015.01.003

Mora, R., Truffello, R., & Oyarzún, G. (2021). Equity and accessibility of cycling infrastructure: An analysis of Santiago de Chile. *Journal of Transport Geography, 91*, 102964. https://doi.org/10.1016/j.jtrangeo.2021.102964

Moreno, C. (2024). *The 15-minute city: The urban planning concept to building sustainable cities*. Wiley.

Moreno, C., Allam, Z., Chabaud, D., Gall, C., & Pratlong, F. (2021). Introducing the "15-Minute City": Sustainability, resilience and place identity in future post-pandemic cities. *Smart Cities, 4*(1), 93–111. https://doi.org/10.3390/smartcities4010006

Mytton, O. T., Townsend, N., Rutter, H., & Foster, C. (2012). Green space and physical activity: An observational study using health survey for England data. *Health & Place, 18*(5), 1034–1041. https://doi.org/10.1016/j.healthplace.2012.06.003

Pozoukidou, G., & Chatziyiannaki, Z. (2021). 15-minute city: Decomposing the new urban planning eutopia. *Sustainability, 13*(2), 928. https://doi.org/10.3390/su13020928

Riggs, W. (2016). Inclusively walkable: Exploring the equity of walkable housing in the San Francisco Bay Area. *Local Environment, 21*(5), 527–554. https://doi.org/10.1080/13549839.2014.1001020

Steuteville, A. D. R. (2023). *Defining the 15-minute city*. https://www.cnu.org

Weng, M., Ding, N., Li, J., Jin, X., Xiao, H., He, Z., & Su, S. (2019). The 15-minute walkable neighborhoods: Measurement, social inequalities and implications for building healthy communities in urban China. *Journal of Transport and Health, 13*, 259–273. https://doi.org/10.1016/j.jth.2019.05.005

The 15-Minute City Concept and Public Health: Quantifying the Link Between Pedestrian Accessibility and Population Well-Being in Maastricht

Clémentine Ravereau, Karima Kourtit, Louafi Bouzouina, and Peter Nijkamp

1 Introduction

There are many pressures on cities today, due to their increasing densification and all the related issues that this implies. Among the issues described by the UN in the Sustainable Development Goals, such as water and energy management, there are also many issues relating to the mobility of people. Despite their dense clustering, city residents exhibit significant daily mobility; urban density alone is not sufficient to reduce the movement of people within cities. These trips are still often made by car, creating severe problems of congestion, air quality and noise. To mitigate the negative externalities due to urban transport, two main axes can be envisaged: changing the dominant mode of transport to more sustainable alternatives and reducing the number of trips. These two axes are combined in the 15-min city concept, created in 2021 by Carlos Moreno. It is an urban planning concept, broader

C. Ravereau
Regional Directorate for the Environment, Planning and Housing in Normandy, Rouen, France

K. Kourtit (✉)
Faculty of Management, Open University of the Netherlands, Heerlen, The Netherlands

EHERO – Erasmus University Rotterdam (EUR), Rotterdam, The Netherlands

University College Wrocław (WSKZ), Wrocław, Poland

L. Bouzouina
LAET, ENTPE, University of Lyon, Vaulx-en-Velin, France
e-mail: louafi.bouzouina@entpe.fr

P. Nijkamp
University College Wrocław (WSKZ), Wrocław, Poland

EHERO – Erasmus University, Rotterdam, The Netherlands

L. Bouzouina et al. (eds.), *15-Minute Cities*, Footprints of Regional Science,
https://doi.org/10.1007/978-3-032-20840-8_15

than a simple question of mobility, whose object is a city within which residents can carry out all activities in less than 15 min (Moreno et al., 2021). Moreno aimed to think of the city as dense neighbourhoods on a human scale, where trips within 15 min should be made by active mode: mainly cycling or walking (Duany & Steuteville, 2021).

The 15-minute city concept has many advantages, including improving quality of life by saving transport time, reducing reliance on cars for everyday mobility and strengthening the social fabric. The fact that all essential services are available within a 15-minute walk of residents' homes encourages them to take the time to talk also to their neighbours, whom they then know better, creating conducive spaces to social interaction (Seaman et al., 2010). Following the 2015 Paris Climate Change Conference (COP21), more cities in Europe and elsewhere are turning to more sustainable city models, such as 15min cities, to promote sustainability and resilience in the face of global challenges. It is also noticeable that political figures are using this concept as a campaign argument, such as Anne Hidalgo as a candidate for the mayor of Paris in 2020. The risk of this political takeover of the concept is its oversimplification. Indeed, it requires a complex and nuanced way of thinking that political discourse does not always allow for, because it needs efficiency and consistency. This popularity can therefore sometimes compromise the very effectiveness of the concept. Today, Moreno continues to defend his concept (Moreno, 2024), which is not without opponents. Indeed, after the successive mobility constraints caused by Covid-19, some population groups simultaneously experienced a heightened desire to travel and a fear of restriction. British MP Nick Fletcher, for example, has said that 15-minute cities would 'take away individual freedoms', and this fear comes at a time when new low-emission zones are being introduced in the UK (Ravenscroft, 2023). There are, of course, many challenges to putting this concept into practice, particularly in already built-up and densely populated cities. Creating compact neighbourhoods requires optimising the use of land as much as possible, and this can be complicated in cities where land availability is limited. What's more, like all recent urban planning movements aimed at reducing urban sprawl, it implies a densification of most neighbourhoods, which at first glance does not appeal to existing residents. Finally, there is a risk of creating the conditions for gentrification of the 15-minute districts, which would exclude low-income households through property prices. This would seriously undermine the social equity that is so fundamental to the project.

The issue of accessibility is a key point in the concept of 15-minute cities, but is the geographical organisation of accessibility sufficient to influence the behaviour, in particular the walking behaviour, of the inhabitants of a city? Researchers have defined concepts such as the Walk Appeal (Mouzon, 2012) to explain that many other factors come into play in the modal choice of walking. This article aims to explore the relationship between the potential accessibility of a city on foot to access a basket of goods and the well-being of the population. The impact of a lack of access to basic services on health and well-being will be assessed. The city of Maastricht was chosen here as a case study, because Dutch cities are cities with a strong and relatively long-standing cycling tradition, which seem fairly close to the ideal of 15-minute cities.

This study will therefore attempt to assign a pedestrian accessibility value to the neighbourhoods of Maastricht based on the number of services to which residents have access within a 15-minute walk, and then compare this value with indicators of health and quality of life in this neighbourhood. This paper will then look at the neighbourhoods with the worst levels of access and try to show the consequences for residents' levels of well-being using data from an atlas of southern Limburg. In particular, we will look at the repercussions of poor access to health services on depressive disorders and on the overall state of health of residents, especially as regular walking has beneficial effects on health.

The methodology is based on calculating isochrones of access to a basket of essential goods. From the isochrone maps, we will derive accessibility scores for each district of Maastricht. Then, linear regression will be used to compare the explanatory nature of these scores with indicators of well-being, to examine the potential benefits and challenges of the 15-minute city concept and to discuss how it can be implemented in practice. Additionally, our study will review existing literature on the topic and draw on case studies from around the world to illustrate key concepts and principles.

2 Overview of Related Works

There is a large body of literature on the concept of accessibility and its importance in urban planning and design. The term accessibility can be broadly defined as the ease with which individuals can reach desired destinations within a given area. Within this broad definition, pedestrian accessibility specifically refers to the ease with which pedestrians can reach destinations within a city (Handy, 1996). Accessibility is one way of responding to mobility issues that does not simply increase the number of journeys, as has been done in most cases to date. Articles focusing on pedestrian accessibility often refer to the need to consider individual and environmental factors that influence pedestrian behaviour, such as demographic characteristics, travel motives, the presence or absence of pavements and pedestrian facilities (Handy, 1996, 2020). It is therefore important to carry out this study with the urban planning context of the area in mind. Public policy solutions to improve pedestrian accessibility may lie in the adjustment or development of existing roads. Furthermore, when we look at pedestrian behaviour, we realise that the amount of time people are prepared to devote to their walking differs according to the activities they have to do. For example, they may spend more time walking to school than walking home from shops with heavy bags to carry (Logan et al., 2022). This data qualifies the 15-minute city approach, which uses a fixed threshold for all types of journey combined. Clearly, pedestrian behaviour depends in part on accessibility to basic services. Walter Christaller (1933) developed the theory of central places to limit travel and offer more services and goods. This theory, although anachronistic to the concept of 15-minute towns and created on a wider scale than that of the city, can help us to think about centralities within city neighbourhoods, to reduce the distances travelled by all the inhabitants of that district.

In the remainder of the study, we can therefore expect to see higher accessibility scores near central locations, particularly near Maastricht city centre. Since the first trials of 15-minute cities, researchers have tried to quantify successes to help planners. They calculated the accessibility of individuals living in the cities in question by defining baskets of essential goods. These baskets of goods group together the activities to which one must have access in less than 15 min, and may differ according to the studies. Indeed, while most studies include shops, health facilities and schools, few include the workplace because it is methodologically complicated to include (Knap et al., 2023). It should be noted that the latter study focused on accessibility by bicycle, which may explain certain differences in approach with our present study. As the authors point out in the limitation and future research section, they propose extending the study to walking accessibility scores for a more nuanced analysis and to include people who are unable to cycle or use a similar mode of transport. According to Maslow (1943), people's basic needs are more or less the same over time, and there is a hierarchy between them, which is explained by the level of motivation to satisfy them. This hierarchy can be represented by a pyramid as depicted in Fig. 1. This paper focuses on the basic needs described in this pyramid and particularly on health, education and food shops. In addition, there is a link between the concept of walkability and the second level of Maslow's pyramid on safety. Encouraging the creation of more walkable areas, which limit the use of cars in our towns and cities, may reduce the risk of accidents. Improving pedestrian access to the basket of goods, therefore, meets both physiological and safety needs. There are wider implications to consider, which go beyond the individual scale. Improving living conditions for residents can have an impact on the attractiveness of urban areas and therefore on land prices. This is a double-edged sword, as it is both good news for cities that invest in pedestrian infrastructure, but also a problem for territorial equity, as gentrification exiles low-income populations from cities.

Fig. 1 Maslow's pyramid of basic needs

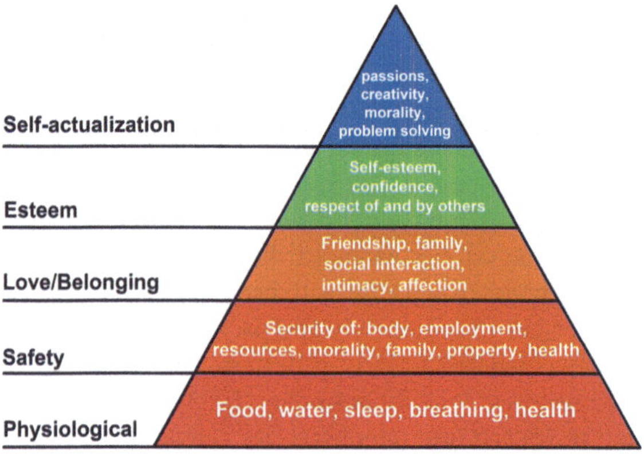

Once this basket of goods has been defined, a first approach consists in looking at accessibility indicators such as the proportion of inhabitants having access to all these services in less than 15 min, or the maximum time it takes for an inhabitant to have access to the furthest of these services (Caubel, 2006). The study by Logan et al. (2022) showed that, taken individually, these calculations can be used to evaluate the progress of a city and to compare cities with each other. They can also inform planners of urban areas where services are lacking, and even which services are lacking (Barbieri et al., 2023). With this information, public authorities can regulate building permits to favour the location of certain services. As a general rule, authorities should also be involved in implementing rent controls and property prices to enable equitable urban development. Previous studies have shown that pedestrian accessibility is particularly important in city centres, where prices are rising the most (Weng et al., 2019).

3 Methodology and Data

To assess urban accessibility, the challenge is to identify the places that should be accessible, because they are part of essential human needs, and then to assign them an area of attraction, from which it is possible to reach them in less than 15 min on foot. By superimposing the layers corresponding to medical services, educational facilities and supermarkets, we can identify the areas of the city that meet the criteria for the 15-minute pedestrian zone and, therefore, a fortiori, the 15-minute city. For neighbourhoods not covered by the isochrones, the option of other soft modes and public transport remains to be considered. Nevertheless, it will probably also be necessary to consider urban redevelopment to allocate more services to these areas. Once accessibility has been assessed, it is compared with the effectiveness of services in different areas. In other words, we examine whether areas with high access also have a better quality of service. For example, are residents in better health if they have good access to healthcare services, or do they perform better at school if they have good access to education services? To locate essential services such as schools, pharmacies, supermarkets and hospitals, we use OpenStreetMap data, which lists all the facilities in a town. This is collaborative data that is continually updated by the community and widely used in urban research because it is easily accessible and very rich.

In the case of the proposed study, the OpenStreetMap data used is located in Maastricht, in the South of the province of Limburg, in the South part of the Netherlands (Fig. 2). The city of Maastricht has a population of around 120,000 (121,151 in 2022 according to Statistica), is crossed by the Meuse, and is made up of five districts, subdivided into 44 neighbourhoods (Table 1). Part of this study consists of assessing the impact of inequalities in access to services in terms of success, using performance indicators, particularly in terms of health status. We have extracted data from the 'Gezondheidsatlas Zuid-Limburg'. This atlas provides indicators on the health and well-being of the citizens of Zuid-Limburg (see Appendix: Fig. 8). The atlas contains ratios of residents who are overweight, suffering from chronic illness, or feeling lonely. But it also includes more subjective indicators

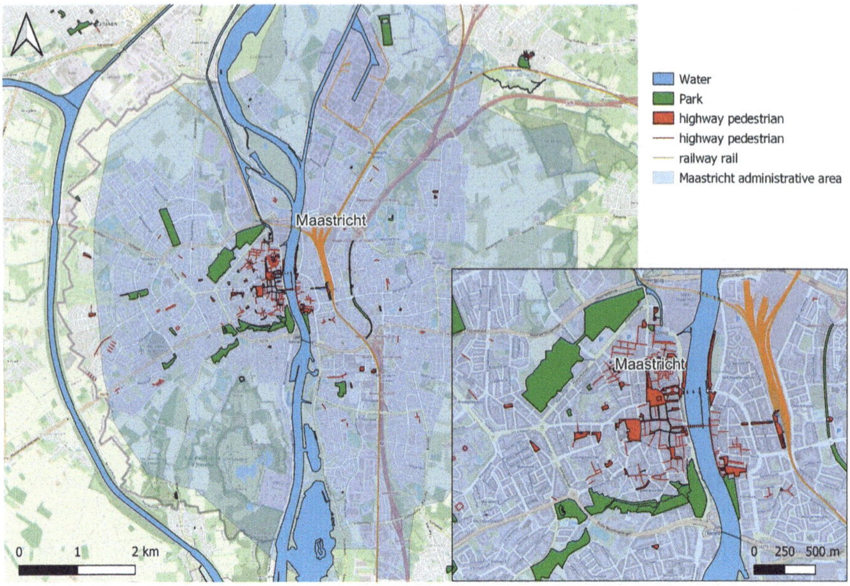

Fig. 2 Topological features and rail transport infrastructure of Maastricht

Table 1 Census of Maastricht districts by borough

Borough	Neighbourhood
Maastricht-Centrum	Binnenstad, Jekerkwartier Kommelkwartier, Statenkwartier Boschstraatkwartier, Sint Maartenspoort Wyck-Céramique
Maastricht-Zuidwest	Villapark, Jekerdal Biesland, Campagne Wolder, Sint-Pieter
Maastricht-Nordwest	Brusselsepoort, Mariaberg Belfort, Pottenberg Malpertuis, Caberg Oud-Caberg, Malberg Dousberg-Hazendans, Daalhof Boschpoort, Bosscherveld Frontenkwartier, Belvédère, Lanakerveld
Maastricht-Noordoost	Beatrixhaven, Borgharen Itteren, Meerssenhoven Wyckerpoort, Wittevrouwenveld Nazareth, Limmel, Amby
Maastricht-Zuidoost	Randwyck, Heugem Heugemerveld, Scharn Heer, De Heeh, Vroendaal

such as the perception of good health, annoyance caused by noise, and satisfaction with the amount of green space in the neighbourhood. Various figures are compared with the geographical reference level. The municipalities are compared with the average for South Limburg, and the neighbourhoods are compared with the average of the municipality to which they belong. To measure the potential accessibility of a service, we use cartography and specifically a QGIS tool called ORS Tools. This is a plugin that defines zones of access to a type of service (geolocated by OpenStreetMap) depending on the mode used. These zones are called isochrones, i.e. geographical zones that represent the equality of travel time from a given point to other points or destinations, in this case, the services are in the basket of goods.

They are displayed on a map or graph showing all the areas that can be reached in the same time from a specific central point. This approach uses the network through Open Route Service Tools, and therefore takes into account natural barriers such as rivers and human barriers such as motorways when calculating paths. In a city like Maastricht, there are many mobility options, from walking to driving, cycling, self-service scooters and public transport. Here, we used the algorithm on the assumption that journeys are made on foot. The algorithm searches for routes accessible to pedestrians and then uses road network data and pedestrian movement profiles to calculate possible routes from the starting point. In our study, we used this plugin with the walking mode and displayed the 5-, 10- and 15-minute isochrones for each service. Also, we obtained a map with all the isochrones; we grouped them together to obtain a single layer covering the entire area from which this type of service can be accessed in less than 15 min. It may be interesting to compare accessibility on foot with that by bike, because the 15-minute city model works thanks to the complementarity of active modes. The accessibility of urban educational amenities can be seen on the following map in Fig. 3.

The map shows that the area covered by the 15-minute cycling isochron is almost equal to the city of Maastricht. This means that the city can almost be considered a 15-minute city from a cycling point of view. However, when compared with the walking isochrones, we can see that there are inequalities of access within the city. It is interesting to extend this comparative analysis now to the other services in the basket of goods and not to restrict it to educational amenities. This remains an area for future investigation. In order to compare the data from the atlas and the isochrones, we need to make the data spatial, so that we can cross-reference them. So, using the OpenStreetMap breakdown of the districts, we outgraded the table with the health indicators and the areas concerned in QGIS. This shows the correlations between poor pedestrian access to certain areas and high rates of chronic disease, see for example Fig. 4.

Other comparison maps are available in the Appendix Fig. 10, and reveal several key findings that are presented in the results section. The maps allow us to understand at a glance the principle of comparison by superimposing layers, but they are difficult to analyse because the health and well-being indicators are on a scale that defies easy representation. The indicators are given for distinct groups of neighbourhoods; we have chosen to adapt to these groupings and to artificially create a score for the accessibility of these neighbourhoods. The accessibility score is based

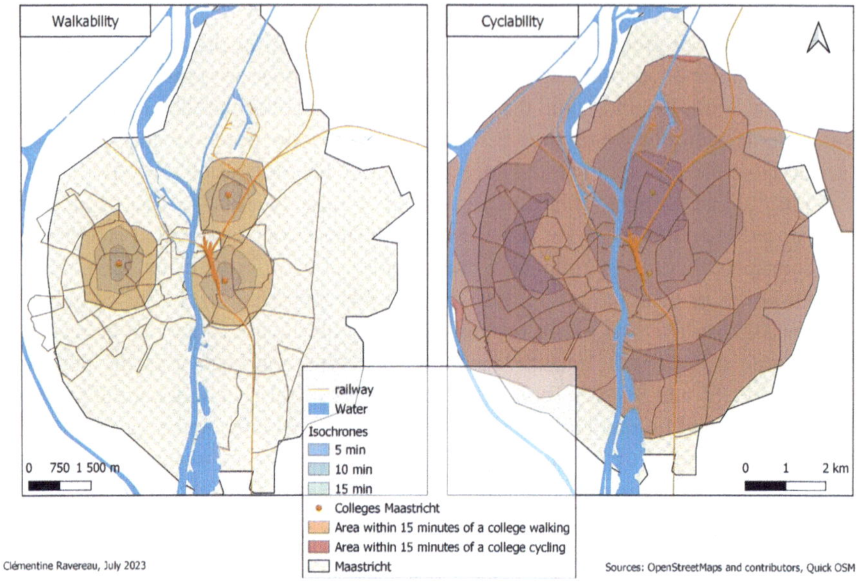

Fig. 3 Comparison of accessibility to the colleges by mode of transport: walking or cycling. Visualisation of the areas from which it is possible to reach them in less than 15 min

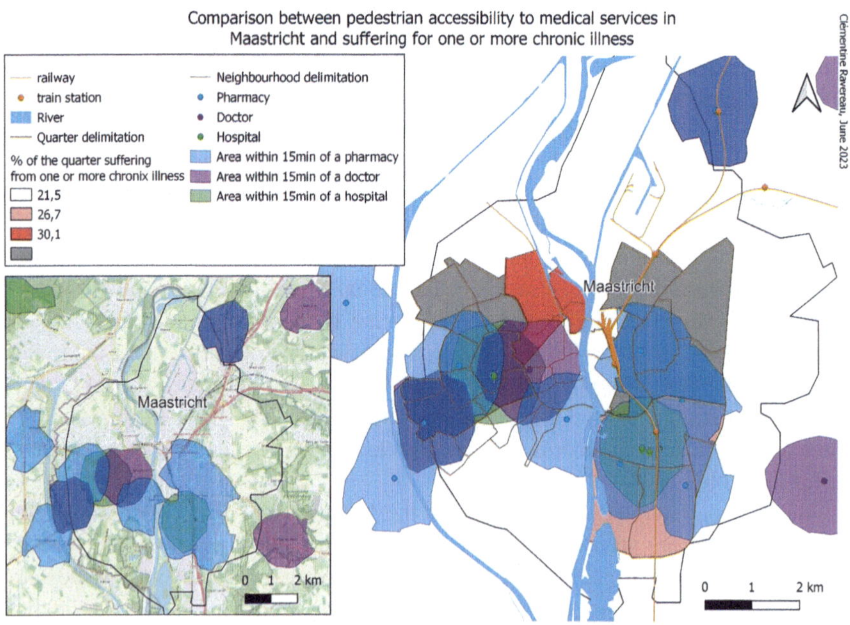

Fig. 4 Example of a comparative map: comparison between pedestrian accessibility to medical services in Maastricht and the percentage of quarters suffering from one or more chronic illnesses

on the presence or absence of essential services isochrones in each neighbourhood. To construct this score, we returned to the neighbourhood scale and, for each service, assigned a value of 0 if the neighbourhood was not covered by the isochrone, 1 if it was fully covered, and 0.5 if it was partially covered. This consistently was applied for all services, mean timed so as to get an overall score for the accessibility of the neighbourhood to the basket of goods. Only with this score calculation were we able to calculate averages for the groups of neighbourhoods in the atlas. Once we have an accessibility score for each of the 12 neighbourhood groupings, we can analyse the links between accessibility and well-being. To do this, we chose to use the multiple linear regression method on RStudio. We tried to explain the indicators correlated with a good pedestrian accessibility score. To assess its individual impact, each indicator was introduced sequentially into the multiple linear regression model. Indicators that diminished the model's overall significance were excluded.

4 Empirical Results

The map in Fig. 5 shows general trends: neighbourhoods located in the North and South are less accessible on foot than others. Pedestrian accessibility shows a gradual decline moving outward. This trend is not surprising, as good walking accessibility is expected in the centre, and, as can be seen on the topographical map, parks

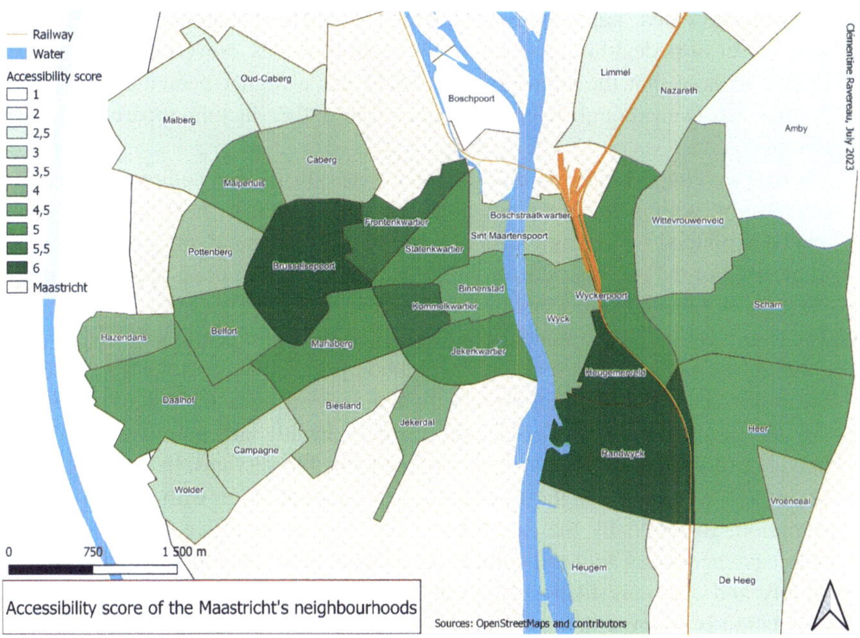

Fig. 5 Map of accessibility scores by neighbourhood

and pedestrian walkways, which are elements that encourage walking, are concentrated in the centre of the city.

A closer inspection shows that the neighbourhoods with the highest scores are not the most central either. These are Brusselsepoort, Randwyck and Heugemerveld, neighbourhoods close to the centre, but not the most central. To understand where these counter-intuitive results come from, we examine the maps in Figs. 6 and 7 that are more specific to a type of service.

We can see that the centre always scores relatively well and that in each of the areas with high scores for health and education, certain neighbourhoods stand out, namely Daalhof and Brusselsepoort for pedestrian access to health and Heugemerveld for pedestrian access to education. The first comparison approach uses the comparative map presented in the methodology section; see also Fig. 10. Unfortunately, few areas are covered by the OpenStreetMap breakdown, which gives incomplete results. Nevertheless, we have already tested comparisons between accessibility on foot and state of health. The results are given in bullet points:

- Firstly, according to the chronic disease indicator, the area in red in the North has the highest rate of chronic disease and is poorly covered by the health service footpaths.
- For residents' self-assessed health status: Where data is available, we can see that the perceptions are highly positive in the centre (see Fig. 6) and that it is better in the South-East than in the north, which is correlated with more pharmacies and the presence of a hospital.
- The feeling of loneliness, the highest level of loneliness, is, as might be expected from the previous maps, in the Maastricht-Noordwest district, which has the worst level of pedestrian accessibility to health services. Nevertheless, it is surprising to note that the centre of Maastricht has a higher percentage than the South-East, even though it is more accessible. Other factors apparently come into play.
- For overweight: We can see that the inhabitants of the city centre are the least overweight, with a 16-percentage point advantage over the next highest, which can be explained by the high level of accessibility in general, not just in the area of health. For the two districts for which we have data, it is once again the North-West district that has the worst health outcome.
- The results on the Dutch standard of physical exercise for health lead to the same conclusions as those on overweight: neighbourhoods that are not easily accessible on foot take less physical exercise, which may increase the risk of obesity.
- For the consumption of alcohol, the primary health indicator works well in neighbourhoods that are well served on foot by health services. In fact, it is in the centre that we find the highest percentage of the population with excessive consumption, followed by the South-East. We may assume that neighbourhoods with easy access on foot to health services have good overall pedestrian accessibility and are more likely to go out drinking. We may also assume that the population is younger in these areas.

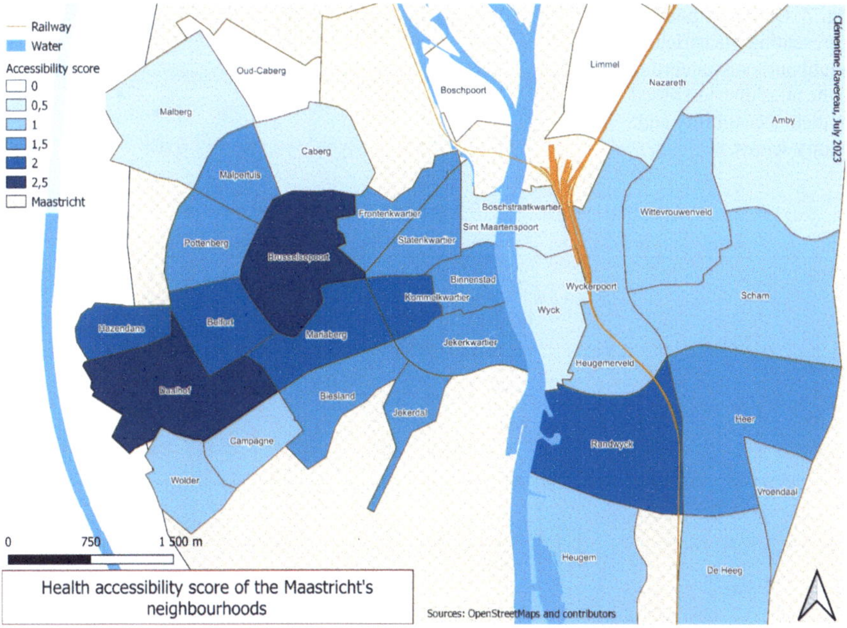

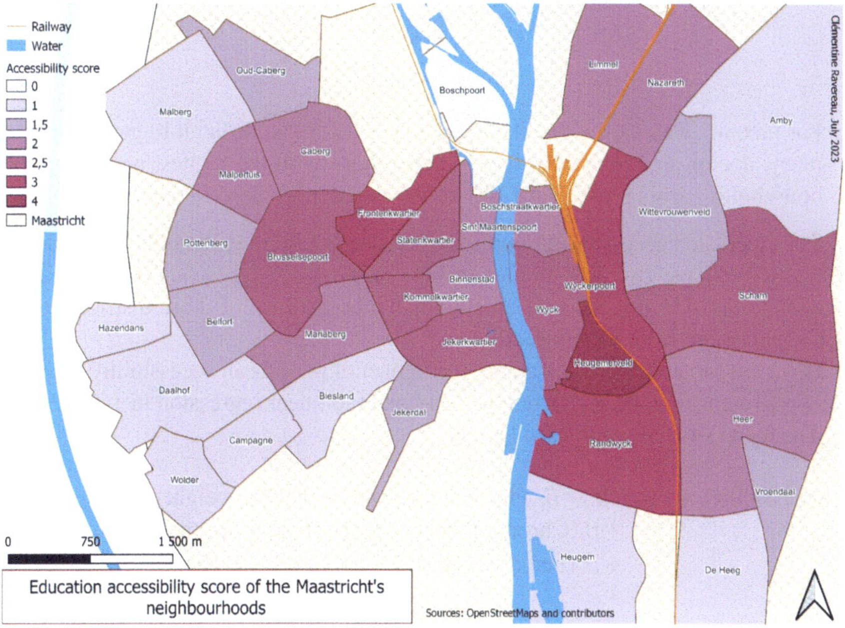

Fig. 6 Maps showing accessibility scores for health (*top*) and education (*bottom*)

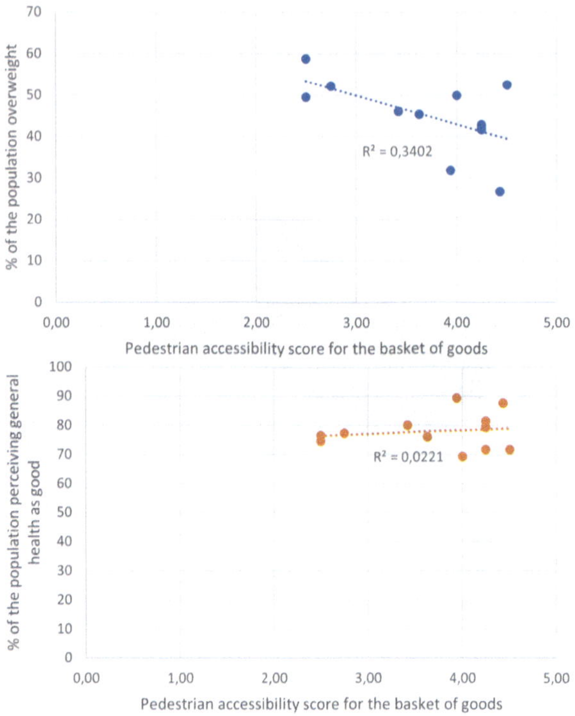

Fig. 7 Clouds of points representing Maastricht's neighbourhoods according to the atlas division based on their accessibility and quality scores

- For income, we can see that the areas where it is most difficult to get to a pharmacy, doctor, or hospital on foot are also those with the highest percentage of households earning less than 120% of the social minimum average.

The calculated accessibility scores are detailed in Figure 9. It would of course be interesting to have also data on the outskirts of the city, as these are the areas least accessible on foot. For the moment, a suitable breakdown of these districts could not be generated from OpenStreetMap.

If we use the above score method, the impact of pedestrian accessibility on people's well-being can be seen from the results of the linear regression in Table 2.

The regression equation used here is:

$$\begin{aligned}\text{Accessibility Score} = \beta_0 &+ \beta_1(\text{Physical Exercise}) + \beta_2(\text{Overweight}) \\ &+ \beta_3(\text{Chronic Illness}) + \beta_4(\text{High-Risk Anxiety/Depression}) \\ &+ \beta_5(\text{Perceives Health Very Good}) + \beta_6(\text{Feels Very Alone}) \\ &+ \varepsilon\end{aligned}$$

We can see that the atlas indicators that are explained by the accessibility score are only those related to physical and mental health. If we look at the results in more detail, we can see that having a good accessibility score contributes to having a rate of individuals who meet the Dutch standards for physical exercise for health.

Table 2 Regression results

Variable	Estimate	Std. error	t-value	p-value
(Intercept)	22.9864	6.4011	3.591	0.0157*
Physical exercise (meets standard)	0.0160	0.0486	0.330	0.7548
Overweight	−0.1448	0.0446	−3.244	0.0228*
Chronic illness(es)	0.0914	0.0930	−0.983	0.3709
High-risk anxiety/depression	−0.0385	0.1319	−0.292	0.7823
Perceives health as very good	−0.1815	0.0812	−2.234	0.0758
Feels very alone	−0.0983	0.0610	−1.612	0.1678
Observations	12			
R-squared	0.7716			
Adjusted R-squared	0.4975			
F-statistic	2.815 ($p = 0.138$)			
Residual std. error	0.5276 ($df = 5$)			

Notes: * $p < 0.05$

Conversely, a poor score for pedestrian accessibility to the basket of goods is correlated with a high rate of overweight, risk of chronic illness, anxiety, depression and loneliness. What is more surprising is that, with a significance of 0.1, good accessibility is detrimental to the perception of good health. If we look at the corresponding scatterplot Fig. 7, we can see that the slope of the trend line remains low, especially compared with the slope of the trend line for the overweight indicator, which is also more significant.

It is interesting to examine whether the statistical links established above make sense in relation to other scientific publications. To this end, a survey carried out in Bristol on 6821 people in 2005 can shed light on the link between overweight and pedestrian accessibility (Coombes et al., 2010). This article focuses on the accessibility of green spaces; it is not limited to pedestrian accessibility but concludes on the basis of proximity to parks, which makes it possible to establish links with pedestrian accessibility. Using logistic regression, this study showed that respondents living closest to a type of green space classified as a formal park were more likely to meet physical activity recommendations and less likely to be overweight or obese. This is consistent with our results.

Empirical links have also already been established between mental well-being and exposure to walkable spaces. Very recently, researchers in Ile de France published a study on the mental well-being of elderly people suffering from momentary depressive symptoms (Fancello et al., 2023). It shows that this well-being improves after 2 h exposure to walkable areas, water features, shops, leisure activities, or cultural sites. These results support the Maastricht findings. Indeed, high levels of pedestrian accessibility may enable residents to spend more time in walkable areas, sometimes on quays or near water features or shops, which increases their well-being and reduces their depressive effects. Finally, an Australian study helps to understand the counter-intuitive results concerning excessive alcohol consumption from our linear regression. This study analysed the link between access to out-of-home drinking establishments and harmful alcohol consumption (Kavanagh et al., 2011). The researchers were able to establish a correlation, thus proving that a large number of outlets were associated with a level of harmful alcohol consumption in this area. These results help to explain the counter-intuitive findings of our study. Indeed, high pedestrian accessibility to a basket of essential goods is probably associated with shopping areas where alcohol outlets are likely to be found. Overall, the study shows that high pedestrian accessibility is a significant predictor of positive health outcomes. This is also the conclusion of a 2018 article that used a sample of 400 individuals in Greece (Panagopoulos et al., 2018). Indeed, the authors explain that improvement of people's health and well-being was identified as an important function of pedestrian zones.

5 Discussion

Based on the results, accessibility in a 15-minute city context is a lever for public action on much broader issues than transport. In Maastricht, linear regression showed that accessibility had a significant impact on overweight and the general perception of a good level of health. These are public health issues.

As we saw in the results section, the perception of good health is not correlated with good accessibility, which is surprising but can be qualified by the small number of observations and related to low significance. We can also imagine that an indicator such as the perception of good health varies greatly according to the socio-economic characteristics of the population, and that residents' access to preventive care is higher among residents in better health, and therefore creates an awareness of their health problems.

Since the Covid-19 pandemic, mental health and the difficulties it poses, particularly for young people, have become a key issue in the public debate. It has been shown that physical mobility constraints and ambient stress have created anxiety and considerably increased mental health problems around the world (Adegboye et al., 2021). Indeed, anxiety and feelings of loneliness are symptoms that have been increasingly highlighted. Public authorities are now paying greater attention to these issues. The results of the linear regression show that good access to essential services is correlated with low rates of people at risk of developing anxiety or depressive disorders, and low rates of people experiencing a strong sense of loneliness. It therefore seems possible to develop policy measures to improve pedestrian accessibility in public health plans targeting mental health. These would be comprehensive measures that would have an impact on mental health and physical health (particularly reducing overweight), as well as opening up neighbourhoods on the outskirts of towns.

It is important not to underestimate the difficulty of implementing such measures for decision-makers, because pedestrian facilities are by nature non-remunerative, unlike motorways, for example, which may have tolls, or self-service bicycles or scooters for active modes, for which one has to pay to hire. This requires a long-term vision of the benefits for the population, which unfortunately cannot be reconciled with political temporariness. This kind of study serves as an argument for those who have the political courage to act for the public good without financial gain. The major limitations of this study lie in the lack of data available for the regressions, as this implies fairly low levels of significance. It might therefore be interesting to create a new database on the level of well-being of the population at the level of Maastricht city neighbourhoods, rather than groupings of neighbourhoods. In addition, this would be an opportunity to include outlying districts such as Sint-Pieter, Belvédère and Vroendaal. Similarly, for the accessibility score, it might be interesting to include restaurants, cinemas, museums, sports and cultural activities in the basket of goods. These studies require financial support, in particular to carry out a survey of a representative sample of the population. Using the same methodology, it would then be possible to establish links between accessibility to the basket of goods on foot and the population's level of well-being, with greater precision, thereby increasing the effectiveness of resulting public policies. As a follow-up, we may extend this analysis by studying the use of bicycles as a means of urban travel, and we could also identify a method for including multimodal travel in the study. This would bring us closer to Carlos Moreno's concept of 15-minute cities.

6 Conclusion

Our study investigated the link between pedestrian accessibility and well-being, focusing on the neighbourhoods of Maastricht, a Dutch city in the Southern part of the country. Pedestrian accessibility is an essential component of the transformation of cities according to the 15 minutes-city model of Carlos Moreno, and to evaluate it from a sustainability perspective.

The study uses the isochrone method for a basket of essential goods. The attribution of accessibility score of all neighbourhoods allows to compare this parameter with the well-being ratios of a health atlas. With a multilinear regression, the research presented here provides evidence of a clear relation between health (physical and mental) and pedestrian accessibility. We can distil from this research that pedestrian access to a basket of essential goods is already a lever for lower obesity rates and lower rates of depression. It is therefore possible to imagine that extending this accessibility to cultural venues, restaurants and sports facilities will further improve the overall health and well-being of the population. These results are important because they give value to walking, which is the most equitable mode of transport since it is accessible to all able-bodied urban residents (Arellana et al., 2021). Its positive externalities have been proven many times over, particularly in terms of improving the health of the population. Yet, this case is still given very little prominence in the urban planning policies of countries in the South. Carrying out studies showing the effectiveness of walking in northern countries could potentially encourage southern countries to orient, in the medium term, their transport policies in this direction. This is a major challenge, given the current and future pressures on cities in the South in terms of sustainability and quality of life.

A limitation of the present study is that the empirical evidence is insufficient to draw quantitative conclusions about the impact of pedestrian accessibility on well-being. We can only draw qualitative inferences about the dynamics of the links between the two parameters. Applying the methods of this study on larger samples could therefore be a fruitful avenue for future studies. This does, however, enhance the value of public policies on pedestrian accessibility by giving them a prominent public health policy dimension.

Appendix

Categorie	Indicator	Average of Maastricht	Maastricht Centrum	Maastricht Noord-O	Maastricht Zuid-Oo	Maastricht Zuid-We	Maastricht Noord-W	Caberg en Malpertuis	Daalhof	De Heeg	[illegible] en Nazareth	Malberg	Pottenberg en Mariaberg	Wyckerpoort en Wittevrouwenveld
	Excessive alcohol consumption	7,2	11.1	5.8	5.7	8.7	6.6	6.4	7.2	7.4	7.1	5.8	8.6	6.2
	Individuals meeting the Dutch standard of physical exercise for health	56,5	65	55	58.2	63.4	51	39.7	44	48.3	59.4	42.9	53.8	56.5
Health	Overweight	40,4	26.8	46.1	42.4	31.9	45.4	49.9	52.5	49.5	52.2	58.8	42.9	41.7
	Suffering from one or more chronic illnesses	26,9	21.5	29.7	26.7	19.8	30.1	29	32	27.9	31.3	38	33.7	30.8
	High risk of anxiety disorders or depression	9,7	8.9	12.5	7.4	5.9	11.1	12.6	11.5	10.5	14.1	13.9	15	14.6
Positiv health	Perceives general health as (very) good	81,3	87.8	80.1	81.6	89.5	76	69.5	71.9	76.7	77.5	74.7	71.9	79.5
	Is concerned about health risks in the residential environment	89,1	93.8	86.7	92.4	90.9	84.8	79.5	84.2	89.8	86.5	84.6	81.6	85.6
	Is severely annoyed by noise	29,4	33.3	30.3	26.9	25.2	29.3	35.2	27.1	28.3	30	26.7	28.6	30.1
Residential quality of the neighbourhood	Finds that there are enough places to meet in the neighbourhood	61,5	61.9	63.9	58.2	72.1	60.2	56.4	57.7	58.4	49	67.2	54.4	64.9
	Finds the area attractive for getting around	75,5	66.9	76.7	80.2	94.9	72	64.9	70.4	78.2	54.2	71.6	65.1	74.8
	Is satisfied with the greenery in the neighbourhood	84,5	73	85	88	96.2	85.9	77.2	86.4	89.7	79.1	82.6	83	80.8
	Households whose income does not exceed 120% of the social minimum	10,2	8.9	12.1	7.8	2.5	13.1	26.2	10	8.6	23	9.4	23.7	15.4
	Feels very alone	14,4	15.8	14.4	10.9	8.5	17.7	23.3	15.1	14.9	21.3	21.5	26.5	15.3

Fig. 8 Extraction of data from the Gezondheidsatlas Zuid-Limburg

Neighbourhoods	Kindergarten	School	Colleges	University	Education	Hospital	Doctors	Pharmacy	Health	Supermarket	Total
Oud-Caberg		1	0,5		1,5				0	1	2,5
Malberg		1			1			0,5	0,5	1	2,5
Caberg		1	1		2	0,5			0,5	1	3,5
Malpertuis		1	1		2	0,5		1	1,5	1	4,5
Pottenberg		1	0,5		1,5	0,5		1	1,5	0,5	3,5
Belfort		1	0,5		1,5	0,5	0,5	1	2	1	4,5
Hazendans		1			1		1	1	2	1	4
Daalhof		1			1	0,5	1	1	2,5	1	4,5
Wolder		1			1		0,5	0,5	1	1	3
Campagne		1			1		0,5	0,5	1	1	3
Brusselsepoort		1	1	0,5	2,5	1	1	0,5	2,5	1	6
Mariaberg		1	0,5	0,5	2	1	0,5	0,5	2	1	5
Biesland		0,5		0,5	1	0,5	0,5	0,5	1,5	1	3,5
Frontenkwartier		1	1	1	3	0,5	1		1,5	1	5,5
Statenkwartier		1	0,5	1	2,5	0,5	1		1,5	1	5
Kommelkwartier		1	0,5	1	2,5	0,5	1	0,5	2	1	5,5
Jekerkwartier	0,5	1		1	2,5		0,5	1	1,5	1	5
Jekerdal		1		0,5	1,5		0,5	1	1,5	1	4
Binnenstad		1		1	2		1	0,5	1,5	1	4,5
Sint Maartenspoort		1	0,5	0,5	2			0,5	0,5	1	3,5
Boschpoort		0			0				0	1	1
Boschstraatkwartier		1		1	2		0,5		0,5	1	3,5
Wyck	0,5	1	1	0,5	3			0,5	0,5	1	4,5
Wyckerpoort	0,5	1	1	0,5	3			1	1	1	5
Heugemerveld	1	1	1	1	4	0,5		0,5	1	1	6
Randwyck	1	1		1	3	1		1	2	1	6
Heugem		0,5		0,5	1	0,5		0,5	1	1	3
De Heeg		0,5		0,5	1	0,5		0,5	1	0,5	2,5
Vroendaal		1		0,5	1,5			1	1	1	3,5
Heer	0,5	1		0,5	2	0,5		1	1,5	1	4,5
Scharn	0,5	1	0,5	0,5	2,5			1	1	1	4,5
Wittevrouwenveld		1	0,5		1,5			1	1	1	3,5
Amby		1			1			0,5	0,5	0,5	2
Nazareth		1	1		2			0,5	0,5	0,5	3
Limmel		1	1		2				0	0,5	2,5

Fig. 9 Accessibility score by neighbourhood

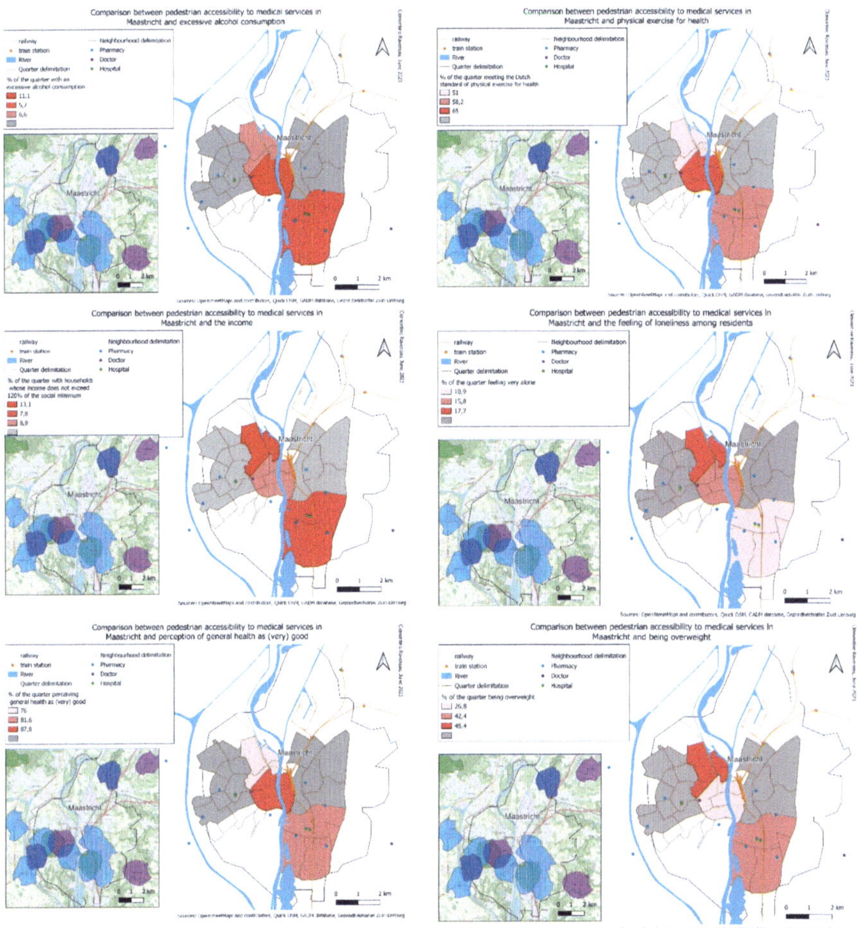

Fig. 10 Maps comparing the walkability of Maastricht districts with indicators of the well-being

References

Adegboye, D., Williams, F., Collishaw, S., Shelton, K., Langley, K., Hobson, C., Burley, D., & van Goozen, S. (2021). Understanding why the covid-19 pandemic-related lockdown increases mental health difficulties in vulnerable young children. *JCPP Advances, 1*(1), e12005.

Arellana, J., Alvarez, V., Oviedo, D., & Guzman, L.-A. (2021). Walk this way: Pedestrian accessibility and equity in barranquilla and soledad, colombia. *Research in Transportation Economics, 86*, 101024.

Barbieri, L., D'Autilia, R., Marrone, P., & Montella, I. (2023). Graph representation of the 15-minute city: A comparison between Rome, London, and Paris. *Sustainability, 15*(4), 3772.

Caubel, D. (2006). *Politique de transports et accès à la ville pour tous ? une méthode d'évaluation appliquée à l'agglomération lyonnaise*. Thèse de doctorat en sciences économique, Université Lyon 2.

Christaller, W. (1933). *Die Zentralen Orte in Süddeutschland* (C. W. Baskin, Trans.). Central Places in Southern Germany.

Coombes, E., Jones, A.-P., & Hillsdon, M. (2010). The relationship of physical activity and overweight to objectively measured green space accessibility and use. *Social Science & Medicine, 70*(6), 816–822.

Duany, A., & Steuteville, R. (2021, February 8). Defining the 15-minute city. Congress for the New Urbanism. https://www.cnu.org/publicsquare/2021/02/08/defining-15-minute-city

Duany, A., & Steuteville, R. (2021). Defining the 15-minute city. *Public Square: A CNU Journal.*

Fancello, G., Vallée, J., Sueur, C., van Lenthe, F.-J., Kestens, Y., Montanari, A., & Chaix, B. (2023). Micro urban spaces and mental well-being: Measuring the exposure to urban landscapes along daily mobility paths and their effects on momentary depressive symptomatology among older population. *Environment International, 178*, 108095.

Handy, S. (1996). Methodologies for exploring the link between urban form and travel behavior. *Transportation Research Part D: Transport and Environment, 1*(2), 151–165.

Handy, S. (2020). Is accessibility an idea whose time has finally come? *Transportation Research Part D-transport and Environment, 83*, 102319.

Kavanagh, A.-M., Kelly, M.-T., Krnjacki, L., Thornton, L., Jolley, D., Subramanian, S. V., Turrell, G., & Bentley, R.-J. (2011). Access to alcohol outlets and harmful alcohol consumption: A multi-level study in Melbourne, Australia. *Addiction, 106*(10), 1772–1779.

Knap, E., Ulak, M.-B., Geurs, K.-T., Mulders, A., & van der Drift, S. (2023). A composite X-minute city cycling accessibility metric and its role in assessing spatial and socioeconomic inequalities—A case study in Utrecht, the Netherlands. *Journal of Urban Mobility, 3*, 100043.

Logan, T.-M., Hobbs, M.-H., Conrow, L.-C., Reid, N.-L., Young, R.-A., & Anderson, M.-W. (2022). The x-minute city: Measuring the 10, 15, 20-minute city and an evaluation of its use for sustainable urban design. *Cities, 131*, 103924.

Maslow, A. H. (1943). A Theory of Human Motivation. *Psychological Review, 2*, 21–28.

Moreno, C. (2024). *The 15-minute city: The urban planning concept to building sustainable cities*. Wiley.

Moreno, C., Allam, Z., Chabaud, D., Gall, C., & Pratlong, F. (2021). Introducing the "15-minute city": Sustainability, resilience and place identity in future post-pandemic cities. *Smart Cities, 4*(1), 93–111. https://doi.org/10.3390/smartcities4010006

Mouzon, S. (2012). *Walk appeal*. Retrieved from https://originalgreen.org/blog/walk-appeal.html?fb_comment_id=fbc_10150916012636891_22379968_10150941931091891#f12adfe3e51721e

Panagopoulos, T., Tampakis, S., Karanikola, P., Karipidou-Kanari, A., & Kantartzis, A. (2018). The usage and perception of pedestrian and cycling streets on residents' well-being in Kalamaria, Greece. *Land, 7*(3), 100.

Ravenscroft, T. (2023). *Uk politician attacks 15-minute city concept in parliament*. Accessed July 31, 2023, from https://www.dezeen.com/2023/02/13/15-minute-city-mp-attack-nick-fletcher/

Seaman, P., Jones, R.-L., & Ellaway, A. (2010). It's not just about the park, it's about integration too: Why people choose to use or not use urban greenspaces. *International Journal of Behavioral Nutrition and Physical Activity, 7*(1), 78. https://doi.org/10.1186/1479-5868-7-78

Weng, M., Ding, N., Li, J., Jin, X., Xiao, H., He, Z., & Su, S. (2019). The 15-minute walkable neighborhoods: Measurement, social inequalities and implications for building healthy communities in urban China. *Journal of Transport and Health, 13*, 259–273.

Mobility and the Digital City

Telework, Mobility, and the Paradox of Proximity in the 15-Minute City

Benjamin Motte-Baumvol

1 Introduction

The 15-minute city has emerged as one of the most influential planning frameworks of the early twenty-first century. It crystallizes long-standing aspirations in urbanism—to build compact, mixed-use, and accessible neighborhoods—while linking them to contemporary concerns about climate change, public health, and resilience. Its guiding principle is straightforward: residents should be able to access essential functions of urban life—work, education, shopping, healthcare, culture, and leisure—within a 15-min walk or cycle from their homes. This vision echoes earlier traditions, from Ebenezer Howard's garden city and Clarence Perry's neighborhood unit to transit-oriented development and walkable urbanism, but goes further by explicitly tying spatial form to sustainability goals. Compactness is presented not merely as an urban lifestyle aspiration but as a necessary condition for reducing emissions, congestion, and social inequalities.

Telework disrupts this framework in fundamental ways. Once a marginal practice, remote work became mainstream during the COVID-19 pandemic, supported by digital infrastructures such as broadband, cloud platforms, and collaborative tools. By decoupling paid work from daily presence at the office, telework loosens the home–work–services nexus that structured twentieth-century metropolitan life. Commuting is no longer a universal necessity, and residential choices are less tightly anchored to the geography of employment. For urban planners, this is a profound shift: the office, long the focal point of commuting and metropolitan structures, has become a variable node—central for some, peripheral for others, and, for certain households, nearly optional.

B. Motte-Baumvol (✉)
SAFIR, Sorbonne University Abu Dhabi, Abu Dhabi, UAE
e-mail: benjamin.mottebaumvol@sorbonne.ae

L. Bouzouina et al. (eds.), *15-Minute Cities*, Footprints of Regional Science,
https://doi.org/10.1007/978-3-032-20840-8_16

This shift gives rise to the paradox of proximity. On the one hand, telework reduces commuting frequency, alleviates peak-hour pressures, and frees time that can be reallocated to local activities, seemingly reinforcing the ethos of the 15-minute city. On the other hand, it often lengthens commuting distances when trips still occur, enables households to relocate farther from employment centers, and generates rebound effects as time saved is redirected to additional nonwork trips. Pre-COVID evidence from England already highlighted this tension: analyses of the UK National Travel Survey showed that hybrid teleworkers often accumulated greater weekly commuting distances than daily commuters, with only high-frequency telework (3 or more days per week) delivering net reductions in distance and emissions (Cerqueira et al., 2020; Cerqueira & Motte-Baumvol, 2022; Motte-Baumvol & Schwanen, 2024; Motte-Baumvol & Schwanen, 2025). In Switzerland, studies revealed comparable patterns. Cross-border evidence from the Franco-Swiss Léman region showed that teleworkers lived farther from their workplaces but partly compensated through greater reliance on public transport (Motte-Baumvol et al., 2025). National survey analyses demonstrated that telework was associated with longer commutes and additional nonwork trips (Melo & de Abreu e Silva, 2017; Cerqueira et al., 2020; Cerqueira & Motte-Baumvol, 2022). These findings underscore that fewer commutes do not always mean shorter commutes, and that sustainability outcomes depend on a combination of work frequency, distance, mode choice, and household routines.

Post-COVID studies further confirm the conditional nature of these outcomes. Comparative surveys conducted in France, the UK, and the United States in 2023 revealed divergent relocation trajectories: outward moves in London, lateral or central relocations in Paris, and relative stability in New York (Motte-Baumvol & Porcher, 2025a, 2025b). A parallel strand of research on the same survey data showed that telework frequency is strongly stratified, concentrated among higher-income professionals in all three cities (Motte-Baumvol et al., 2024, under review). National evidence from France indicates additional complexity: telework days are unevenly distributed during the week, with Fridays and Mondays disproportionately favored, generating new temporal pressures on transport systems and workplace organization (Belton Chevallier et al., 2024). Collectively, these results show that telework reshapes urban mobility in diverse ways: reducing commute frequency but lengthening distances, loosening the home–work bond but reinforcing inequalities, enabling temporal flexibility but destabilizing spatial planning frameworks such as the 15-minute city.

The present chapter builds on this decade of research led by Benjamin Motte-Baumvol and collaborators. Across multiple national and metropolitan contexts, this work documents how digital flexibility reshapes commuting, relocation, and sustainability. It provides the empirical foundation for three core arguments. First, proximity in the digital age must be reframed as hybrid, encompassing both physical accessibility to nondigitizable activities and digital accessibility to remote work and services. Second, the environmental promise of telework is conditional: only high-frequency remote work consistently reduces emissions, and outcomes vary sharply by context. Third, reconciling telework with the 15-minute city requires integrated policies that reinforce neighborhood services, sustain public transport, and expand access to digital flexibility without deepening sociospatial inequalities.

The structure is as follows: Section 2 presents the conceptual background of proximity and digital flexibility. Section 3 examines evidence on telework and commuting distances, followed by Sect. 4 on residential relocation. Section 5 compares metropolitan patterns in Paris, London, New York, and Switzerland. Section 6 discusses strategies for reconciling telework with the 15-minute city, and Sect. 7 concludes by reframing the paradox of proximity as both a challenge and an opportunity for urban planning.

Research Trajectory on Telework and Mobility

The arguments in this chapter build on a decade of research led by Benjamin Motte-Baumvol and collaborators, examining how telework reshapes commuting, residential mobility, and sustainability across diverse contexts. Together, these studies document the paradox of proximity: fewer commutes but longer distances, conditional environmental gains, and socially selective access to flexibility.

Early analyses of the UK National Travel Survey showed that telework does not automatically reduce emissions. Hybrid workers—teleworking 1 or 2 days per week—often accumulated more weekly commuting distance than daily commuters, with consistent reductions only achieved at 3 or more days of remote work (Cerqueira et al., 2020). A complementary study revealed that telework also reshapes household mobility, as time saved from commuting was frequently reallocated to shopping, escorting children, or leisure trips, offsetting potential environmental benefits (Cerqueira & Motte-Baumvol, 2022). Further research emphasized weekly patterns, showing that hybrid teleworkers concentrated longer commutes into fewer days, raising weekly totals compared to daily commuters (Motte-Baumvol & Schwanen, 2024). A dedicated analysis of Fridays added a temporal dimension: commuting declined sharply on this day, but nonwork mobility rose, illustrating that environmental impacts depend on the rhythms of the week as much as on commuting distance (Motte-Baumvol et al., 2024).

Parallel work in Switzerland brought additional nuance. A study of the Franco-Swiss Léman region, where cross-border commuting is widespread, revealed that teleworkers tended to live farther from their workplaces, often on the French side of the border, where housing is more affordable. Environmental outcomes were mediated less by reductions in travel demand than by modal shifts, as teleworkers relied more on public transport and less on cars (Motte-Baumvol et al., 2025). An ad hoc national survey further demonstrated that telework did not systematically reduce daily mobility: nonwork trips were often more frequent, and residual commuting persisted even on supposed telework days (Motte-Baumvol et al., 2022). A subsequent analysis of the same survey highlighted the links between telework, residential relocation, and job change, showing that longer commutes among teleworkers reflected selection effects—telework being more common among workers already living at a distance—while actual moves remained limited (Motte-Baumvol et al., 2023).

Comparative surveys after COVID-19 extended these insights to global metropolitan contexts. Research in France, the UK, and the United States showed that teleworkers were more likely to relocate after 2020, but outcomes diverged: outward moves in London, lateral or central relocations in Paris, and relative stability in New York (Motte-Baumvol & Porcher, 2025a, 2025b). Ongoing analyses of Paris, London, and New York also demonstrate that telework remains socially stratified, concentrated among higher-income, highly educated professionals, thereby reinforcing inequalities in residential choice and mobility opportunities (Motte-Baumvol et al., 2024, under review).

Taken together, these studies establish a coherent research trajectory. They show that telework reduces commuting frequency but often lengthens distances; it generates rebound effects in nonwork mobility; it interacts with relocation differently across institutional and metropolitan contexts; and it remains unevenly distributed across social groups. This body of work provides the empirical foundation for the present chapter, situating the paradox of proximity at the heart of debates on reconciling telework with the 15-minute city.

Defining Telework

The term *telework* has been used in policy and research for several decades, yet its meaning often overlaps with related notions such as *remote work*, *work from home* (WFH), and *hybrid work*. Clarifying these distinctions is essential for analytical precision.

Telework refers broadly to the performance of paid work away from the employer's premises, enabled by information and communication technologies. Its defining feature is the spatial decoupling of the worker from the office, with digital tools sustaining professional tasks. In this sense, telework is broader than WFH, encompassing coworking hubs, satellite offices, and other remote sites.

Work from Home (WFH) is a narrower category, referring specifically to telework carried out at the worker's residence. Before COVID-19, WFH was the primary way telework was measured in statistical surveys, because the home remained the dominant remote location. In 2019, INSEE reported that only about 7% of French employees mainly worked from home, compared with 12% in the UK (ONS) and 24% of employed people in the USA who did some work at home on an average day (BLS).

Hybrid work denotes a temporal arrangement in which workers divide their time between the office and a remote site. Since 2020, hybrid patterns have become dominant in many countries. By 2021–2023, most USA and UK teleworkers reported 2–3 days of remote work per week, while in France, adoption remained more limited and socially selective, concentrated in metropolitan areas and higher-skilled occupations.

Remote work is a broader and less precise umbrella concept. It includes telework but also more geographically untethered forms such as *digital nomadism*, where paid work is decoupled not only from the office but also from fixed residence.

The COVID-19 pandemic transformed both the prevalence and perception of telework. During the strictest lockdowns in 2020, levels peaked at around 60% of full-time workers in the USA, 46% in the UK, and over 40% in France. These figures fell once restrictions were lifted but stabilized well above pre-pandemic levels. By 2023, around one-third of employees in the USA and the UK, and about one-quarter in France, continued to telework regularly, typically in hybrid form.

For this chapter, telework is defined *operationally* as working from home at least 1 day per week, with hybrid work understood as any arrangement combining telework days and office presence. WFH is treated as a subcategory of telework, while remote work is recognized as a broader umbrella that extends beyond conventional labor markets. This operational definition aligns with the empirical analyses presented throughout the chapter, ensuring consistency across national and metropolitan comparisons.

2 Conceptual Background: Proximity and Digital Flexibility

The principle of proximity has long been central to urban planning. From Howard's garden city and Perry's neighborhood unit to transit-oriented development and compact city strategies, planners have consistently sought to minimize spatial and temporal distances between homes, workplaces, and services. The 15-minute city represents the latest articulation of this tradition, emphasizing compactness, mixed land use, and localized access to everyday functions. By organizing daily life around short travel times, it promises lower carbon emissions, healthier lifestyles, and greater social cohesion. In time-geographic terms, it operationalizes Hägerstrand's time–space prism, seeking to expand individuals' opportunities by reducing constraints of distance and travel time.

Yet proximity is not only a spatial condition; it is also shaped by technological and social transformations. The spread of information and communication technologies (ICTs) has progressively loosened the binding of work and services to specific places. E-commerce, e-government, and online education already reconfigured access to goods and services before the pandemic. The massive expansion of telework during and after COVID-19 extended this flexibility to paid employment, decoupling professional activity from daily copresence at the workplace. In doing so, digital tools destabilize the classical logic of accessibility planning: the office, long a fixed anchor of urban and regional structures, becomes a variable node.

Clarifying terminology is essential. In this chapter, *telework is defined operationally as working from home at least 1 day per week, enabled by ICTs*. Work from Home (WFH) is treated as a subcategory, while *hybrid work* refers to arrangements combining office and remote days. The number of days matters: hybrid patterns reduce commuting frequency but often increase commuting distances on in-office days. Broader notions of *remote work* sometimes include digital nomadism or itinerant practices entirely detached from metropolitan labor markets, but here the focus remains on regular telework as observed in statistical surveys and ad hoc studies.

Levels of telework varied considerably before the pandemic. In 2019, about 7% of French employees regularly worked from home, compared with 12% in the UK and roughly a quarter of US workers who did some work at home on an average day. During the strictest lockdowns of 2020, these shares soared to over 40% in France, nearly 60% in the USA, and 46% in the UK. Since 2021, telework has stabilized above prepandemic levels, with roughly one-third of workers in the USA and UK, and one-quarter in France, teleworking regularly—most commonly in hybrid form, typically 2 or 3 days per week.

These shifts have profound implications for the concept of proximity. Earlier planning frameworks assumed that reducing travel required bringing home and work closer together. Digital connectivity now allows individuals to separate residential choice from workplace location. The result is a redefinition of proximity in hybrid terms: physical accessibility remains crucial for nondigitizable activities such as schooling, healthcare, and leisure, while digital accessibility enables professional and social interactions to occur independently of spatial distance.

At the same time, the 15-minute city itself should not be seen as an uncontested or self-evident solution. As Schwanen (2024) has argued, its claims rest on fragile empirical foundations. While compact, mixed-use design is associated with lower car use, evidence that localized living automatically strengthens social ties, reduces inequality, or produces broad sustainability gains is far less robust. Moreover, the 15-minute city downplays the role of political economy, property markets, and governance capacity in shaping outcomes. In this sense, telework and the 15-minute city face parallel critiques: both promise to transform urban mobility and accessibility, but both depend on institutional, economic, and social conditions that may amplify or limit their effects.

Empirical research underscores this conditionality. Studies in England show that only high-frequency telework—3 days or more per week—consistently reduces commuting-related emissions, while 1- or 2-day arrangements often increase total travel demand (Cerqueira et al., 2020; Motte-Baumvol & Schwanen, 2024). Comparative surveys in France, the UK, and the USA revealed divergent relocation logics: outward dispersal in London, central or lateral moves in Paris, and stability in New York (Motte-Baumvol & Porcher, 2025a, 2025b). Analyses of the Franco-Swiss Léman region further demonstrated that teleworkers often live farther from workplaces but reduce car use indirectly, mediated by modal shifts rather than reduced mobility overall (Motte-Baumvol et al., 2025).

Taken together, these findings illustrate a paradoxical reconfiguration of proximity: temporal sovereignty and spatial dispersion coexisting within the same practice. For planners, the challenge is not to choose between physical and digital forms of accessibility, but to reconcile them. The 15-minute city can remain relevant if reframed as a hybrid model—anchoring everyday life in local services while accommodating digital flexibility—so that telework becomes a lever for sustainability rather than a source of dispersal and inequality.

3 Telework and Commuting Distances: Evidence from Pre- and Post-COVID Studies

One of the most persistent assumptions about telework is that fewer commutes automatically translate into less travel. Yet empirical evidence across diverse contexts shows this relationship is far from straightforward. Teleworkers frequently live farther from their workplaces than nonteleworkers; they commute less often, but when they do, their journeys are often longer. This paradox complicates the environmental promise of telework.

Pre-COVID Evidence Studies using the UK National Travel Survey highlighted this tension well before the pandemic. Teleworkers were found to have longer commutes than their office-based counterparts (Budnitz et al., 2020; Caldarola & Sorrell, 2022). Analyses coauthored by Motte-Baumvol showed that hybrid workers teleworking 1 or 2 days per week often accumulated *more* weekly commuting kilometers than daily commuters, with only those working remotely 3 days or more achieving net reductions in distance and emissions (Cerqueira et al., 2020; Motte-Baumvol & Schwanen, 2024). Similar patterns appeared in the United States and the Netherlands, where telework supported tolerance of longer commutes, since residential choices were guided more by housing affordability and amenities than by workplace proximity (Zhu, 2013; de Vos et al., 2018).

Swiss Evidence Analyses from Switzerland deepen this picture. In the Franco-Swiss Léman region, teleworkers tended to live farther from their workplaces, especially on the French side of the border, yet they were more likely to rely on public transport than nonteleworkers (Motte-Baumvol et al., 2025). Reductions in car use were thus indirect, mediated by modal shifts rather than by telework itself. A national survey further highlighted that telework often failed to reduce total mobility: many teleworkers engaged in additional nonwork trips and sometimes even commuted on supposed telework days, producing a "mobility paradox" where flexibility reshaped rather than diminished demand (Motte-Baumvol et al., 2022). Another national analysis underscored the role of selection: teleworkers tended already to live farther from their jobs, while actual relocations remained limited (Motte-Baumvol et al., 2023). These findings caution against treating telework as a uniform driver of commuting change, emphasizing the importance of institutional and spatial contexts.

Post-COVID Shifts The pandemic massively expanded telework and made hybrid arrangements mainstream. Comparative surveys in France, the UK, and the United States in 2023 showed that teleworkers were more likely to have relocated since 2020, but outcomes diverged: London saw outward relocations into the commuter belt, Paris experienced lateral or central moves within a constrained metropolitan fabric, and New York remained relatively stable, with high-frequency telework concentrated near the core (Motte-Baumvol & Porcher, 2025a, 2025b). These contrasting patterns reveal how telework reshapes commuting geographies differently depending on labor markets, planning regimes, and housing systems.

Conditional Sustainability A consistent finding emerges across cases: the environmental benefits of telework are conditional. In England, only high-frequency teleworkers (3 or more days per week) achieved clear reductions in CO_2 emissions, while 1- or 2-day arrangements often increased emissions due to longer commutes on in-office days. In Switzerland, rebound effects further complicated the picture, as time saved from commuting was often reallocated to other activities—shopping, leisure, childcare—offsetting potential gains (Cerqueira & Motte-Baumvol, 2022).

Synthesis Telework reduces the *frequency* of commuting but not necessarily its *spatial* or *environmental* footprint. For the 15-minute city, this means that proximity cannot be assessed only in terms of trip counts. Commuting distances, modal choices, and nonwork mobilities are equally critical dimensions. The paradox of proximity thus highlights that the key question is not simply *how often* people travel, but also *how far*, *by what mode*, and *for what purposes*.

4 Residential Relocation and Urban Form

Beyond its effects on commuting distances, telework also reshapes residential choices. By loosening the daily obligation to live near the workplace, it alters how households weigh trade-offs between housing, amenities, and accessibility. Yet the evidence shows that these outcomes are highly uneven, mediated by housing markets, institutional frameworks, and labor market structures.

Pre-COVID Evidence Even before the pandemic, research pointed to the potential of telework to reconfigure residential geographies. In the United States, teleworkers were disproportionately located in suburban and exurban areas, attracted by larger homes and better amenities (Zhu, 2013). Ory and Mokhtarian (2006) showed that teleworkers might move outward, capitalizing on reduced commuting frequency, or inward, seeking urban amenities while limiting commuting to a few days per week. European studies echoed this ambivalence: in Portugal and the

Netherlands, telework enabled some dispersal but did not systematically generate suburban sprawl (de Abreu e Silva & Melo, 2018; de Vos et al., 2018).

Swiss Evidence Studies led by Motte-Baumvol and collaborators provide more granular insights into how national and regional contexts mediate residential outcomes. In the Franco-Swiss Léman region, where cross-border commuting is widespread, teleworkers were more likely to live farther from their workplaces, often relocating to the French side of the border to benefit from cheaper housing (Motte-Baumvol et al., 2025). These moves reflected not only reduced commuting obligations, but also wage differentials, housing costs, and divergent planning systems. The environmental balance was mixed: commuting frequency declined, but longer distances offset potential gains, and modal choice (greater reliance on public transport) became decisive.

Analyses of Swiss national survey data refined this picture. One study showed that longer commutes among teleworkers were not primarily the result of relocation, but of *selection effects*: workers already living at a distance were more likely to telework (Motte-Baumvol et al., 2023). Another highlighted a "mobility paradox": even on supposed work-from-home days, teleworkers frequently made additional trips for shopping, childcare, or leisure, and sometimes commuted nonetheless (Motte-Baumvol et al., 2022). These results underline that telework changes the *form* rather than the *volume* of mobility, and that relocation effects cannot be understood in isolation from broader daily activity patterns.

French Post-COVID Evidence National survey data collected in France after COVID-19 provide additional nuance. Motte-Baumvol et al. (2024) mapped the actual and ideal weekly distribution of remote work, showing that while many employees expressed a preference for more frequent telework, realized patterns stabilized around 2–3 days per week in large metropolitan areas. These findings confirm that telework is now institutionalized in hybrid form, but also socially selective: uptake was highest among higher-skilled professionals in central urban regions. Residential implications followed this stratification. Telework provided some households with new flexibility to adjust their location, but for many others, constraints linked to affordability, housing supply, or occupational ineligibility limited relocation.

Post-COVID Relocations Comparative surveys across France, the UK, and the United States in 2023 showed that teleworkers were indeed more likely to relocate since 2020, but with divergent outcomes (Motte-Baumvol & Porcher, 2025a, 2025b). In London, telework supported outward relocations into the commuter belt, consistent with high housing costs and flexible labor markets. In Paris, relocations were more lateral or central, reflecting compact urban form and restrictive planning regimes. In New York, high-frequency teleworkers largely remained near the core, stabilizing concentration rather than encouraging dispersal. These contrasts demonstrate that telework does not dictate relocation outcomes; it interacts with housing supply, planning institutions, and metropolitan form.

Inequalities in Mobility Evidence from Switzerland, France, and cross-national comparisons underscores that not all workers can seize the residential flexibility telework potentially affords. Relocations are most common among higher-income, highly educated professionals with both occupational autonomy and financial resources. For others, housing affordability constraints, family obligations, or the absence of telework eligibility limit mobility. Telework, therefore, risks reinforcing sociospatial inequalities: some households gain new locational freedoms, while others remain locked into constrained geographies.

Rebound and Trade-Offs Even when relocations occur, outcomes are not always those anticipated. Households moving outward may save on housing costs but face longer commutes on office days, eroding environmental benefits. Conversely, some moves are driven less by workplace distance than by access to schools, amenities, or quality-of-life factors, with telework acting as an enabling condition rather than a primary cause. These trade-offs highlight the need for careful interpretation of telework's role in residential change.

Synthesis Overall, telework facilitates a loosening of the home–work nexus, but its residential effects remain strongly mediated by structural and contextual conditions. In London, it tends to encourage outward moves; in Paris, it produces limited dispersal within a constrained urban fabric; in New York, it stabilizes centrality; in Switzerland, it reshapes cross-border living while revealing the importance of selection effects and rebound mobilities; and in France more broadly, it has institutionalized hybrid practices without unleashing a wholesale residential exodus. For the 15-minute city, this means that planning must consider not only commuting but also residential mobility, ensuring that digital flexibility does not exacerbate inequalities or undermine sustainability.

5 Comparative Metropolitan Patterns: Paris, London and the UK, New York, and Switzerland

The geography of telework is highly uneven across metropolitan contexts. Evidence from Paris, London, and New York demonstrates how differences in urban form, labor markets, and institutional frameworks shape not only the frequency of telework but also its spatial consequences. These international comparisons can be complemented by insights from the UK as a whole and from Switzerland, including both the cross-border dynamics of the Léman region and national survey evidence, as well as recent survey work in France that highlights post-COVID stabilization of hybrid models. Together, these cases underscore the diversity of outcomes and the context-dependency of telework's spatial effects.

London and the Wider UK In London, telework adoption correlates strongly with distance from the central business district (CBD). High-frequency remote work is most common in outer London and the commuter belt, where high housing costs push households outward and where professional and managerial jobs compatible with telework are widespread. This geography translates into longer commuting distances on in-office days. Wider UK analyses reinforce the pattern: Motte-Baumvol and Schwanen (2024) showed that hybrid teleworkers often accumulate higher weekly commuting distances than daily commuters, since reductions in frequency are offset by longer average trip lengths. The series of UK studies using National Travel Survey data further established that only when telework reaches 3 or more days per week do net reductions in emissions materialize (Cerqueira et al., 2020; Motte-Baumvol et al., 2019, 2021, 2023, 2024). For policymakers, this creates a dual challenge: risks of reinforcing car dependence in peri-urban zones, and the erosion of the financial base of transit systems reliant on peak-hour commuting.

New York The New York metropolitan area exhibits a contrasting profile. High-frequency telework is concentrated close to the core, particularly in Manhattan and adjacent boroughs, among professional and executive workers in finance, law, and advanced services. These sectors remain anchored in the CBD, with strong agglomeration economies sustaining concentration. As a result, telework has not generated widespread outward relocations but has instead stabilized metropolitan centrality. The challenge is less sprawl than the financial viability of dense transit: fewer commuting trips reduce ridership and revenue in a system heavily dependent on high passenger volumes. Here, telework highlights the fragility of transit funding models that hinge on daily flows.

Paris and Metropolitan France Paris presents yet another configuration. Levels of telework adoption remain lower than in London or New York, reflecting institutional rigidities, occupational structures, and a narrower pool of eligible jobs. When relocations occur, they are more often lateral within the metropolitan fabric or toward central neighborhoods rather than outward. The compactness of the Paris region and restrictive planning policies constrain suburban expansion. Motte-Baumvol et al.' (2024) post-COVID French survey revealed that while many workers expressed a preference for more frequent telework, realized patterns stabilized around 2–3 days per week, primarily among higher-skilled professionals in metropolitan cores. This stratification means that telework reinforces sociospatial divides: higher-income workers disproportionately benefit from flexible arrangements, while others remain tied to fixed schedules and long commutes.

Switzerland and the Léman Region Evidence from Switzerland adds a distinctive dimension. In the Franco-Swiss Léman region, teleworkers were more likely to reside farther from their workplaces, often across the border in France, where housing was cheaper (Motte-Baumvol et al., 2025). Yet environmental outcomes were mediated by modal shifts: while commuting distances lengthened, reliance on public transport increased, reducing car dependence. At the national scale, survey evidence showed that telework often produced additional nonwork trips and even occasional commuting on telework days, a "mobility paradox" where flexibility reshaped rather than reduced mobility demand (Motte-Baumvol et al., 2022). Further analyses highlighted selection effects: teleworkers were already more likely to live farther away, rather than relocation being widespread (Motte-Baumvol et al., 2023). Telework in Switzerland thus combines tolerance of long distances with rebound mobilities and cross-border logics, mediated by strong public transport infrastructures.

Comparative Lessons Taken together, these cases underline that telework cannot be understood as a universal driver of either dispersal or concentration. Its spatial consequences are strongly context-dependent. In London and the wider UK, telework reinforces outward moves and long-distance commuting. In New York, it consolidates CBD-oriented labor markets while destabilizing transit finances. In Paris and metropolitan France, it accentuates social selectivity more than spatial dispersal, stabilizing hybrid arrangements among elite groups. In Switzerland, it lengthens residential distances but also generates rebound trips and modal shifts that reduce car use. For the 15-minute city, the lesson is clear: telework reshapes proximity in highly contingent ways. Policy responses must therefore be territorially specific—supporting sustainable mobility and coworking in London's peripheries, safeguarding transit stability in New York, widening access to flexible work in Paris, and leveraging modal and cross-border dynamics in Switzerland.

6 Reconciling Telework with the 15-Minute City

The paradox of proximity—fewer commutes but longer distances, reduced frequency but greater dispersal—represents a central challenge for reconciling telework with the 15-minute city. A framework designed around localized living and colocated daily functions is destabilized when distance increases, even as trip frequency declines. Yet telework and the 15-minute city are not inherently contradictory. The question is whether the framework, widely celebrated but also criticized for its idealism, can be reworked to account for digital forms of accessibility and the inequalities they entail.

Hybrid Definitions of Proximity At its core, the 15-minute city rests on a physicalist logic: everyday needs should be met within a short radius of home. Telework disrupts this logic by loosening the home–work tie and allowing professional activity to occur without daily presence in the office. Empirical evidence across the UK, France, the USA, and Switzerland demonstrates this duality: reduced frequency of trips but, often, greater commuting distances when they occur (Cerqueira et al., 2020; Motte-Baumvol & Schwanen, 2024; Motte-Baumvol et al., 2021, 2022, 2023; Motte-Baumvol & Porcher, 2025a, 2025b). Schwanen (2024) critiques the 15-minute city precisely for its reliance on a singular, local definition of accessibility that underplays digital mediation. The evidence here confirms that proximity today must be reframed in hybrid terms, where digital and physical dimensions coexist.

Conditional Environmental Outcomes The 15-minute city assumes that spatial proximity produces automatic sustainability gains. Telework challenges this premise. Findings from England show that only high-frequency telework—3 days or more per week—consistently reduces emissions, while hybrid arrangements often raise overall travel demand (Cerqueira et al., 2020; Motte-Baumvol & Schwanen, 2024). Cross-national surveys reveal divergent residential logics: outward dispersal in London, lateral or central relocations in Paris, and core stability in New York (Motte-Baumvol & Porcher, 2025a, 2025b). In Switzerland, the picture is even more nuanced: cross-border teleworkers live farther away but reduce car use through modal shifts (Motte-Baumvol et al., 2025), while national surveys highlight rebound trips and selection effects (Motte-Baumvol et al., 2022, 2023). These findings confirm Schwanen's (2024) argument that the 15-minute city rests on an overly linear link between proximity and sustainability, when in practice, outcomes are highly contingent on context, frequency, and modal systems.

Rebound Effects and Nonwork Mobility Time saved from commuting is not automatically reinvested in local life. English and Swiss evidence shows that teleworkers often make additional shopping, leisure, or childcare trips on work-from-home days, offsetting potential environmental gains (Cerqueira & Motte-Baumvol, 2022; Motte-Baumvol et al., 2022). Comparative surveys reveal that while some households do reinforce neighborhood activity, others use telework to justify dispersed residential strategies. Here again, Schwanen's critique is instructive: the 15-minute city tends to assume benign uses of local accessibility, but telework demonstrates that temporal sovereignty can generate both neighborhood vitality and new forms of mobility demand.

Equity and Stratification One of Schwanen's (2024) key criticisms is that the 15-minute city risks overlooking inequalities in who can actually live a "15-minute life." Telework amplifies this blind spot. Across contexts, access to remote work is strongly stratified by income, education, and occupation. Evidence from Paris, London, and New York confirms that high-income professionals disproportionately benefit from telework flexibility, while others remain tied to long, daily commutes (Motte-Baumvol et al., under review). In Switzerland, national surveys show that teleworkers already tend to live farther away, highlighting how structural inequalities shape who can use telework to their advantage (Motte-Baumvol et al., 2023). Without redistributive interventions, telework risks reinforcing rather than reducing sociospatial inequalities—precisely the critique levied against the universalist aspirations of the 15-minute city.

Policy Directions Reconciling telework with the 15-minute city requires responding to both the empirical evidence and the conceptual critiques:

- *Neighborhood Services and Third Places*: Investments in local amenities, coworking hubs, and third places can anchor telework in neighborhood life, addressing Schwanen's concern that the 15-minute city risks idealizing proximity without creating conditions for it to thrive.

- *Sustainable Suburban Mobility*: In peri-urban contexts such as London, extending active and public transport is essential to prevent telework-driven dispersal from entrenching car dependence.
- *Transit Funding in Dense Cores*: In cities like New York, flatter ridership curves challenge revenue models; new mechanisms are needed to maintain the dense transit networks on which localized accessibility depends.
- *Equity and Access*: In Paris and elsewhere, widening telework access beyond privileged groups is critical if the 15-minute city is to remain a socially just framework.
- *Cross-Border Coordination*: In regions like Léman, telework interacts with wage differentials, housing costs, and transport systems across jurisdictions; only coordinated policies can reconcile digital flexibility with sustainability.

Toward Integrated Planning Telework thus tests the conceptual limits of the 15-minute city. It exposes the fragility of its linear assumptions about proximity and sustainability, while underscoring the necessity of addressing inequalities. Yet the two frameworks can be reconciled if proximity is redefined in hybrid terms: physical anchors for nondigitizable activities, complemented by digital accessibility that expands temporal and spatial flexibility. The lesson is clear: telework should not be seen solely as a challenge to proximity, nor the 15-minute city as a rigid physicalist model. Instead, together they point toward a new planning paradigm—an integrated vision of accessibility that accepts the coexistence of locality and dispersal, immobility and mobility, digital and physical proximities.

7 Conclusion

Telework has unsettled long-standing planning assumptions about the role of spatial proximity. Whereas the 15-minute city emphasizes localized access to daily functions, telework loosens the home–work nexus: it reduces commuting frequency but often lengthens commuting distances and reshapes residential patterns. This paradox of proximity complicates efforts to cut emissions, sustain transit systems, and reinforce neighborhood-based accessibility.

Comparative evidence underscores the diversity of outcomes. In London, telework encourages outward relocations and longer commutes; in New York, it consolidates CBD-oriented labor markets while destabilizing transit revenues; in Paris, it amplifies social selectivity rather than spatial dispersal; and in the Léman region, it combines longer commuting distances with modal shifts away from the car. Across these contexts, only high-frequency telework consistently reduces commuting-related emissions, while hybrid arrangements often erode gains through longer trips and rebound mobilities.

For the 15-minute city, the implication is that proximity must be reframed in hybrid terms. Physical accessibility to essential services remains indispensable, but digital proximity has become an equally structuring force of urban life. Reconciling the two requires territorially specific policies: reinforcing neighborhood services and coworking hubs, extending sustainable mobility infrastructures into peri-urban zones, sustaining dense transit networks in urban cores, and widening access to telework beyond privileged groups.

Telework neither guarantees nor undermines the viability of the 15-minute city. It demands integrated strategies that align digital flexibility with sustainable and equitable urban living. Understood in this way, the paradox of proximity is less a contradiction than an invitation to innovate—redefining accessibility for a digital urban age in which mobility and immobility, locality and dispersal, must be managed together. In this sense, telework confirms Schwanen's (2024) critique: the 15-minute city cannot be treated as a universal blueprint, but its adaptability lies precisely in its capacity to evolve into a hybrid framework that integrates both digital and physical proximities.

References

Belton Chevallier, L., Aguilera, A., Viana Cerqueira, E., Pigalle, E., & Motte-Baumvol, B. (2024). Mapping out the current and ideal weekly distribution of remote work in France after COVID-19. *International Journal of Sustainable Transportation, 18*(11), 962–971. https://doi.org/10.1080/15568318.2024.2425726

Budnitz, H., Tranos, E., & Chapman, L. (2020). Telecommuting and other trips: An English case study. *Journal of Transport Geography, 85*, 102713. https://doi.org/10.1016/j.jtrangeo.2020.102713

Caldarola, B., & Sorrell, S. (2022). Do teleworkers travel less? Evidence from the English National Travel Survey. *Transportation Research Part A: Policy and Practice, 159*, 282–303. https://doi.org/10.1016/j.tra.2022.03.026

Cerqueira, E. D. V., & Motte-Baumvol, B. (2022). Flexible work arrangements and household-related journeys. Who takes the lead in dual-earner heterosexual couples? *Travel Behaviour and Society, 26*, 240–249. https://doi.org/10.1016/j.tbs.2021.11.003

Cerqueira, E. D. V., Motte-Baumvol, B., Chevallier, L. B., & Bonin, O. (2020). Does working from home reduce CO2 emissions? An analysis of travel patterns as dictated by workplaces. *Transportation Research Part D: Transport and Environment, 83*, 102338. https://doi.org/10.1016/j.trd.2020.102338

de Abreu e Silva, J., & Melo, P. C. (2018). Home telework, travel behavior, and land-use patterns: A path analysis of British single-worker households. *Journal of Transport and Land Use, 11*(1). https://doi.org/10.5198/jtlu.2018.1134

de Vos, D., Meijers, E., & van Ham, M. (2018). Working from home and the willingness to accept a longer commute. *The Annals of Regional Science, 61*(2), 375–398. https://doi.org/10.1007/s00168-018-0873-6

Melo, P. C., & de Abreu e Silva, J. (2017). Home telework and household commuting patterns in Great Britain. *Transportation Research Part A: Policy and Practice, 103*, 1–24. https://doi.org/10.1016/j.tra.2017.05.011

Motte-Baumvol, B., & Porcher, T. (2025a). *Telework and the reconfiguration of urban space: Intra-metropolitan relocation patterns in Paris, London, and New York after COVID-19.* Under review.

Motte-Baumvol, B., & Porcher, T. (2025b). Assessing the link between COVID-19-induced telework adoption and residential relocation in France, the UK, and the US. *Cities, 163*, 106037. https://doi.org/10.1016/j.cities.2025.106037

Motte-Baumvol, B., & Schwanen, T. (2024). Telework, travel times, and peak hour avoidance in England: An overview using travel times across five weekdays. *Travel Behaviour and Society, 34*, 100668. https://doi.org/10.1016/j.tbs.2023.100668

Motte-Baumvol, B., & Schwanen, T. (2025). *What Shapes Work-From-Home Frequency? Spatial configurations and cross-national variations in Paris, London, and New York.* Under review.

Motte-Baumvol, B., Schwanen, T., & Bonin, O. (2024). Telework and the day-to-day variability of travel behaviour: The specificities of Fridays. *Transportation Research Part D: Transport and Environment, 132*, 104245. https://doi.org/10.1016/j.trd.2024.104245

Motte-Baumvol, B., Ravalet, E., & Rérat, P. (2025). *Why do teleworkers live further from their workplace than non-teleworkers?* Under review.

Ory, D. T., & Mokhtarian, P. L. (2006). Which came first, the telecommuting or the residential relocation? An empirical analysis of causality. *Urban Geography, 27*(7), 590–609. https://doi.org/10.2747/0272-3638.27.7.590

Schwanen, T. (2024). The 15-minute city: A solution to saving our time and our planet: Carlos Moreno. Hoboken, NJ: Wiley, 2024. *The AAG Review of Books, 12*(4), 32–34. https://doi.org/10.1080/2325548X.2024.2389788

Zhu, P. (2013). Telecommuting, household commute and location choice. *Urban Studies, 50*(12), 2441–2459. https://doi.org/10.1177/0042098012474520

Teleworking and the 15-Minute City: Rethinking Lifestyles Between Proximity and Hypermobility

Laura Hostettler Macias, Emmanuel Ravalet, and Patrick Rérat

1 Introduction

The idea of the *15-minute City* (15mC) was first conceptualized by Moreno as an urban model designed to enhance sustainability while simultaneously improving quality of life by emphasizing the need for proximity, where daily needs can be met by a short walk or a bike ride, and where mobility, but especially car-dependent mobility, is reduced (Moreno et al., 2021).

How can teleworking influence mobility practices, the role of mobility and the model of the 15mC? The COVID-19 pandemic induced an important learning effect in the ways to work remotely. In an effort to contain the virus, millions of jobs were temporarily relocated from the centralized offices to the homes of workers, who, due to the nature of their jobs, were able to telework. This shift produced an unprecedented reduction in daily commuting. After the pandemic, telework (TW) has declined but to a much higher level than before the pandemic and has led to a reconfiguration of work geographies (Hostettler Macias et al., 2025; Sweet et al., 2023; Luca et al., 2025).

The digitalization of work and the rise of teleworking have a substitution effect that can be traced back to the idea of Zelinsky's model of mobility transition. Already in the 1970s, Zelinsky suggested that advances in transportation could reduce internal migration, while simultaneously extending commuting distances. With the subsequent development of information and communication technologies (ICTs), these transitions began reshaping the labor market and work practices, including telework. As a result, daily mobility could decline, as individuals "use the flow of information rather than engage in a territorial shift" (Zelinsky, 1971: 222). In other words, ICTs enable digital mobility instead of actual physical mobility.

L. Hostettler Macias (✉) · E. Ravalet · P. Rérat
IGD, OUVEMA, University of Lausanne, Lausanne, Switzerland
e-mail: laura.hostettlermacias@unil.ch

L. Bouzouina et al. (eds.), *15-Minute Cities*, Footprints of Regional Science,
https://doi.org/10.1007/978-3-032-20840-8_17

Furthermore, this work practice has also had an impact on the role of the home and its immediate surroundings as multifunctional spaces where work, leisure, and family life increasingly converged. With this adaptation and with the reduction of having to commute, "[…] living and working remotely within a 15-Minute neighbourhood equipped with high-quality proximity services can be seen as an ideal condition for improving quality of life" (Nuvolati & Romens, 2025: 219–220). Elldér (2022) presents evidence from Sweden that teleworkers tend to walk more than non-teleworkers. They also walk more when doing service-related errands but cycle less when commuting to work. In the context of the 15mC concept, this suggests that remote work may encourage local, walkable trips for daily needs.

So, telework not only has an impact on how and where we work, but also on how our mobility patterns are impacted by changing the geographies of work.

From this perspective, telework can be viewed as a potential enabler of the 15mC, as it aligns the geography of work with the geography of local life, thereby opening opportunities to reduce structural reliance on long-distance commuting. Moreover, by decentralizing work practices, teleworking holds the potential to rebalance mobility patterns more broadly, supporting shifts toward active modes of travel, strengthening local economies through increased neighborhood consumption, and fostering new spatial routines that are more closely tied to proximity than to distance (Di Marino et al., 2022).

Yet this optimistic view is complicated. Teleworking is not a uniform practice; it ranges from occasional home-based days to full remote employment, with each form producing distinct relationships to urban space. Moreover, rebound effects can be observed, including long-distance commutes on non-telework days, the rise of multi-residential arrangements, and intensified leisure travel. These dynamics point to a complex interaction between proximity and hypermobility that challenges simplistic assumptions about telework as an automatic sustainability solution in the sense of 15mC. With hypermobility, we refer to the intensification of mobility practices, encompassing, for example, increased travel frequency, longer commuting distances, and the spatial extension of daily activities, enabled by advancements in transport and communication systems.

It is important to note here that mobility and lifestyle are deeply interconnected. Lifestyle can be defined through the patterns of people's behavior, their values, and also their choices; it is the expression of identities and shows also how daily life is organized. It encompasses both objective conditions (such as income, household structure, and occupation) and subjective orientations (such as attitudes, preferences, and values) that shape how people live, work, and move through space (Lanzendorf, 2002). Furthermore, people's mobility practices both shape and reflect how they live, work, and relate to place. For example, choices about where to live, how to commute, and how often to run errands locally or travel longer distances all form part of a person's mobility patterns which, needless to say, are influenced by personal values, household circumstances, income, access to different transport modes, and the urban environment itself (Lanzendorf, 2002).

So, to capture the complexity of telework, mobility practices, and the concept of 15mC, it is necessary to move beyond measuring the sheer number of commuting trips and instead consider commuting distances, residential (re)locations, and the broader spectrum of spatial practices that telework enables. Teleworking reshapes proximities not only around the home but also around secondary residences. Such shifts demand a more nuanced understanding of how telework reorganizes lifestyles, reshapes urban–peri-urban relations, and redefines the balance between proximity and mobility or in some cases hypermobility. In this chapter, we want to explore the following research questions:

> What are the effects of teleworking (TW) on the geography of lifestyles—considering spatial practices, residential choices, and relationships to urban and peri-urban environments?

> How do different forms of telework mediate the tension between proximity and hypermobility?

Our findings will highlight the complex relationship between telework, commuting practices, and the concepts of proximity in the sense of the 15mC and hypermobility. By combining qualitative and quantitative data analysis, we are able to identify patterns that demonstrate both the potential and the limitations of telework in relation to spatial organization and daily mobility. In the following section, we will introduce our methodology and the context of the case study.

2 Methodology and Context

This study employed a mixed methods approach, integrating both quantitative and qualitative data collection and analysis. The quantitative component provided breadth and generalizability, while the qualitative component offered depth and contextual understanding. Together, they produced a more comprehensive picture than would have been possible through a single methodological lens.

The quantitative part was based on a large-scale survey carried out in the spring of 2022, during the post-crisis period, across the entire Swiss territory. To ensure representativeness, the survey was conducted in the three main national languages and targeted a working population reflective of Swiss employees. Moreover, data collection was carried out through a professional polling institute, employing a probabilistic sampling method to enhance the validity of the results. The questionnaire, which contained 55 questions and took approximately 15 min to complete, was designed to capture a wide range of information on telework, commuting, and workplace practices. In addition to standard questions on work arrangements, respondents were asked to provide a detailed record of the trips they had made on their most recent teleworking day and on the most recent day spent at their regular workplace. A total of 5100 Swiss workers completed the survey. The dataset was subsequently cleaned and weighted to align with the sociodemographic and geographic characteristics of the Swiss population, drawing on reference data from the Swiss Labour Force Survey (SLFS). This ensured that the results were both robust and representative of national patterns.

The qualitative part consisted of 30 biographical in-depth interviews conducted between 2022 and 2023. The interviews followed a semi-structured guide that addressed employment trajectories, telework arrangements, commuting patterns, housing choices, and familial responsibilities. Interviews typically lasted between 45 min and 1 h, although several extended further, providing the opportunity to explore participants' experiences in considerable depth. The biographical format proved especially valuable, as it illuminated how individuals made sense of telework and navigated its implications, embedding professional practices within broader residential, relational, and family contexts. The interviews were transcribed verbatim and analyzed using the software MAXQDA. The thematic analysis guided the identification of recurring patterns, meanings, and thematic dimensions across the dataset through an iterative process of coding, refinement, and interpretation.

The interview sample represented a diverse group of Swiss teleworkers. Participants ranged in age from 27 to 63 years and included 14 women and 16 men. They lived in varied household arrangements, including single households, cohabiting partnerships, families with one to three children, and situations where individuals were living again with their parents. While most participants were based in urban settings, six lived in more periurban or rural contexts. In terms of professional background, they were predominantly employed in the tertiary sector, in fields such as architecture, engineering, consulting, academia, IT services, and library science. Many worked part-time on a voluntary basis, often as a result of family responsibilities or lifestyle choices.

The integration of the quantitative and qualitative components took place at both the design and interpretation stages. The survey results established broad, generalizable patterns in telework and commuting behavior across Switzerland, while the interviews provided insights into the lived experiences underlying these patterns. Through this combination, the study was able to triangulate findings, enhance validity, and enrich statistical evidence with contextualized narratives, thereby offering a holistic understanding of telework in the Swiss context.

2.1 Context of Switzerland

Switzerland provides a particularly relevant context for examining the relationship between the 15mC concept and teleworking practices. The country's urban system is characterized by a highly polycentric structure. A network of cities embedded in urban regions combines high population density with strong economic activity.

A defining feature of Switzerland's spatial organization is its highly efficient public transport network. The national rail system, integrated with regional and local services, ensures that even villages and peri-urban areas are well connected to the major urban centers. This accessibility has traditionally supported long-distance commuting, with a significant proportion of the workforce travelling daily across boundaries of cantons.

A further mention is that the country has a predominantly tertiary economy, with 77.8%[1] of the workforce employed in services. Many of these occupations are well-suited to teleworking, as was demonstrated during the COVID-19 pandemic when large segments of employees shifted to remote work arrangements. The proportion of people who telework reached approximately 45.3% in 2022 (Hostettler Macias et al., 2025).

Building on the methodological approach described above within the context of Switzerland, the following section examines the empirical results, highlighting key trends in teleworking practices.

3 Results

3.1 *Four Types of Teleworkers Based on Their Mobility Practices*

The qualitative fieldwork allowed us to distinguish four types of teleworkers, differentiated the groups by commuting times and associated mobility practices. The first group, referred to as *short-distance commuters*, reside within a 15-min radius of their conventional workplace and typically reach it on foot, by bicycle, or an electric bicycle. This group operates within a framework that most closely approximates the 15mC model, both on teleworking days and on the days they work in their conventional workplace. Our analysis is restricted to the domains of housing and work, which are particularly affected by teleworking practices. Telework, in their case, reinforces already localized routines and consolidates a local lifestyle. One of the characteristics is that they all live in urban centers, which was found in the quantitative data as well, which will be presented in more detail further down in the results section.

The subsequent sections introduce examples that exemplify findings from the qualitative data with the example of B. He is a software developer at the Swiss Railways; he lives alone in a three-room apartment, and he has a separate room as an office available.

B's office is just a 5-min bike ride from his apartment; so through his short commute, he already had a proximity lifestyle. He began teleworking during the pandemic in March 2020, having rarely worked from home before. He is originally from Zurich, where his family still lives, but he has his social setting in Bern. In line with the concept of the 15mC, his weekly routine primarily involves walking to sports activities or cycling to social engagements beyond work. For him, teleworking is not about reducing commuting time but, in his words:

[1] In the year 2024 see: https://www.bfs.admin.ch/bfs/en/home/statistics/work-income/employment-working-hours/labour-force-characteristics/economic-sector.html

> Home office is a blessing, because it means not having to work every day in an open space office with 50 other people.

Another short commuter example is T. He is a communication manager, and thus he is able to telework a lot. He shares a flat with a friend, and they have a big living room where he occasionally works from, as his roommate also works from home from time to time. However, T said that when he needs to concentrate, he works at his desk, which is in the same room as his bed. He already teleworked before the pandemic. From his apartment to the office space, it takes him about 7 min by bike. He works from the official office about one time a week; he told that for the exchanges it helps to be present in the office. However, his job involves a lot of storytelling, so having to be creative, which T finds it easier to be inspired at home. Since he lives in the city center, his daily mobility is proximity-based; he walks or cycles to the places he needs to go. Sometimes during the winter, when biking becomes harder, he takes the bus to play tennis.

The second group, *mid-distance commuters*, travel less than 30 min to work by any mode of transport, excluding the short commuters. While their commutes are relatively manageable, telework offers partial relief, reducing the frequency of commutes but not eliminating them. On teleworking days, the principles of proximity are more visible, whereas on days working from their conventional workplace, their commute can involve a car drive or public transportation, but what can be observed is that telework can concentrate activities around the home.

L works at the University of Bern; she lives with her husband in a small town in the agglomeration of Bern. Her commute consists of an 18-min train ride, adding the quick bike ride to the train station from home, and then a short walk from the train station to the main building of the University of Bern. She has about a 30-min commute. Her husband also works in the city of Bern, so they sometimes commute together. He is a dentist so he cannot telework. She teleworks from home about 2 days out of the week. She recounted:

> Working at the office I walk a lot around the neighborhood having to go from one University building to another and going to grab lunch somewhere close by. But on these days, it also means that once I get home after my commute there is not much time left to work out. When I am teleworking from home, I do feel a stronger need to move my body because of the lack of movement throughout the day. So, after sending the last email of the day I can easily work out before dinner.

A also works in Bern at an NGO, she lives with her husband and her 2-year-old son as well in a small town in the agglomeration of Bern. They live next to the train station, where the train ride takes about 20 min to Bern, and then she has to take a bus, which takes about another 10 min to get to the office. She works part-time and teleworks 2 days a week being 1 day at the office in Bern. She said that with a small child, her mobility is much more centered around their home. Her son goes to a daycare 4 min from their house, and her husband works at a local shop in the village. A told that being able to telework makes the organization of daily life much easier.

The third group consists of *long-distance commuters*, who spend 30 min or more on commuting, regardless of the mode of transport. For these individuals, telework represents a significant shift, as it substantially reduces the burden of long commutes.

R is a project manager for the Swiss parliament, and he has been working at his current job for 5 years now. He lives in a small town and works in Bern. He works 4 days from home and is 1 day at his office in Bern. Although the train ride itself takes only about 20 min, his total commute, including the trip to the station from home and the walk or bus ride from the station to the office, can easily stretch to 50 min or more. For R, this daily travel made up a significant part of his day, time that could have been spent otherwise. Before the pandemic, he went to the office every day, making the routine a constant presence in both his life and his family's. Now, with teleworking, the shift has been profound. When he finishes work and shuts down his computer, he is immediately there, fully present and ready to spend time with his family.

> This is my ideal situation, as it gives me time to be at home with my almost 7-month-old, which is such a precious time. In my work breaks I can see her, I come down from our home office and have a quick coffee or we have lunch together. Before the work meeting on teams [ICT tool], I just quickly go back to the office and shut the door

M lives alone in a studio apartment in Bern and works at the Swiss Radio and Tv station (SRF) in Zurich. Her commute is very long from door to door; she has about 1 h and 50 min to sometimes 2 h. She has to take the bus to the train station in Bern, then take the train to Zurich, and then take another bus. She told:

> On the days that I go to Zurich I always work on the train, so those days are usually long days were I sometimes work over-time. And on my teleworking days because I am already at home, after turning off my computer I can go grocery shopping or work out all before dinner. I have much more time for myself and my hobbies

The fourth group, *multilocals*, includes individuals who rent a pied-à-terre near their workplace or maintain a second home in which they telework at least several times a month. This group illustrates the ways in which telework enables multi-scalar geographies of work and life. Their practices oscillate between proximity, when temporarily embedded in one locality, and hypermobility, through sustained circulation between their residences.

G has been working as a journalist for the past 14 years, primarily writing for a sports magazine. She lives with her husband in Zurich, and her office is in a municipality in the canton of Schwyz, resulting in a commute of approximately 1 h and 10 min. Holding a bike GA, she often brings her bicycle on the train to expedite the final leg of her journey from the station to the office. She has been teleworking for at least 10 years, noting that her profession is highly compatible with this practice. G and her husband have a secondary residence in the mountains in Lenk. She reported spending significant time there, even teleworking from there. The proximity to nature and the opportunity for various sports activities in the mountains are very dear to her. When asked whether she would consider relocating permanently to Lenk, she explained that her additional role as a swimming instructor in Zurich and her husband's position at ETH Zurich, which relies heavily on university infrastructure, keep them based in the city of Zurich.

S permanently lives in a small mountain village with her husband. She works as a library and document assistant in Bern, and her husband works at their village cable car company. Her commute is very long. F, she must take the cable car, which takes about 10 min, then she takes the train for 2 h and 10 min, and, lastly, she has a walk of about 7 min to arrive at the library. She works part-time, and when she has to be in Bern, she stays 1–2 nights per week at her sister's apartment. The rest of her week, she works from home in her living room, although she told me that it is not an ideal situation as she finds it sometimes hard to concentrate when her husband is at home. She said that without telework it would be hard to continue working in Bern, as it would mean she would have to live almost a double life, coming back to their apartment and seeing her husband only on the weekends.

> I always joke with my sister that I am her parasite. But we both love seeing each other and when I am staying at her place we often go to the cinema, concerts, or art exhibitions, as there is nothing like that in our small village.

3.2 *Where the 15-Minute City Meets Telework: Quantitative Insights Groups*

The quantitative analysis provides a complementary perspective by indicating the prevalence of these groups within the wider teleworking population. We want to now present the distribution, which confirms the diversity of arrangements while underscoring two important findings (Fig. 1). First, the proportion of teleworkers

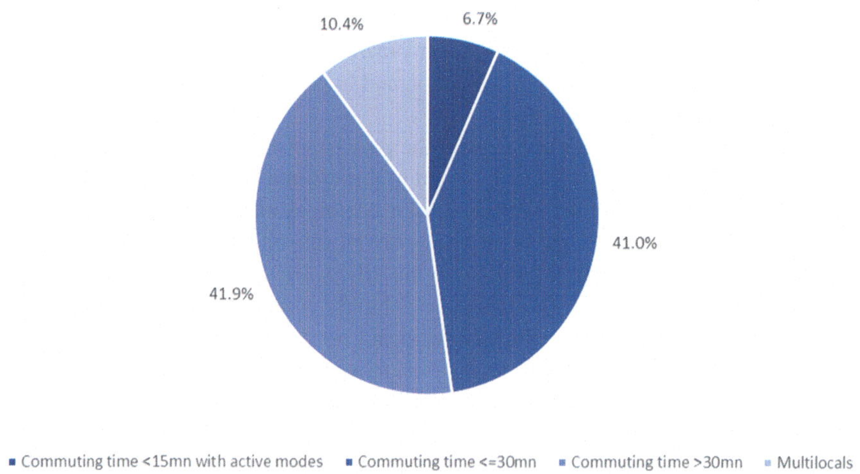

Fig. 1 Percentage of the four teleworker groups

who reside within a 15-min walk or cycle of their workplace is low, limiting the direct relevance of the 15mC model to only a minority.

Second, 10.4% of teleworkers fall into the multilocal category, a surprisingly high proportion that highlights the extent to which telework sustains practices of multilocality. While these practices may temporarily align with the 15mC in one location, they more fundamentally reflect the emergence of lifestyles structured around circulation across multiple places, extending beyond the model's proximity-based rationale and enabling mobility patterns aligned more with hypermobility practices.

Table 1 highlights the relationship between proximity and hypermobility in the context of teleworking. The results indicate that individuals residing closer to their workplace are more likely to be recent adopters of telework, suggesting that the appropriation of telework during the health crisis engaged groups with previously limited inclination toward this mode of working. In contrast, longer teleworking experience is more prevalent among individuals living at greater distances from their workplace, a tendency that is particularly pronounced among multilocation workers. Nonetheless, the observed differences between these groups do not reach statistical significance. From the perspective of the 15mC, these findings suggest that teleworking may reduce the functional importance of physical proximity to the workplace while simultaneously reinforcing the value of local accessibility for everyday needs, thereby reshaping the relationship between residential location, work practices, and urban mobility.

And finally, in geographical terms, we observe in the following results that the residential localization according to the teleworker groups shows how the impact of teleworking potentially supports and creates new lifestyles.

Table 2 further contextualizes these patterns by showing that *multilocals* are disproportionately urban, whereas *mid-* and *long-distance commuters* are more commonly located in rural or peri-urban areas. Taken together, the results from Tables 1 and 2 indicate that the experience of teleworking intersects with broader spatial structures of settlement: urban multilocals combine flexibility of place with access to dense local amenities, while rural and peri-urban commuters rely more heavily on long-distance mobility to maintain their work–life arrangements.

Lastly, from the survey, we can observe that *short-distance commuters* who engage in telework often remain at home for the majority of their teleworking days,

Table 1 Experience in teleworking in relation to the four teleworking types

	Recent adopters of telework (%)	Experienced teleworkers (%)	(%)
Commuting time <15 min with active modes	66.7	33.3	100.0
Commuting time <=30 min	59.7	40.3	100.0
Commuting time >30 min	56.5	43.5	100.0
Multilocals	54.1	45.9	100.0
Total	58.2	41.8	100.0

Table 2 The four teleworking groups and in what kind of area their residence is located

	Residential locations					
	Medium size and large centers (%)	Small centers (%)	Metropolitan suburbs (%)	Peri-urban suburbs (%)	Rural spaces (%)	Total (%)
Commuting time <15 min with active modes	36.9	12.3	30.8	9.2	10.8	100.0
Commuting time <=30 min	26.0	8.7	36.8	14.1	14.4	100.0
Commuting time >30 min	26.9	7.0	34.9	18.3	12.8	100.0
Multilocals	54.5	4.0	26.3	6.1	9.1	100.0
Total	30.1	7.8	34.5	14.7	12.9	100.0

spending even more time in their domestic environment than other categories of teleworkers. For this group, teleworking reinforces local living, where work, consumption, and daily life are concentrated within the immediate household. In contrast, *multilocal commuters* experience teleworking days differently. Although they are generally less mobile on those days compared to their regular routines, their reduced mobility is frequently connected to the strategic use of teleworking as an opportunity to move between different residences. This duality illustrates how teleworking not only reshapes patterns of daily mobility but also highlights the differentiated ways in which proximity, distance, and multilocal living are negotiated depending on the commuter profile.

3.3 Summarizing and Exploring the Potential Effects of Telework on the 15-Minute City

3.3.1 Short-Distance Commuter

In the example of B, we see that teleworking has a minimal impact in terms of mobility, as his short commute and reliance on walking and cycling for social and recreational purposes remained largely unchanged. But what can be observed is that teleworking reinforces his already proximity-based lifestyle. For individuals who already live in environments where most essential services are close at hand, the shift to remote work collapses professional activity into the home and sometimes eliminates the need for daily commuting altogether.

3.3.2 Mid-Distance Commuter

What can be observed is that despite the reduction in commuting frequency, *mid-distance commuters* still rely on public transportation or car drives to overcome the distance between their home and workplace on office days. Their mobility patterns alternate between the daily commute to their workplace and localized movement during telework days. We see that in this group, already the 15mC or proximity-oriented urban design only partially applies because their weekly routines include significant travel to workplaces. In these first two lifestyle groups, teleworking serves as an enabler of convenience, reducing travel demands and providing additional time for personal and family-related activities, such as resting or caring for children.

3.3.3 Long-Distance Commuters

Considering also in this group the impact of teleworking on mobility patterns, we can observe that, on days when commuting remains necessary, hypermobility continues to dominate, leaving this group largely disconnected from the 15mC model, or only experienced during teleworking days. Therefore, telework introduces hybrid mobility patterns into their daily practices: proximity and localization when working remotely, combined with hypermobility when working at their conventional workplace. While the frequency of commuting decreases among teleworkers, their overall time–distance budget remains above average, indicating that teleworking may serve as a compensatory mechanism that renders long commutes more manageable. We can observe the emergence of a dual-center arrangement in which work is divided between the conventional workplace and teleworking from their home. This creates a hybrid situation where teleworking does not eliminate the commute entirely but reshapes its role and frequency. The home becomes a secondary place of professional activity, allowing workers to reduce the number of long journeys while still maintaining ties to the main workplace. Teleworking functions as a mechanism that enables residential anchoring, linked to factors such as community attachment, family networks, schooling, and home ownership, while maintaining continuity in one's professional trajectory.

3.3.4 Multilocals

In the example of G, the ability to telework contributes to her mobility pattern of hypermobility. Traveling back and forth between her office, the secondary home and Zurich, she exhibits the highest level of mobility among the four groups studied. We can observe that the relevance of the 15mC is therefore intermittent and contingent, as it applies to specific contexts of residence; moreover, it can appear on teleworking days while being undermined by the broader pattern of multilocality. We can observe here the emergence of a triple-center or multiple-center arrangement where work is done from different places, resulting in a hypermobility pattern. By reducing spatial and temporal constraints, teleworking enables multilocal living, allowing individuals to balance work and personal life across different places.

In conclusion, these groups illustrate the divergent lifestyle effects of telework. For *short commuters*, and to a certain extent *mid-distance commuters*, telework enhances localization by concentrating daily practices within a short spatial radius. Conversely, for *long commuters* and *multilocals*, telework contributes to the maintenance, and in some cases the intensification, of hypermobility. For *long commuters*, telework allows them to continue sustaining demanding commuting patterns with reduced frequency, while *multilocals* employ telework to distribute their activities across multiple locations, sometimes on a multicity scale.

4 Discussion: Does Telework Foster the Emergence of 15-Minute Cities

Even though its magnitude is lower than during the pandemic, teleworking is a growing trend that influences the way individuals live, work, and organize their lifestyle in time and space. The effects of teleworking on everyday and residential mobility—and ultimately on the model of the 15mC—will greatly depend on how people consider and use the time they save by not commuting as well as the number of days they telework (or as enabled by their employees).

Our results show that teleworking could have various and even antagonistic effects. Will teleworking be used to foster proximity, localized lifestyles, and a slowing down of life rhythms? Or will it be used to enable hypermobility, such as long-distance commuting or multilocal living?

In advance, it is noteworthy that, across all the teleworker groups, the principles of the 15mC are largely realized on teleworking days. However, a closer examination on the weekly scale reveals distinct mobility practices among the different groups. This diversity of teleworkers and the impacts on their mobility will be present in the following section.

Teleworking may first foster a localized lifestyle, as in the case of short-distance commuters. Teleworking leads to a rediscovery of proximity and an organization of the lifestyle near the place of residence, at least if local services and opportunities are provided. The time saved may be used for short-distance trips (e.g., escorting children, shopping) or for other home-based activities (as shown by the share of people not leaving their home). Short-distance commuters are overrepresented in urban areas and teleworking is a way to develop a local anchorage and connections with the neighboring areas. The percentage of those teleworkers is small in comparison to the other groups, but they could enhance the residential attractiveness of urban settings from small to big cities (Rérat, 2012b, 2019).

In the remaining groups, the daily organization may be based on proximity on teleworked days but not on the days people commute to their conventional workplace. Teleworking could be used to mitigate mobility and to adopt a quieter life rhythm less based on commuting on a weekly scale. However, teleworking could also be used to diminish the spatial constraints in the choice of a dwelling or a job as the improvement in classic transport infrastructure has done. Teleworking—as in the last decades—could foster longer distances between the places of everyday life and lead to more urban sprawl. It could also reinforce residential anchorage and the decline of internal migration over the last decades (observed in Switzerland as in many Western countries), and their replacement by long-distance commuting (Rérat, 2014). An additional trigger could be the need or wish to have a dedicated room for teleworking in the dwelling that would favor (re)locating in less dense and more remote areas where housing prices are much lower than in centers.

The model of the 15mC aims to foster lifestyles less (car-based) mobility-dependent by providing functions (dwelling, work, health, leisure, education, and shopping) in a radius reachable on foot or by bike. Based on our results, we identify three additional planning principles that must go along with the 15mC. First, on a local scale, there is a need of mitigating the rise in housing space consumption and of avoiding that teleworking put an additional pressure on the housing market and lead to more urban sprawl. Coworking spaces could be implemented at the scale of neighborhoods. Second, urban sprawl has to be regulated at a regional scale. Densification and compact city policies have to provide enough dwellings in central and well-connected areas, and enhance the quality of life (Rérat, 2012a). Third, 15mC—and notably in small towns around bigger cities—have to be connected by public transport or cycling highways to avoid car dependence.

On a scientific level, more research will be needed to analyze how teleworking and the time saved by avoiding commuting trips are used notably through a life course and a lifestyle perspective (Hostettler Macias et al., 2022). This would enable a more precise assessment of the effects of teleworking and the role of proximity on residential and everyday mobility, such as the way teleworking is organized within couples, whether it influences residential relocation and housing consumption over the life course, or to which extent that the proximity enabled by teleworking is compensated by other forms of mobility (e.g., working on vacation).

5 Limitations

Despite the strengths of the dataset and the analytical approach, several limitations should be acknowledged. First, the empirical data provide information only about the individual respondent and not about other members of the household. This restriction makes it difficult to capture intra-household mobility dynamics or spatial decision-making processes. For example, a household may include one individual who teleworks and therefore has a short commuting distance, while their partner or another household member undertakes long-distance commuting. Such variations could significantly influence overall mobility patterns, residential choices, and daily spatial organization, but remain unobserved in the current dataset.

And secondly, the available data are confined to information concerning housing and workplace locations. Consequently, the spatial scope of the analysis does not extend to other activity locations, such as leisure, shopping, or education, which also play an important role in shaping everyday mobility and spatial practices. This limitation restricts the ability to fully assess the broader impacts of teleworking on total travel demand and spatial behavior.

References

Di Marino, M., Tomaz, E., Henriques, C., & Chavoshi, S. H. (2022). The 15-minute city concept and new working spaces: A planning perspective from Oslo and Lisbon. *European Planning Studies, 31*(3), 598–620. https://doi.org/10.1080/09654313.2022.2082837

Elldér, E. (2022). Active travel and telework in Sweden: Teleworkers walk more, but cycle less. *Transportation Research Part D: Transport and Environment, 109*, 103362. https://doi.org/10.1016/j.trd.2022.103362

Hostettler Macias, L., Ravalet, E., & Rérat, P. (2022). Potential rebound effects of teleworking on residential and daily mobility. *Geography Compass, 16*(9), e12657. https://doi.org/10.1111/gec3.12657

Hostettler Macias, L., Ravalet, E., & Rérat, P. (2025). How does telework impact daily and residential mobilities: New geographies of working and living in Switzerland. *Applied Geography, 178*, 103591. https://doi.org/10.1016/j.apgeog.2025.103591

Lanzendorf, M. (2002). Mobility styles and travel behavior: Application of a lifestyle approach to leisure travel. *Transportation Research Record, 1807*(1), 163–173. https://doi.org/10.3141/1807-20

Luca, D., Özgüzel, C., & Wei, Z. (2025). The new geography of remote jobs in Europe. *Regional Studies, 59*(1), 2352526. https://doi.org/10.1080/00343404.2024.2352526

Moreno, C., Allam, Z., Chabaud, D., Gall, C., & Pratlong, F. (2021). Introducing the "15-Minute City": Sustainability, resilience and place identity in future post-pandemic cities. *Smart Cities, 4*(1), 93–111. https://doi.org/10.3390/smartcities4010006

Nuvolati, G., & Romens, A. I. (2025). Remote working and using the services of the 15-minute city. An analytical model based on data collected in Milan. *Sociologica, 19*(2), 217–236. https://doi.org/10.6092/issn.1971-8853/20720

Rérat, P. (2012a). Gentrifiers and their choice of housing: Characteristics of the households living in new developments in Swiss cities. *Environment and Planning A, 44*(1), 221–236. https://doi.org/10.1068/a4496

Rérat, P. (2012b). Housing, the compact city and sustainable development: Some insights from recent urban trends in Switzerland. *International Journal of Housing Policy, 12*(2), 115–136. https://doi.org/10.1080/14616718.2012.681570

Rérat, P. (2014). The selective migration of young graduates: Which of them return to their rural home region and which do not? *Journal of Rural Studies, 35*, 123–132. https://doi.org/10.1016/j.jrurstud.2014.04.009

Rérat, P. (2019). The return of cities: The trajectory of Swiss cities from demographic loss to reurbanization. *European Planning Studies, 27*(2), 355–376. https://doi.org/10.1080/09654313.2018.1546832

Sweet, M., Van Loon, R., & Burke, M. (2023). What might working from home mean for the geography of work? *Journal of Transport Geography, 108*, 103596. https://doi.org/10.1016/j.jtrangeo.2023.103596

Zelinsky, W. (1971). The hypothesis of the mobility transition. *Geographical Review, 61*(2), 219–249. https://doi.org/10.2307/213996

Parking Availability Prediction Using Deep Learning Approaches: Temporal Convolutional Networks (TCN) and TimesNet

Hana Gharrad, Satria Bagus Wicaksono, Ansar Yasar, Iñaki Cejudo Fresnadillo, Josep Maria Salanova Grau, and Evdokimos Konstantinidis

1 Introduction

In many cities, the time-consuming search for a parking spot is a major barrier to accessibility. This phenomenon, known as cruising for parking, creates congestion, increases emissions, and directly contradicts the efficiency goals of a 15-minute city. Predicting the availability of parking spaces is a critical challenge in urban mobility. In case of nonavailability of parking slots, most drivers will cruise roads at low speed, creating more congestion and more emissions. In addition, erratic driver behavior, such as frequent lane changes during parking space searches, increases the risk of traffic accidents, particularly when compounded by stress. The use of machine learning techniques in the prediction of available parking spaces has emerged in recent years as a more accurate method than pure statistical methods.

H. Gharrad (✉)
UHasselt, Transportation Research Institute, Hasselt, Belgium

Urban Planning, Economics and Transport Laboratory (LAET – ENTPE), University of Lyon, CNRS, Vaulx-en-Velin, Lyon, France
e-mail: hana.gharrad@entpe.fr

S. B. Wicaksono · A. Yasar
UHasselt, Transportation Research Institute, Hasselt, Belgium

I. C. Fresnadillo
Fundación Vicomtech, Basque Research and Technology Alliance (BRTA), Donostia-San Sebastián, Spain

J. M. S. Grau
Centre for Research and Technology Hellas (CERTH), Hellenic Institute of Transport, Thermi, Thessaloniki, Greece

E. Konstantinidis
Lab of Medical Physics and Digital Innovation, Aristotle University of Thessaloniki, Thessaloniki, Greece

L. Bouzouina et al. (eds.), *15-Minute Cities*, Footprints of Regional Science,
https://doi.org/10.1007/978-3-032-20840-8_18

Other recent studies focused on comparing the performance of these different methods. Accurate short-term prediction of parking availability can optimize parking turnover, particularly in the context of autonomous vehicles, and open the opportunities for the implementation of dynamic on-demand parking rates.

Many factors could influence the availability of parking slots, including the arrival time, the location of the parking, the weather conditions, the density of population in the parking area, etc. The performance of the prediction model depends on the size and quality of the data and variables recorded and provided, the selected model and its structure and parameters, and the time of the prediction. In this chapter, we evaluate the performance of Temporal Convolutional Networks (TCN) and Time Series Analysis with 2D Tensor Transformation (TimesNet) on the prediction of available parking slots at a given time. We use a medium-size parking dataset of 16 parking lots located in Spain, recorded over 6 months. We evaluate the two models based on the coefficient of determination R^2 and the Mean Absolute Error (MAE). In the second section, we start by reviewing existing works that addressed parking prediction using machine learning techniques. In the third section, we present the data used, and we explore seasonality and the stationarity of the data. In the fourth section, we present the methodology and the network structure used. In the fifth section, we present the analysis and the findings. The last section concludes the study and summarizes the relevant findings.

2 Related Works

In Ye et al. (2020) the authors propose a short-term prediction of available parking space based on Gradient Boosting Decision Tree (GBDT) and Wavelet Neural Network (WNN) and an improved WNN algorithm with (WA) decomposition and Particle Swarm Optimization (PSO). These models were tested on large-scale parking data to dynamically predict available parking space (APS) and reveal its spatial-temporal characteristics in a central city in three closely located areas in Zhejiang province, China. The authors focus on daily and weekly patterns and on 15-minute forecast intervals since it is most appropriate for drivers' travel plans and flexible parking management. The proposed model, Wavelet Analysis-Particle Swarm Optimization-Wavelet Neural Network (WA-PSO-WNN) further enhances PSO-WNN by incorporating Wavelet Analysis (WA) decomposition and reconstruction. Compared to the standalone WNN model, WA-PSO-WNN increased accuracy by an average of 80.86% (weekday) and 71.37% (weekend), indicating a three to five times improvement. Despite its longer training times, the suggested WA-PSO-WNN model, combining wavelet analysis for data pre-processing and particle swarm optimization for initial weight determination, demonstrated superior prediction accuracy and stability for short-term APS forecasting, especially when dealing with nonstationary time series fluctuations. In Provoost et al. (2019), the authors developed a real-time prediction model for parking area states (occupancy, in-flux, and out-flux) up to 60 min ahead, using publicly available historic and real-time data sources from the city of Arnhem, the Netherlands. These data sources include parking, traffic, and weather information collected from August 2017 to April 2019. Two models were tested and evaluated, a Feed-Forward Neural Network (FFNN) and a Random Forest (RF) model. These models were compared against a naïve, seasonal random walk model using Mean Absolute Scaled Error (MASE). From the reported results, the FFNN model provided the best accuracy. This model outperformed the RF approach on all assessed performance measures, although the differences were small. Both the FFNN and RF models significantly outperformed the base model. On the other hand, the RF model was slightly faster for predictions (1.32 s) than the FFNN (1.57 s). In addition to the temporal pattern, parking occupancy pattern could also vary based on additional factors such as the geographical location of the parking space, its proximity to services and facilities, and parking policies. These factors could play an important role in identifying the occupancy pattern. In Parmar et al. (2021), the authors propose a model to analyze the influence of car users' socioeconomic and travel characteristics on parking duration using Artificial Neural Network (ANN) and Local Interpretable Model-agnostic Explanations (LIME). Economic parameters, particularly parking fee per hour, profession, and income, were found to be the most important predictors for parking duration. Instead of predicting the number of parking slots. The authors in Bassetti et al. (2022) suggested estimating the probability of finding a parking slot for a given road segment. Three classes were designed: Green (60% parking probability), Yellow (30% parking probability), and Red (8% parking probability). The data used for training and evaluating the models was primarily generated synthetically by a Cruising for Parking Simulator (CPS). The simulator simulated 40 users moving for 200 weeks in the San Giovanni city area of Rome, Italy. This process collected information about users' parking, un-parking, and cruising actions over road segments at different time slots. From this simulated trip data, ten features were extracted for each road segment at a given time slot. These features were designed to exploit information that would be available from smartphone data, such as total parking mean, total cruising mean, in-trip parking, cruising, circling mean, in-day parking, cruising mean, and distance features. Four supervised machine learning models were trained and compared: K-Nearest Neighbors (KNN), Gradient Boosting (GB), Random Forest (RF), and ANN. All models achieved over 90% accuracy proofing that the developed machine learning models were highly efficient in classifying parking availability at the street segment level. The integration of different data sources could expand the prediction capability of the model and its accuracy. In Inam et al. (2022), the authors suggested to build an end-to-end machine learning pipeline for on-street parking prediction using multisource data and investigate the impact of multisource data integration on prediction accuracy. The authors utilized a publicly available open data from the city of Melbourne, Australia and integrated four datasets, which are parking occupancy, pedestrian, traffic, and weather data from 1 January 2017 to 31 December 2017. The authors trained and compared Seven machine learning and deep learning models, which are Multilayer Perceptron (MLP), Random Forest (RF), Decision Trees (DT), K-Nearest Neighbors (KNN), Gradient Boosting (GB), Adaptive Boosting (AB), and linear Support Vector Machine (SVC). For the evaluation, the authors used metrics such as accuracy, precision, recall, F1-score, and AUC. The best-performing model was RF, which outperformed other models, achieving an average accuracy of 81% and an AUC of 0.81. This model was selected for the On-Street Parking (ONSP) model and deployed for real-time prediction. From the results, the integration of multisource data significantly improved prediction accuracy by 10% compared to models using only occupancy data.

TCN is a relatively new type of neural network, distinguished by its architecture composed entirely of convolutional layers. It has demonstrated superior performance in various tasks, including audio synthesis and language modelling at both character and word levels. TCNs are particularly adept at processing sequential or serialized data. A key characteristic of TCN is its ability to predict future data points solely based on past observations (Liu et al., 2019). This reliance on historical data makes TCNs highly suitable for time series prediction like APS. Recently, Wu et al. (2023) proposed TimesNet with a modular architecture aiming to capture the temporal patterns derived from different periods. TimesNet can identify the existence of multiple periodic patterns within time series data. It employs a modular architecture that transforms one-dimensional (1D) time series into a 2D representation. This transformation allows the data to simultaneously reveal characteristics of both intra-period and inter-period variations (Zuo et al., 2023). Consequently, TimesNet efficiently extracts these multi-periodic features by effectively capturing the temporal 2D variations, leading to notable generality and enhanced performance in time series prediction. In this study, we focus on the evaluation of the TCNs and TimesNet on predicting the availability of parking slots in the selected parking spaces and reveal daily and weekly patterns based only on occupancy data.

3 Data Collection

The data was collected from a sensor-based solution from the 30th of July 2024 to the 8th January 2025. The locations of the parking spaces are presented in Fig. 1. For July and January, the monthly data was partially collected. While for the months from August to December, the data was fully collected. In total, the data include 137 days with 12979 records for each parking. The data collected in July was removed (1 day). The dataset includes information related to the capacity of each parking space, the timestamp, and the number of free parking slots recorded. The features of date, hour, day, month, day of the week, and whether it is a weekend were added as predictive variables. In the raw data, the timestamps were not perfectly aligned to a 15-minute interval due to system or measurement delays. Stagnant records were identified and marked to apply local and global polynomial interpolation. For each parking, the number of daily records is between 94 and 96 records. To explore the collected dataset, the variation in the number of free parking slots was evaluated in Table 1. The parking with zero variability in one of the months was removed from the dataset since this may indicate a serious issue with the hardware or networking settings in the used device. The recording of the number of free slots is measured at the same time for all 16 parking spaces.

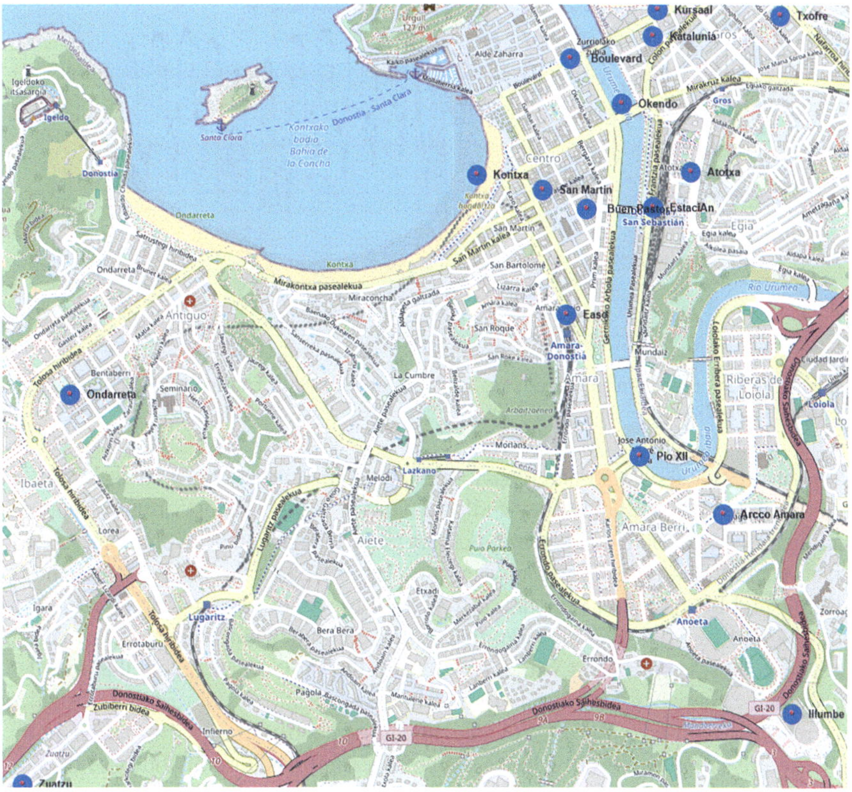

Fig. 1 Geographical location of the selected parking spaces

Table 1 Variability in the number of free slots for each month for different parking spaces

Parking	August	September	October	November	December	January
Arcco	21.16	25.89	47.22	39.77	54.95	0.0
Atotxa	25.27	45.16	55.6	56.79	61.72	36.88
Boulevard	99.82	110.08	106.51	102.75	114.24	76.69
Buen Pastor	103.02	115.06	111.07	85.11	115.56	47.64
Easo	41.34	43.17	47.58	42.83	51.71	28.72
Estación	74.23	105.14	63.97	70.6	76.06	44.9
Illumbe	0.0	0.0	0.0	0.0	0.0	0.0
Katalunia	96.85	115.7	117.0	115.69	135.26	78.94
Kontxa	193.54	191.41	182.28	166.43	196.93	83.07
Kursaal	66.67	71.85	54.06	50.22	58.95	44.62
Okendo	186.33	196.08	179.97	167.46	206.28	84.99
Ondarreta	0.0	86.35	0.0	23.25	0.0	0.0
Pio XII	0.0	0.0	0.0	0.0	0.0	0.0
San Martin	86.73	92.27	54.74	55.35	0.0	0.0
Txofre	25.84	36.58	35.84	33.77	77.02	65.87
Zurriola	0.0	0.0	0.0	0.0	0.0	0.0

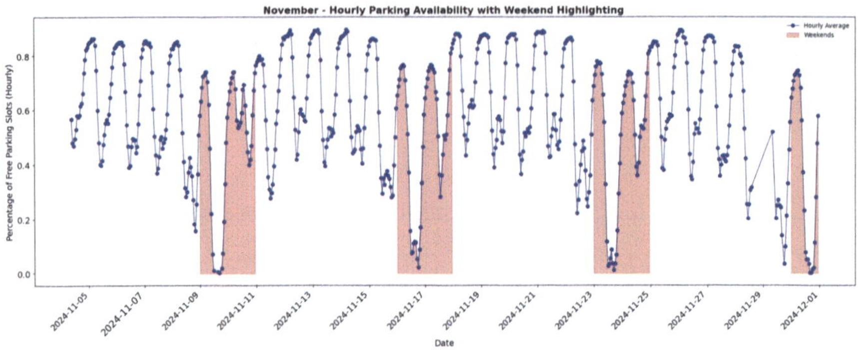

Fig. 2 Variation of the number of free slots in the parking Boulevard during November (Weekend highlighted in *red*)

To ensure that the data is stationary (constant statistical properties over time), we conducted the Augmented Dickey-Fuller test (ADF Test). The test strongly supports that the time series data contains predictable patterns that can be captured by appropriate machine learning. For Boulevard parking, for example, the ADF statistic of -13.56 (highly significant) for first differences, of -5.35 (strong) for seasonal differences and of -18.98 (very significant) for combined differences. The pattern across weeks was consistent with 70–80% stability on weekends, and the bounded daily changes between -0.15 and $+0.15$, which shows structured variation. Detailed information for these tests is provided in the Appendix (Table 4). Occasionally, intervals are skipped, possibly due to system interruptions, unavailability of data, or network delays. Data were collected at 15-minute intervals. The 15-minute interval for parking slots forecasting proves to be more suitable for drivers' travel planning while offering greater flexibility for parking management operations.

In this study, we use only daily and weekly temporal patterns to predict the number of parking slots available in a parking space. Although the integration of other dimensions and variables such as geographical and demographic properties can enhance the prediction performance of the model. We focus more on comparing the capabilities of TCN and TimesNet in identifying heterogeneous time patterns. Figure 2 shows the variation of the availability of parking slots (as a percentage) in the Parking Boulevard during the month of November. From initial observations, it is evident that the data are characterized by strong daily and weekly periodic trends.

For further exploration, hourly aggregated data shows regularity. The tests of seasonality (repeating pattern) in the data, we used Fisher's g Test and the ANOVA (for daily period, 24-h cycle). Both tests confirmed that the periodicity is statistically significant.

4 Methodology

Based on the data exploration analysis: The parking data was first filtered based on the variance in the number of free slots per month. The parking with zero variance indicates an issue with the sensors and hardware used (Table 1). Data associated with these parking places were removed. For the TCNs, the used model is structured in three temporal blocks. These blocks are chained with a doubling dilation rate to capture both short-term and long-term dependencies without using large kernels (Fig. 3).

For TimesNet, the fundamental building block is the TimesBlock, which begins by using a Fast Fourier Transform (FFT) to analyze the input sequence and identify its top-k most dominant periods ($k = 5$). For each period, the model reshapes the 1D time series into 2D tensors, where one dimension is the period length and the other is the number of periods. This 2D tensor is then processed by a sophisticated Inception Block, which applies multiple parallel convolutional kernels (the number of kernels used is 3) of different sizes to effectively capture patterns at various scales within the 2D structure, followed by a GELU activation (Fig. 4).

The collected data and realized analysis were uploaded and hosted on a node under the RAISE platform. The RAISE platform establishes the foundational infrastructure for a distributed crowdsourced data processing system. The main objective of this platform is the transition from simply making data openly available to enabling its effective processing. Fundamentally, RAISE aims to implement a mechanism where computational algorithms are deployed directly to the datasets, rather than requiring the data to be moved to the algorithms. The used data and analysis are stored under Research Analysis Identifier (2025).

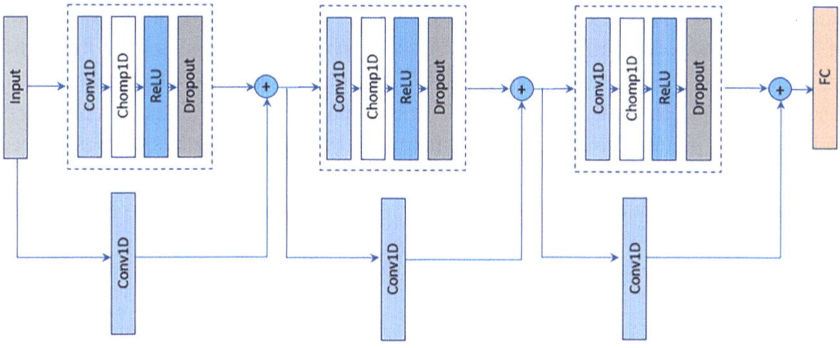

Fig. 3 Architecture of the used TCN network

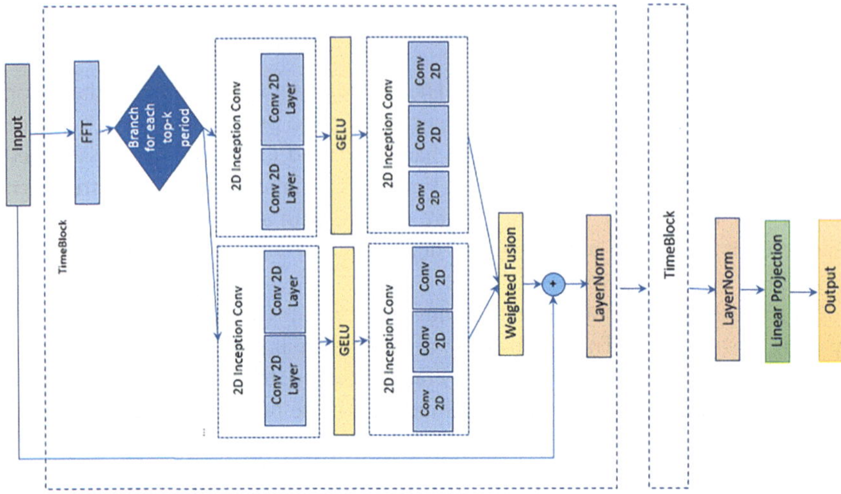

Fig. 4 Architecture of the used TimesNet network

5 Analysis and Findings

After collecting, transforming the data, and preparing the models, we ran the experiment by training the TCN and TimesNet models for parking occupancy prediction. We train a separate model for each parking space instead of a single model across all spaces. The rationale is to capture more precise patterns for each slot in our prediction. Subsequently, we conduct prediction analysis on each parking slot and use MSE and R^2 score to compare the model's performance. Data collected during the month of November was dedicated to testing, while all the other data was used for the training. The selected sequence was 700 for TCN, aiming to capture immediate temporal relationships, patterns spanning several hours and daily (around 100 records per day), and multi-day patterns. While TimesNet uses a 500 sequence length, since our task is time series prediction, we first reshape the data into sequences. For the TCN model, the sequence is defined as:

$$X = \{T_1, T_2, \ldots, T_n\}, Y = \{T_n\}$$

In this setting, the model predicts the last value of the sequence, T_N, given the sequence $\{T1, T2, \ldots, T_N\}$. This formulation is consistent with the TCN architecture, which processes sequences and adjusts the prediction horizon accordingly. For the TimesNet architecture, the sequence is structured as:

$$X = \{T_1, T_2, \ldots, T_n\}, Y = \{T_{n-\text{label.length}} \ldots, T_{n+1}\}$$

Here, TimesNet is used to predict T_{n+1}. To align with its encoder–decoder design, we define overlapping sequences with a length equal to the label length. This overlap ensures compatibility with the TimesNet structure, although the final prediction is made only on the nonoverlapping value at the end of the target sequence.

Training is done in the specified environment, with experiments run on an NVIDIA RTX 2080 with 12 GB RAM, under Ubuntu 20.2. The training parameters for both models are summarized in Table 2. For the TimesNet model, we also implement an early stopping mechanism, which terminates training if there is no improvement in the loss after a fixed number of epochs. This is introduced due to the longer training time required by TimesNet compared to TCN. The training is performed by splitting the dataset into training, testing, and evaluation sets. The training and testing sets are used to train the model, while the evaluation set is used to assess its performance.

After training the models, we evaluate the prediction results of TimesNet and TCN using MAE and R^2 scores. Our goal is to assess both the prediction accuracy (MAE) and the ability of the models to explain variance in the data (R^2). Table 3 presents the overall evaluation of both TCN and TimesNet. It can be observed that TimesNet outperforms TCN in both MAE and R^2. The results indicate that TimesNet achieves more accurate predictions across parking locations with less variability, while also explaining the variance in the data better than TCN.

Additionally, Fig. 5 shows that most TimesNet predictions achieve both high accuracy (low MAE) and stronger explanatory power (higher R^2). As expected, higher MAE values are usually associated with lower R^2 scores. However, in some parking locations, we observe cases of low MAE but also low R^2, which suggests that although the absolute prediction error is small, the model fails to capture fluctuations in the data and underestimates its variance. Figures 6 and 7 present the prediction results for a sample parking location. Overall, both models exhibit limitations in capturing sudden peaks and drops. Although TimesNet achieves higher performance, it does not perform very well for all parking locations. In particular, both

Table 2 Training parameters for TCN and TimesNet models

Parameter	TCN	TimesNet
Optimizer	Rmsprop	Adam
Learning rate	0.001	0.001
Number of Epochs	50	50
Criterion	MSE	MSE
Number of Sequences	700	500
Label Length	–	199

Table 3 Model performance comparison for MAE and R^2

Metric	Model	Mean	Std.	Min.	Max.
MAE	TCN	0.0977	0.0235	0.0735	0.1407
MAE	TimesNet	**0.0618**	**0.0159**	**0.0403**	**0.0898**
R^2	TCN	0.5105	0.2495	−0.0255	0.8077
R^2	TimesNet	**0.7535**	**0.1984**	**0.1685**	**0.8649**

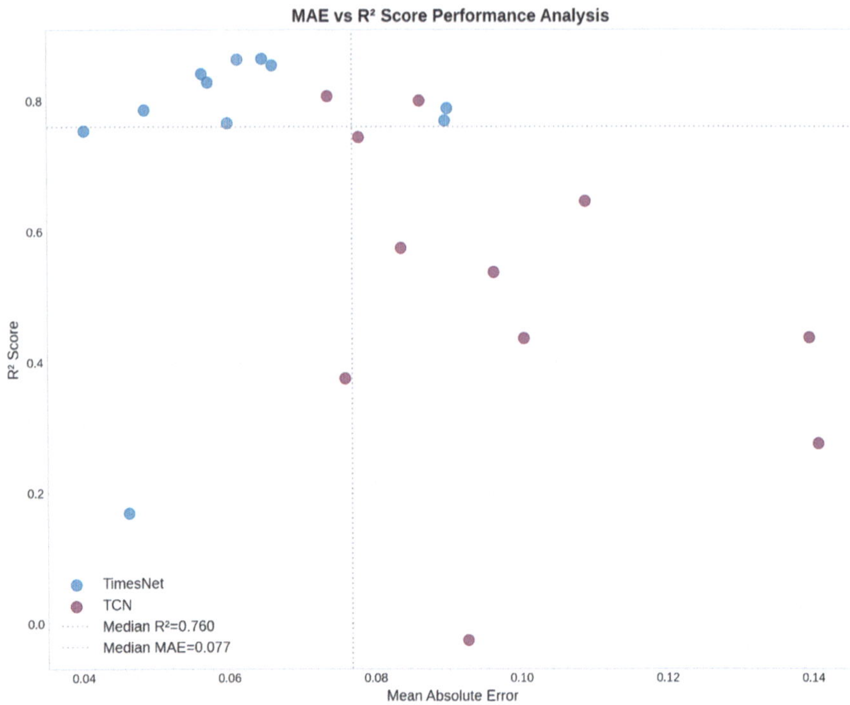

Fig. 5 MAE vs R^2 analysis

TimesNet and TCN encounter difficulties in the Txofre parking space, which shows stronger fluctuations in occupancy patterns.

6 Conclusion

An accurate prediction of the availability of parking slots and the redirection of the drivers to parking spaces with higher predicted availability can reduce cruising, improve the experience of the driver, and reduce congestion. Although the prediction of the availability of parking slot require the integration of other factors like urban setting and geospatial information from other sources. However, the ability to capture both short and long temporal relationships depends on the structure and the processing step of the trained model. This study shows that TimesNet can provide

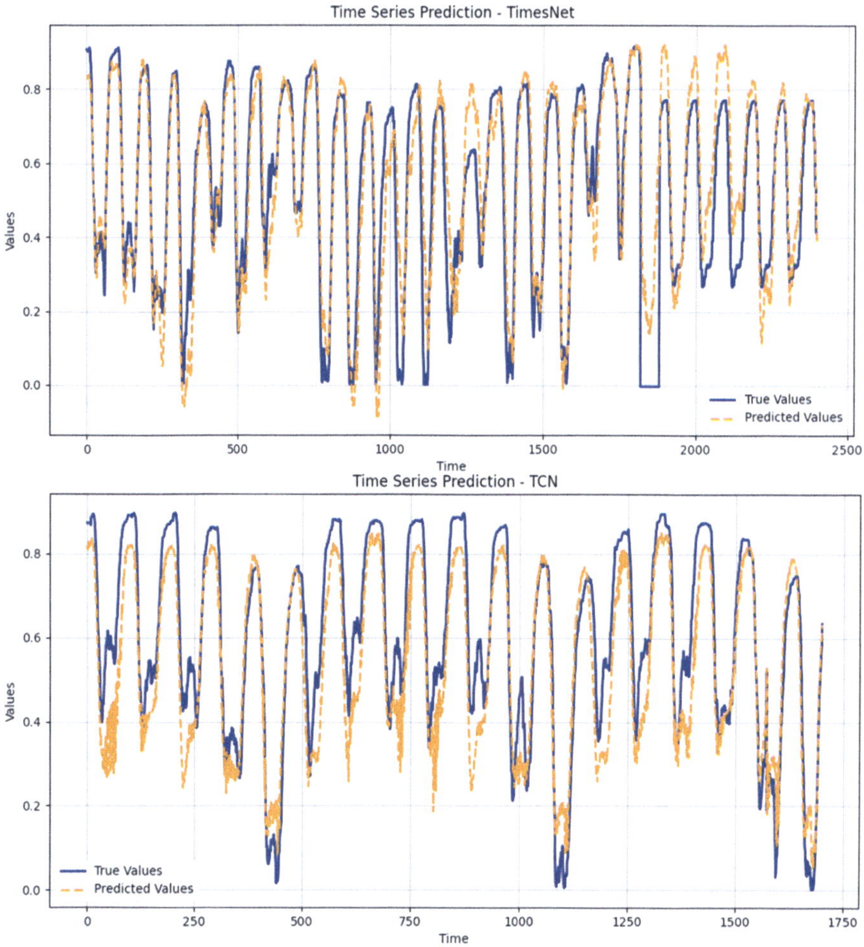

Fig. 6 Prediction results of TCN and TimesNet models for the parking Boulevard

promising predictions, which can be further improved with additional integration of data sources. Improved prediction models can be used to adjust parking prices based on predicted demand (higher prices in high-demand zones to encourage turnover) and provide guidelines and recommendations for urban developers to convert underutilized parking spaces into bike lanes, parklets, or pedestrian plaza which support the 15-minute city's walkable/bikeable infrastructure. The prediction models could also be used as a decision support tool to extend or reduce parking hours (for

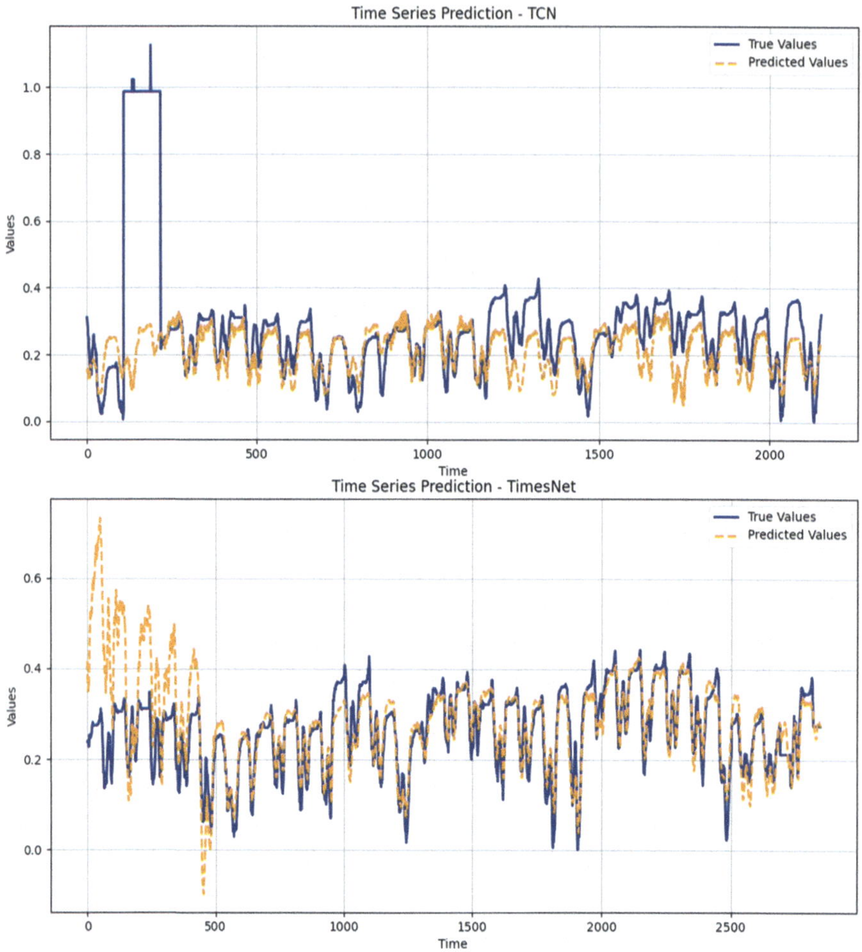

Fig. 7 Prediction results of TCN and TimesNet models for the parking Txofre

residents and visitors) and re-size the parking zones based on the actual demand, which at the same time make the parking system fairer and more balanced for different group of users.

Acknowledgment This work was conducted as a mobility pilot and was supported by the project RAISE (Research Analysis Identifier System), funded by the European Union Grant Agreement *No 101058479*.

Appendix

Table 4 Stationary test conducted per parking space

Parking	ADF[1]	ADF[2]	ADF[3]	Weekend stability	Bounded changes	Max daily change	Combined Diff. stationary
Txofre	−5.41	−5.36	−4.92	69.53	50.77	0.49	Stationary
Boulevard	−13.56	−4.74	−7.29	52.72	55.20	0.45	Stationary
Okendo	−5.57	−4.14	−7.52	71.56	74.85	0.25	Stationary
Easo	−5.24	−4.31	−5.01	72.83	48.69	0.51	Stationary
Atotxa	−5.93	−4.21	−7.14	85.58	79.54	0.20	Stationary
Estación	−6.59	−4.04	−5.97	71.79	75.76	0.24	Stationary
Kontxa	−5.74	−5.19	−6.87	63.92	66.46	0.34	Stationary
Arcco	−6.65	−6.49	−4.62	87.91	79.13	0.21	Stationary
Katalunia	−6.52	−4.17	−6.74	59.22	67.94	0.32	Stationary
Buen Pastor	−12.60	−10.70	−9.52	59.16	50.32	0.50	Stationary
Kursaal	−6.55	−4.97	−5.16	64.58	72.90	0.27	Stationary

References

Bassetti, E., et al. (2022). Exploiting user behavior to predict parking availability through machine learning. *Smart Cities, 5*(4), 1243–1266. https://doi.org/10.3390/smartcities5040064

Inam, S., et al. (2022). Multisource data integration and comparative analysis of machine learning models for on-street parking prediction. *Sustainability, 14*(12), 7317. https://doi.org/10.3390/su14127317

Liu, Y., et al. (2019). *Time series prediction based on temporal convolutional network.* 2019 IEEE/ACIS 18th International Conference on Computer and Information Science (ICIS), pp. 300–305. https://doi.org/10.1109/ICIS46139.2019.8940265

Parmar, J., Das, P., & Dave, S. M. (2021). A machine learning approach for modelling parking duration in urban land-use. *Physica A: Statistical Mechanics and Its Applications, 572*, 125873. https://doi.org/10.1016/j.physa.2021.125873

Provoost, J., et al. (2019). Short term prediction of parking area states using real time data and machine learning techniques. arXiv preprint. *arXiv*:1911.13178. https://doi.org/10.48550/arXiv.1911.13178

Research Analysis Identifier. (2025). *Experiment results on 15/09/2025*. Satria Baguswicaksono processed the dataset with PID 21.T15999/raise/96. https://doi.org/21.T15999/raise/96

Wu, H., et al. (2023). TimesNet: Temporal 2D-variation modeling for general time series analysis. *arXiv*:2210.02186 [cs.LG]. https://arxiv.org/abs/2210.02186

Ye, X., et al. (2020). Short-term prediction of available parking space based on machine learning approaches. *IEEE Access, 8*, 174530–174541.

Zuo, C., et al. (2023). An ensemble framework for short-term load fore-casting based on TimesNet and TCN. *Energies, 16*(14), 5330. https://doi.org/10.3390/en16145330

The manufacturer's authorised representative in the EU is Springer Nature Customer Service Centre GmbH, Europaplatz 3, 69115 Heidelberg, Germany. If you have any concerns regarding our products, please contact ProductSafety@springernature.com

Printed and bound by CPI Group (UK) Ltd, Croydon, CR0 4YY
21/09/2026
02219986-0013